CONCEPTS IN SMART SOCIETIES

Next-generation of Human Resources and Technologies

Editors

Chaudhery Mustansar Hussain
Department of Chemistry and Environmental Science,
New Jersey Institute of Technology, Newark, USA

Antonella Petrillo
Department of Engineering
University of Naples Parthenope, Napoli, Italy

Shahid Ul Islam
Department of Biological and Agricultural Engineering,
University of California Davis, Davis, USA

CRC Press
Taylor & Francis Group
Boca Raton London New York

CRC Press is an imprint of the
Taylor & Francis Group, an **informa** business

A SCIENCE PUBLISHERS BOOK

First edition published 2024
by CRC Press
2385 NW Executive Center Drive, Suite 320, Boca Raton FL 33431

and by CRC Press
4 Park Square, Milton Park, Abingdon, Oxon, OX14 4RN

Library of Congress Cataloging-in-Publication Data (applied for)

ISBN: 978-1-032-17034-3 (hbk)
ISBN: 978-1-032-17036-7 (pbk)
ISBN: 978-1-003-25150-7 (ebk)

DOI: 10.1201/9781003251507

Typeset in Palatino Linotype
by Radiant Productions

Preface

Welcome to an extraordinary time, a time where ecological and digital transition has taken over and revolutionized our lives in every aspect. We live in a smart society, a society in which digital solutions have radically transformed the way we interact, work and live.

The smart society is a phenomenon that has developed explosively in recent decades. The world has become an interconnected place, where information and communications flow rapidly across global computer networks. Our homes, our offices, our cities and even our bodies have become intelligent, equipped with sensors and devices that allow us to monitor and control every aspect of our lives.

This book is a journey into the universe of the smart society. We will explore the challenges and opportunities this new era presents us, analyzing how technology has affected the way we work, communicate, consume and relate.

But smart society is not just about technology and sustainability. The smart society is a complex phenomenon that also involves social, economic, political and ethical issues. The book explores the implications of this transformations.

This book does not intend to provide definitive answers, but to stimulate reflection and discussion on what it means to live in a smart society.

The smart society is here to stay, and we must be prepared to face its consequences. We must be aware of the challenges it poses and the opportunities it offers. We must be ready to adapt and innovate, to find creative solutions to the problems that will come our way.

We are on the eve of an unprecedented revolution, in which technology is changing the world in ways we never imagined. We hope this book inspires you to explore, question, and participate in this amazing transformation. The smart society is here, and the future is in our hands.

Editors

Chaudhery Mustansar Hussain
Department of Chemistry and Environmental Science
New Jersey Institute of Technology, Newark, N J 07102, USA
chaudhery.m.hussain@njit.edu

Antonella Petrillo
Department of Engineering
University of Naples Parthenope, CDN 80143 Napoli, ITALY
antonella.petrillo@uniparthenope.it

Shahid Ul Islam
Department of Biological and Agricultural Engineering
University of California Davis, United States
shads.jmi@gmail.com

Contents

PART IV: Super Smart Society and Sustainability

Introduction

We live in a society driven by rapid, unexpected changes. Less than 10 years ago the concept of the fourth industrial revolution was introduced: a more aware and oriented reality towards the Smart Society. By this we mean novelties in production technologies, enabling Information Technology (IT) services and greater attention to energy consumption. Today, we are discussing the fifth stage in the evolution of society.

Metaverse, connected mobility, robotics, bioeconomy, decarbonization, ecological transition, and digitalization will be the central themes that will have the greatest impact on society and the industries of the future. Focusing on sustainability through technological innovation is the key indication for building a society 5.0.

The concept of the Super Smart Company—or Society 5.0—was born in 2016, thanks to a Japanese research conducted by Hitachi and the University of Tokyo. It indicates the ideal form for the society of the future in which intelligent systems, exploiting technological platforms to their full potential, are able to process huge amounts of data and analyse complex scenarios. In this way, human beings would be constantly supported by the virtual, within increasingly interconnected societies: A digital transformation supported by Artificial Intelligence (AI). The original definition literally quotes: Through an initiative that merges physical space (real world) and cyberspace by exploiting ICT to its full potential, we propose an ideal form for the society of the future: a Super Smart Society that will bring well-being to the people. At the time they did not yet call it metaverse, but it is clear that the connection between technology and the activities we carry out every day is now so deeply rooted, to the point that in many situations it is hard to even point out an effective difference between the digital and the real sphere. In fact, Society 5.0 identifies the use of technology as an enabler for the design and adoption of highly impactful and innovative solutions, for the benefit of sustainable development centered on people.

Society 5.0 will be able to contribute to the achievement of the - Sustainable Development Goals (SDGs), agreed by the UN and ideally achievable by 2030. The question is: How? One possibile answer could be that Society 5.0 will facilitate the sustainable and resilient development of cities and regions through the efficient management of the water system, energy consumption, and the construction of circular production and consumption models. In particular, digital acceleration can be precious in protecting environmental sustainability: the digitization of processes, data collection, and processing –together with the connection infrastructures –allows the prevention of any environmental damage and the safety of infrastructures. Thus, digitization becomes the vector of an economic-social model with a human-centric vision.

In this regard, it should be noted that the fundamental principles of Company 5.0 have been a fundamental source of inspiration for the definition of Industry 5.0. While, the Society 5.0 stands as a real evolution of the previous Information Society, Industry 5.0 does not aim to overcome the technological paradigm of Industry 4.0 but to make its applications compatible with the sustainability and inclusion criteria necessary for giving rise to a truly human-centered socio-economic system. Society 5.0 and Industry 5.0 also have a fundamental point in common, which could be summarized in a famous quote by Albert Einstein: There is no challenge without a crisis. The crisis is the greatest blessing for people and nations, because the crisis brings progress. In the crisis is inventiveness, discoveries, and great strategies.

Thus, this book aims to outline strategic lines and suggest future directions for the development of the super smart society which pursues responsibility and sustainability by including technologies and skills. This book is intended to be a useful resource for anyone who deals with innovation and digitalization. Furthermore, we hope that this book will provide useful resources ideas, techniques, and methods for further research on these issues. Special thanks to all the authors who contributed to the success of the project. As editors of this book, we profusely much thank the authors who accepted to contribute with their invaluable research and the referees who reviewed these papers for their effort, time, and invaluable suggestions. Our special thanks to the editorial team Vijay Primlani, Jyotsna Jangra, and Raju Primlani for their precious support and their team for this opportunity to serve as guest editors.

Enjoy the book!

Editors

Chaudhery Mustansar Hussain
Department of Chemistry and Environmental Science
New Jersey Institute of Technology, Newark, N J 07102, USA
chaudhery.m.hussain@njit.edu

Antonella Petrillo
Department of Engineering
University of Naples Parthenope, CDN 80143 Napoli, ITALY
antonella.petrillo@uniparthenope.it

Shahid Ul Islam, PhD
Department of Biological and Agricultural Engineering
University of California Davis, United States
shads.jmi@gmail.com

PART I

Concept of Super Smart Society

1

Modern Society
Perspective and Development

*Baffo Ilaria** and *Travaglioni Marta*

1. Introduction

Big Data Analytics, Artificial Intelligence (AI), Internet of Things (IoT) are some of the technologies widely used in everyday reality. Private and professional life is saturated with digital data and information technologies (ITs) through which ideas are developed and shared. Due to new technologies, in the last ten years our lives have been transformed, with the advent of the smartphone, new ways of shopping, new ways of working, etc. Digital technology has transformed an industrial society focused on production into one where information is at the center of society. On 22 January 2016, the government of Japan published the 5th Science and Technology Basic Plan (Cabinet Office, 2016a). The plan proposes the idea of "Society 5.0", based on a vision of a future society driven by scientific and technological innovation. The intention behind this concept is described as follows: *Through an initiative merging the physical space (real world) and cyberspace by leveraging ICT to its fullest, we are proposing an ideal form of our future society: a 'super-smart society' that will bring wealth to the people. The series of initiatives geared toward realizing this ideal society are now being further*

Department of Economics Engineering Society and Business Organization (DEIM), University of Tuscia, Largo dell'Università s.n.c., Loc. Riello, Viterbo, 01100, Italy.
    Email: m.travaglioni@gmail.com
* Corresponding author: Ilaria.baffo@unitus.it

deepened and intensively promoted as Society 5.0. (Cabinet Office, 2016a). Society 5.0 is so called to indicate the new super smart society created by transformations led by scientific and technological innovation, after hunter-gatherer society, agricultural society, industrial society, and information society, as shown in Fig. 1.1.

	🏃 Society 1.0	🌿 Society 2.0	🏛 Society 3.0	🖥 Society 4.0	🏃 Society 5.0
Economic and Social Innovation by Deeping of Society 5.0					
Period	Birth of human beings	13,000 BC	End of 18th century	Latter half of 20th century	From 21st century
Society	Hunter-gatherer Coexistence with nature	Agrarian Development of irrigation techniques	Industrial Invention of steam locomotives	Information	Super smart
Productive Approach	Capture/Gather	Manufacture	Mechanization Start of mass production	ICT	Merging of cyberspace and physical space
Material	Stone - Soil	Metal	Plastic	Semiconductor	Material 5.0
Transport	Foot	Ox - Horse	Motor car – Boat – Plane	Multimobility	Autonomous driving
Form of settlement	Nomadic, small settlement	Fortified city Firm establishment of settlements	Linear industrial city	Network city	Autonomous decentralized city
City ideals	Viability	Defensiveness	Functionality	Profitability	Humanity

Fig. 1.1 Contextualizing Society 5.0 (Graphic elaboration of the authors).

In 2016, *Comprehensive Strategy on Science, Technology and Innovation for 2016* (Cabinet Office, 2016b) was released and the following year, the 2017 edition of its comprehensive strategy was published (Cabinet Office, 2017). In this edition, Society 5.0 is further described as follows: *Society 5.0, the vision of future society toward which the Fifth Basic Plan proposes that we should aspire, will be a human-centered society that, through the high degree of merging between cyberspace and physical space, will be able to balance economic advancement with the resolution of social problems by providing goods and services that granularly address manifold latent needs regardless of locale, age, sex, or language to ensure that all citizens can lead high-quality, lives full of comfort and vitality* (Cabinet Office, 2017). Japan is one of the nations with the greatest technological development geared toward social welfare. It is, particularly, about balancing economic and technological progress with the resolution of social issues (Žižek et al., 2021; Gurjanov et al., 2020). Therefore, the objective of Society 5.0 is to create a people-centric society, where cyberspace and physical space are integrated. The Society 5.0 concept developed by Japan addresses the economy and citizens, promoting the idea of a Smart Society, where information technologies outline the profile of a new super-intelligent society (Haque et al., 2021). Digital transformation will once again radically change many aspects of society, affecting private life, public administration, industrial structure, and employment (Nunes et al., 2021; Palumbo et al., 2021). Society 5.0 is envisioned as a society in which anyone can create value (Bibri and Krogstie, 2017), consistent with the future sustainable strategies developed with the 17 United Nations Sustainable Development Goals (SDGs). The goals of Society 5.0 are also SDGs of the 2030 Agenda (Falanga et al., 2021; Israilidis et al., 2019). Therefore, Society 5.0 can be regarded as a means by which the SDGs can be achieved. From this perspective, Industry 4.0 can promote sustainable innovation (Yigitcanlar et al., 2018; Yigitcanlar et al., 2021; Yigitcanlar and Cugurullo, 2020; Yigitcanlar et al., 2020; Yigitcanlar et al., 2017). In other words, Society 5.0 is a model for communicating the government's vision of a future society to industry and public.

2. Society 5.0 as a New Frontier of Human Evolution

The health emergency from Covid-19 and the Russia-Ukraine conflict have caused a series of consequences, including energy crisis, difficulties in finding raw materials and their increase in prices. However, these problems do not seem to be holding back the technological advancement that has begun with Industry 4.0. In fact, technological advancement is rapidly leading towards a new and revolutionary paradigm of society and fast. Big Data regulates almost every aspect of an individual's life, and collective consciences are increasingly leaning towards economic models focused on environmental sustainability. Society has yet to address the issues of the present, where the productive approach and the way in which data is currently stored, processed, and shared is still critical. In fact, humanity has evolved to reach the Society 4.0 model (Information Society), where information and communication technology (ICT) platforms (such as, e.g., the Cloud) are used as a profit-based manufacturing approach. Therefore, the operation of the process is still very dependent on the end user, who uses ICT systems to collect information previously entered manually, re-elaborates them, and shares them externally just as manually. Since the 4.0 model is severely limited by what people can do, as data production increases globally, finding the necessary information and analyzing its risks becoming a complex task. The new model of Society 5.0 is a new phase in human evolution, in which some issues related to the information society are filled using AI, which acts as a link between the physical world and the virtual world. In the Information Society (Society 4.0), the common practice is to collect information through the network and have it analyzed by humans. On the contrary, in Society 5.0 people, things, and systems are all connected in cyberspace, and the optimal results achieved by AI (exceeding the capabilities of human beings) are returned to physical space.

This process brings new value to industry and society in ways that were not possible before.

2.1 Toward a people-centric society

Society 5.0 will become a People-centric Society. In this vision, Society 5.0 is a society that can balance economic progress with solving social issues, ensuring that all citizens can lead a high-quality life full of comfort and vitality. However, balance economic development, social problem solving, and quality of life is difficult, therefore achieving Society 5.0 is a challenge. In fact, if economic growth, the society of mass production and consumption are pursued, damage to the planet could occur during the process. On the contrary, if environmental welfare is the main objective, consequences on the quality of life and economic well-being could result. For example, if humanity minimized energy consumption, the quality of life would decrease, and the economy would grind to a halt. Society 5.0 is an attempt to overcome this seemingly unresolved issue. Solving social problems without sacrificing quality of life is difficult for another reason. It requires balancing what is best for society with what is best for the individual. This challenge is related to how we understand quality of life and social welfare. There are many different definitions and measures of well-being, but it is not possible to quantify it in most cases. Therefore, to date the vision of Society 5.0 is based on two types of relationships: the relationship between technology and society and the relationship between individuals and society mediated by technology.

Society 5.0 identifies technology as an enabler for the design and adoption of highly impactful and innovative solutions that benefit people-centered sustainable development. Technological and social evolution in Society 5.0 is based on a cooperative approach, bringing together all the traditional innovation players (institutions, research centers, private actors, and civil society) and integrating new principles that are equality, equity, solidarity, sustainability, inclusion, and change. The cooperative approach enables the full and conflict-free achievement of economic development and the resolution of issues and challenges that threaten sustainability. The problems and challenges are, for example, the decline in the birth rate combined with the increase in the old-age rate, the consequent reduction in the labor force and the increase in security costs, and the environmental footprint in terms of resource consumption and the alteration of overall balances with anthropogenic activity. According to the United Nations, life expectancy at birth is set to increase, extending average life as early as 2030. At the same time, the fertility rate is set to decline, reaching just below the replacement level needed for demographic stability in 2050. Therefore, modern societies must begin to question the strategies to ensure healthy aging and fulfilling life while maintaining social and economic balance. In this regard, innovation has already achieved good results in agri-food, construction, transportation, and medicine sectors, etc. For example, in medical and healthcare, technologies such as AI, 3D printing, Virtual and Augmented Reality, nanotechnology, and robotics are strongly contributing to the transition to an increasingly predictive, preventive, personalized, precise, and patient-centric model of healthcare.

Ensuring a healthy and long-lived life implies ensuring a healthier and more welcoming world, starting with curbing global warming. In fact, the climate emergency has become an increasingly pressing issue, as the planet is warming at an unprecedented rate on geological scales. Extreme climatic phenomena increase in intensity and frequency, resulting in economic, social, and environmental damage. To ensure the protection of the most fragile ecosystems, society and economic prosperity, the green transition can no longer be postponed and must be on a global scale. In this context, the realization by 2050 of "Net Zero" (planet-neutral) economies and societies is crucial, as the current unprecedented concentration of CO_2 and climate-changing gas emissions is the direct cause of temperature rise on the Planet, with almost unanimous assent from the scientific community. The recent intergovernmental panel on climate change (IPCC) report, released in August 2021, contains some clear indications of the risks and costs of climate change that have already occurred and those that are expected in the coming decades (Fig. 1.2). The main conclusions are important and serve as a final warning to governments, institutions, businesses, and citizens who have failed to implement concrete and effective measures to reduce climate gas emissions over the past 30 years, despite numerous commitments.

2.2 *Society 5.0 and sustainable development*

The digital transformation envisioned by Society 5.0 promises to be a major weapon in the fight against climate change through its potential to enable energy efficiency, new organizational models, and less impactful consumption styles on the planet. Digitization of processes, collection, and processing of large amount of data, and connecting infrastructure create an increasingly interrelated ecosystem that generates benefits on economic productivity, and environmental sustainability. Dematerializing, measuring, and improving production and consumption processes generate greater economic welfare with lower environmental impact and positive social inclusion impacts. Therefore, these actions represent the convergence of the two transformations taking place globally, the green and the digital ones (Fig. 1.3).

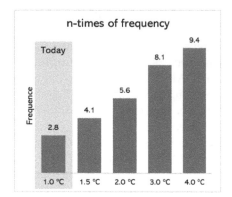
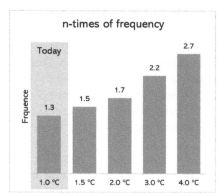
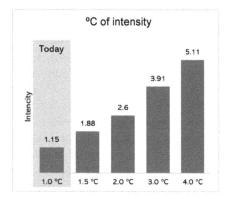
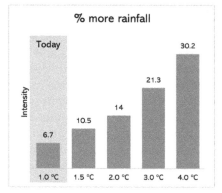

Fig. 1.2 Increased frequency and intensity for extreme weather events that would occur every 10 years in the various scenarios of temperature increase. Median scenario versus baseline 1850–1900. (Graphic elaboration of the authors) (Source: IPCC, 2021).

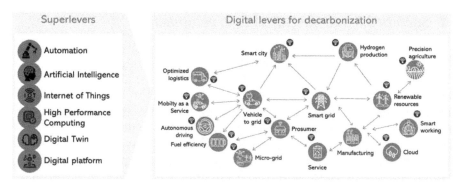

Fig. 1.3 Digital transformation in energy infrastructures.

For the resolution of these problems, the concept of Society 5.0 promotes concrete efforts around three pillars:

1. The transition towards the "Society 5.0" model and Productivity Revolution, through IoT, Big Data and AI technologies.
2. The creation of resilient, environmentally friendly, and attractive communities through Future City Initiatives to achieve the United Nations SDGs.
3. Empowerment of future generations and women through a revolution in human resource development to make the most of rich creative and communication skills, focusing on the gender goals of the SDGs.

In the Society 5.0 approach, a real paradigm shift is promoted (Fig. 1.4). Traditionally, technology and innovation were responsible for social evolution. In the Society 5.0 vision, digitization becomes a tool for differentiating and meeting society's needs by providing the necessary products and services in the quantities required, in the ways and at the times people need them. In this way, Society 5.0 contributes to the achievement of the UN SDGs. Specifically, nine different areas can be identified in which Society 5.0 can help achieve the SDGs.

- Cities and regions. The development of sustainable and resilient urban realities affects several SDGs, such as the efficient management of the water system, energy consumption, and more generally the building of sustainable production and consumption models. As the global urbanization rate set to grow by 2050, Society 5.0 makes a key contribution by ensuring increasingly effective and timely services to citizens and increasingly sustainable development solutions.

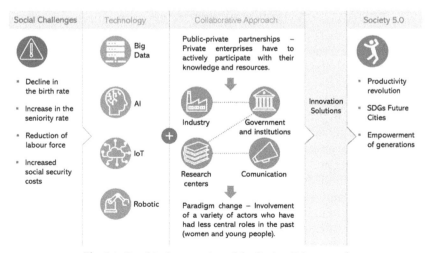

Fig. 1.4 Graphical processing of the Society 5.0 approach.

- Energy. The availability of sustainable, secure, and competitive energy is a key factor in the prosperity of modern society and future generations. Digital technologies make a valuable contribution to the challenges of the energy sector along the entire value chain by enabling the gradual decarbonization of the sector. In fact, the decarbonization process introduces many system challenges, such as managing an increasing share of non-programmable sources and the gradual down-streaming of production with the transformation of consumers into active players in the supply chain.

- Prevention and mitigation of natural disasters. The intensification of extreme natural events requires the rapid identification and implementation of solutions to mitigate climate change. Society 5.0 contributes to reduce impacts and risk to individual safety, property, and people's lives by preventing, monitoring, and securing infrastructure.

- Health and medicine. The health sector is under the combined pressures of an aging population and public spending constraints, and the Covid-19 pandemic. Therefore, it requires a comprehensive rethink aimed at ensuring smart and universal care. The combination of digital technologies (IoT and Advanced Data Analytics) enables the creation of innovative and cost-effective healthcare services.

- Agriculture and Food. New technologies enable sustainable solutions for the entire supply chain, from production to distribution of food, through processing and promoting conscious consumption. In particular, the best impacts in agricultural production can be achieved through digital technologies and data to implement solutions that increase production efficiency, reduce stress on soils and waste of natural resources, avoid waste, and make production more sustainable and healthier.

- Logistics solutions. By facilitating the flow of goods and commodities, logistics plays an essential role in economic growth, providing the infrastructure that supports the performance of all productive and economic activities. Cutting-edge technological solutions improve real-time monitoring and control, with increasingly accurate demand forecasting and service delivery through data analytics. These innovations can be integrated into efficient models based on automated solutions (e.g., autonomous driving, drones, and robots) that can increase the overall efficiency of the industry.

- Manufacturing and services. Industrialization processes can be refined, from design and development to logistics, to make them equitable, responsible, and sustainable. Current technologies allow to adopt organizational models and production logics that maximize efficiency, sustainability (including circularity) and productivity, making available products that are sustainable, safe, better able to meet people's needs and increase the level of competitiveness.

- Finance. The synergy of IoT, Machine Learning and AI provides the predictive capabilities that allow to create personalized services, more informed decision-making, and risk mitigation. The use of these technologies, digital currencies and blockchain systems is aimed at improving user-experiences, delivering higher-value services, and achieving a higher level of transparency and security.
- Public services. A rethinking of public service delivery and management systems inspired by the principles of interoperability will enable a timely and increasingly appropriate response to citizens' needs. Society 5.0 aims to facilitate the exchange of information between various local and national authorities, fostering the generation of new creative solutions to optimize processes by calibrating them to the new needs of the community.

2.3 Industry 4.0 and Society 5.0: Aims and common issues

In November 2011, the German Federal Government published "High-Tech Strategy 2020 Action Plan for Germany", which outlined a high-tech strategic initiative called Industry 4.0. This vision preceded the Society 5.0, as proposed in 2016 by the 5th Science and Technology Basic Plan. The goals of Industry 4.0 have been outlined in the German Federal Government's High-Tech Strategy 2020 Action Plan for Germany, the German equivalent of Japan's Science and Technology Basic Plan. Compared to Society 5.0 (outlined by the 5th Science and Technology Basic Plan), Industry 4.0 shares some common goals. Both paradigms focus on the use of technology, including IoT, AI, and Big Data Analytics. Similarly, both involve a top-down, state-driven approach with collaboration between industry, academia, and the government sector. However, there are some differences. Industry 4.0 advocates smart factories, while Society 5.0 calls for a super smart society. Although both paradigms support the implementation of cyber-physical systems (CPS), the scope of implementation differs. In Industry 4.0, CPS is implemented in the manufacturing environment, while in Society 5.0, it is deployed throughout the company. Further differences are found in the measurement of outcomes. Industry 4.0 aims to create new value and minimize production costs. The outcomes in Industry 4.0 allow relatively simple and clear performance metrics. In contrast, Society 5.0 aims to create a super smart society where metrics are much more complex. According to Comprehensive Strategy on Science, Technology, and Innovation for 2017, success should be measured by how much society can *balance economic advancement with the resolution of social problems by providing goods and services that granularly address manifold latent needs regardless of locale, age, sex, or language to ensure that all citizens can lead high-quality, lives full of comfort and vitality* (Cabinet Office, 2017). Technological innovations can have different future effects. Industry 4.0 is based on a manufacturing-centered industrial revolution but does not provide guidance on the impacts it has on the public. Society 5.0 puts people at the center and thus focuses strongly on the impact that technology has on people to create a better society. In this case, Society 5.0 provides guidance through a scenario of reform aimed at generating an inclusive society that meets people's different needs and preferences. Another important difference is that Industry 4.0 has emphasized which technologies to implement. Thus, there is a choice of which technologies to implement to achieve certain predetermined goals. In contrast, in Society 5.0 there is no choice about the type of technology, effectively forcing the company to implement AI-based technology that can connect physical space with virtual space. To sum up, in Industry 4.0 the generation of knowledge and intelligence is done by humans with the help of technology; in Society 5.0 the generation of knowledge and intelligence will come from machines through AI at the service of people. Figure 1.5 shows the main principles of Industry 4.0 and Society 5.0.

The problems that countries face affect the dimensions of sustainability (environmental, social, and economic) in a very complex way, as improving one dimension can often undermine the others. For example, if social spending is stitched back together, it would benefit the nation fiscally, but would bring serious problems in medical and healthcare settings. Consequently, to solve the problem of balancing dimensions and creating a people-centered society, clarifying the target metrics of this type of society and the roles that policy and technology should play in achieving them is a necessary action. Industry 4.0 has a vision of smart factories; therefore, it has identified the manufacturing sector as the main physical space (real world). Instead, it identified the CPS-centered cyber architecture as cyberspace; here information is integrated horizontally across different sectors and vertically within production systems. As evolution, the physical space (real world) that Society 5.0 identifies is society; regarding cyberspace, the CPS-centered cyber architecture is identified, as per the vision of Industry 4.0. However, in Society 5.0, information is integrated horizontally across service sectors

	INDUSTY 4.0 (Germany)	SOCIETY 5.0 (Japan)
Design	• High-Tech Strategy 2020 Action Plan for Germany (BMBF, 2011) • Recommendations for implementing the strategic initiative Industry 4.0 (Industrie 4.0 Working Group, 2013)	• 5th Science and Technology Basic Plan (released 2016) • Comprehensive Strategy on Science, Technology and Innovation for 2017 (released 2017)
Objectives and Scope	• Smart factories • Focuses on manufacturing	• Super-smart society • Society as a whole
Key issues	• Cyber-physical systems (CPS) • Internet of Things (IoT) • Mass customization	• High-level convergence of cyberspace and physical space • Balancing economic development with resolution of social issues • People-Centric society

Fig. 1.5 Industry 4.0 vs. Society 5.0.

(e.g., energy, transportation) and vertically within systems that track the history and attributes of each service user (such as medical information, consumption behavior, and educational history). In this context, strong and robust information security must be ensured to enable its use. Both Society 5.0 and Industry 4.0 reflect Japan's and Germany's responses to global initiatives, and both make a statement to the international community. In addition, both paradigms aim to integrate information across industries or sectors, and both address the same challenges, seeking to achieve a global cyber architecture as a secure environment for creative activities. To realize the people-centric society vision of Society 5.0, the needs of society must be balanced with the needs of the individual and how the balancing is implemented. The balancing must be optimal, and if this issue is not solved, no progress can be made. In this context, policy and technology must coordinate with each other; in this way everyone understands how each policy proposal or technology development fits into and contributes to Society 5.0. Otherwise, technology will advance independently of policy and policy will proceed in an uncoordinated way with technology, without understanding how technology and policy fit into the larger framework of Society 5.0. In other words, Society 5.0 revolutionizes industry through technology integration as it does with Industry 4.0, but it also seeks to revolutionize public living spaces or people's habits. Further progress needs to be made in promoting the initiatives implemented in smart cities. In addition, the policies needed to optimize society and solve social problems must be properly connected to the technology needed to provide high-quality social services that enable the public to live comfortable lives.

3. Technological Pillar of Super Smart Society

Society 5.0 defines technology as a key element in the integration of the physical and virtual worlds. Although the goals of Society 5.0 have not yet been fully achieved, citizens' lives are already characterized by the synergy between the real world and the digital world. This is true in both the private and professional dimensions, with the use of smartphones, tablets, PC, etc. Entertainment, payments, health, everything is increasingly passing through digital devices, and with the pandemic, the trend has strengthened further, making habits and lifestyles evolve. The increase in the use of devices is so significant that the synergy between real and virtual world has become very strong and it is difficult to establish a boundary between the two worlds. Considering the technologies that enable integration between the physical and virtual worlds, it is useful to focus attention on three key technological categories:

1. The device that produces data.
2. The connectivity that enables the transmission of data.

3. The digital infrastructure, such as storage and computational capacity, that are built to handle huge amounts of data.

It is important to emphasize a common trait of these technology categories, which is their relationship to the data that are produced annually and their availability in high volumes that have never been recorded before. The growth of data produced by devices is set to increase, and the recent report Global DataSphere Forecast, 2021–2025 (Reinsel and Rydning, 2021) showed that growth in the volume of data produced annually between 2021 and 2022 occurred at a compound annual rate (CAGR) of 23%. In Fig. 1.6, the volume of data created and replicated globally with time interval between 2010 and 2025 is represented and measured in zettabytes.

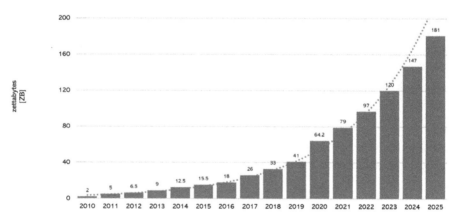

Volume of Data Created and Replicated Globally

Fig. 1.6 Volume of data created and replicated globally, 2010–2025. (Graphic re-working of the authors) (Source: International Data Corporation, 2022).

3.1 Device proliferation and data production

One reason why the explosion in data generation has occurred is the increasing prevalence of suitable tools to produce it. The tools involve society in terms of private use and professional use. There is not always a boundary between private and professional use. In fact, some devices are suitable for joint private and professional use. For example, smartphones, tablets, and PC can be used indiscriminately for both private and professional activities. Going into device details, IoT-based devices (wearables, connected home appliances, and virtual assistants) and non-IoT-based devices (smartphones, tablets, and PC) are considered. Despite the wide use of non-IoT devices, IoT-based devices have reached and surpassed non-IoT devices as of 2019, and their diffusion is increasingly massive (Wang, 2020), as shown in Fig. 1.7.

This proliferation of devices has a positive impact on the social dimension of sustainability. However, it raises important environmental issues. In fact, the growth of devices has important implications on the environmental footprint, with reference to the lifecycle of the device: from the extraction of raw materials that in some cases very valuable, the processing of raw materials, the energy consumption to produce them and to keep them in operation, until the end-of-life management.

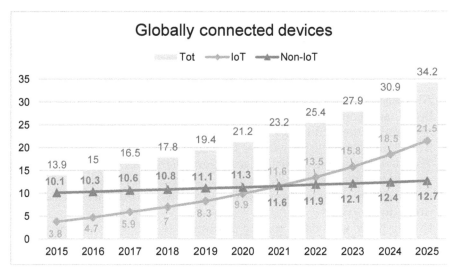

Fig. 1.7 Total number of devices connected globally, 2015–2025 (Graphic reworking of the authors) (Source: Indeema, 2022)

3.2 *Connectivity for data transmission*

Connectivity is another crucial aspect that enables the construction of the Smart Society. Connectivity is the possibility that devices communicate with each other and with data centers to allow data to be exchanged with performance consistent with their use. The theme of connectivity can be declined in two directions: coverage and performance of broadband lines. Considering the mobile coverage, according to The Mobile Report 2022 (GMSA, 2022) the penetration rate in 2021 was 67% of the world's population (percentage that equals 5.3 billion mobile subscribers), and it is expected to grow to 70% in 2025 which means it will reach 5.7 billion subscribers. In terms of performance, the next five years will coincide with the emergence of 5G technology. This technology will be expected to account for at least a quarter of connections in 2025 worldwide. In addition, peaks close to or above 50% are expected in Europe, the United States, and China. The deployment of 5G technology will also have important implications in terms of connection speed. This is relevant in all areas where low latency is of paramount importance, such as autonomous driving, robotics, etc. (Siriwardhana et al., 2021). Mobile connectivity is now pervasive, spreading in a penetrating way, to prevail and dominate in society. It is also used to give connectivity to the population and businesses especially in sparsely populated areas. Other telecommunications technologies are undergoing expansive phases and concern, e.g., satellite technologies, terrestrial fiber optics, submarine cables, etc. These technologies can improve society from a technological point of view, ensuring process control in the industrial environment but also improving the quality of life in the daily reality of people. The ITU/UNESCO Broadband Commission (Citaristi, 2022), the ITU has set goals to be achieved by 2025 to ensure greater global broadband deployment (del Portillo et al., 2021). These goals are as follows:

- All countries will need to have a National Plan for the adoption of broadband connectivity. Therefore, policy will have to devote special attention to the technology.
- Broadband connectivity will need to achieve 75% penetration globally. This percentage will be distributed appropriately according to the type of country in which it will be implemented. Therefore, it will have to be 65% in developing countries and 35% in countries considered underdeveloped.
- Pricing is a key issue to pay attention to. To enable broadband connectivity services, prices for basic services will have to remain cheap and not exceed 2% of monthly Gross National Income (GNI) in developing countries.
- Some small and medium-sized enterprises (SMEs) are still not well connected. Therefore, the lack of connection of more than 50% of SMEs by sector will have to be solved.
- In relation to the SDGs, gender equality should be ensured on all targets under the UN plan.
- 60% of youth and adults will need to have a minimum level of digital proficiency. Digital technologies interact and relate to many transversals and widely applicable skills. For example, how we build knowledge by analyzing information; how we become social actors and express our opinions; how we develop and progress as individuals. The development of digital proficiency is a part of this broader mosaic of skills development, and it is increasingly important.
- 40% of the world's population will have to use digital financial services. Digital financial services have the potential to reduce costs by maximizing economies of scale, to increase the speed, security, and

transparency of transactions, and to enable more personalized financial services that serve the poor.

3.3 *Digital infrastructures for data management*

For devices to be used by taking advantage of broadband connection services, an infrastructure network is required. In terms of digital infrastructure, several phenomena are happening simultaneously. Regarding super computers (HPC - High Performance Computing), after a period of strong growth, 2021 was a year of relative stability. However, it is a prelude to the unfolding of the European strategy on pre-exascale HPCs and the forthcoming commissioning of the first exascale-class HPCs. Under the EU initiative, eight HPCs are being developed, shown in Fig. 1.8.

In fact, the HPC computing power of China, the United States, and Japan has remained just over 70% of the global HPC power, with Italy maintaining the 11th position, thanks to the stable presence of ENI's HPC steadily in the global top ten. Computational capacity is the basis of an efficient infrastructure. Enterprises are increasingly faced with the need to manage large amounts of data, often distributed in very different domains, such as internal and third-party managed datacenters, Remote Office Branches, Public Clouds, and terminals, as shown in Fig. 1.9.

Therefore, the need to provide greater computational capacity leads to continued investment in IT infrastructure (including more traditional "on premise" infrastructure) with particularly strong growth in the "as a Service" model. In particular, the growth of the Cloud in its various forms continues to be significant, both in Infrastructure as a Service (IaaS) and Platform as a Service (PaaS). In IaaS, the effect of scale combined with the ability to use multiple cloud providers has become the main driver. While in the case of PaaS, innovation in services is the key driver to sustain its

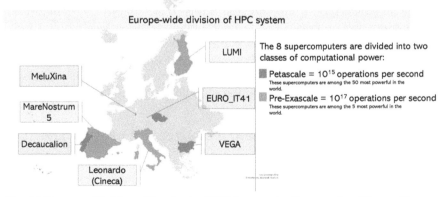

Fig. 1.8 European Division of Developing HPC Systems, by Computational Power Classes.

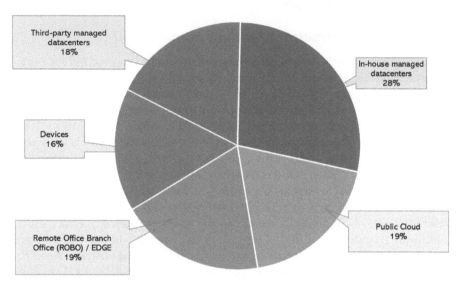

Distribution of Data among Various Storage Systems

Fig. 1.9 Percentage distribution of data at the enterprise level, between the various storage systems. (Graphic elaboration of the authors.) (Source: IDC Cloud Data Storage & Infrastructure Trends Survey, Seagate Technology, 2021)

growth. Finally, a point should also be made about Edge Computing. Edge Computing is a distributed computing model in which data processing occurs as close as possible to where the data is generated, improving response times, and saving on bandwidth. Therefore, the rise of IoT devices will result in the greatest amount of data (which is equivalent to 75% in 2025), being generated outside of large, centralized datacenter (Gartner, 2021). Figure 1.10 shows general data from enterprises by place of creation and processing between 2018 and 2025.

The explosion of data and the tools that enable its generation, transmission, and processing underlie the many technological applications that enable the development of a data-driven society: from precision medicine to domestic or industrial robotics, from smart grids to autonomous driving, from smart cities to decentralized finance to digital twins, all examples that put together outline the so-called Super Smart Society.

4. The Fundamental Role and Requirement of Artificial Intelligence

The strategy for AI is based on trust as a precondition for ensuring an anthropocentric approach of Society 5.0. AI is not a technology for its own

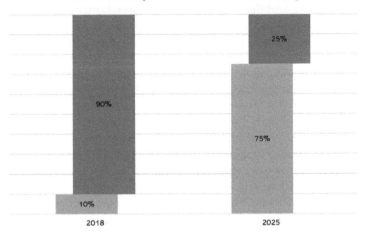

Data Generated by Place of Creation and Processing

■ Non-IoT ■ IoT

Fig. 1.10 General data from enterprises by place of creation and processing (percentage values), 2018 and 2025 (Graphic elaboration of the authors) (Source: Gartner, 2021).

sake but is a tool in the service of people whose goal is to improve the well-being of human beings. Therefore, it is necessary to ensure the reliability of AI. The values on which our societies are based must be fully integrated into how AI is developed. Today's society, and western culture, particularly, is founded on the values of respect for human dignity, freedom, democracy, equality, the rule of law and respect for human rights, including the rights of people belonging to minorities. These values are common in a society characterized by pluralism, nondiscrimination, tolerance, justice, solidarity and equality between women and men. With reference to the European Union, it has a solid regulatory framework designed to become the international standard for anthropocentric AI. The General Data Protection Regulation ensures a high level of protection for personal data and provides for the implementation of measures to ensure data protection by design and by default (Regulation (EU) 2016/679). The regulation removes barriers to the free movement of non-personal data and ensures the processing of all categories of data anywhere in Europe. The cybersecurity regulation contributes to building confidence in the online world, and the proposed electronic privacy (e-privacy) regulation also has the same goal. In general, AI poses new challenges because it enables machines to learn, make decisions, and execute them without human intervention. Before long, this mode of operation will become the norm for many kinds of goods and services, from smartphones to automated and self-driving cars, robots, and online applications. However, the decisions made by the algorithms could be based on incomplete and therefore unreliable data, tampered with because of cyberattacks, tainted by bias, or simply incorrect. Therefore, uncritically applying technology as it is developed would lead to problematic results and reluctance on the part of citizens to accept or use it. On the contrary, AI technology should be developed in a way that puts the human being at the center and allows it to win the public's trust. Accordingly, AI applications should comply with the law, observe ethical principles, and ensure that their practical implementations do not result in unintended harm. Diversity in terms of gender, race or ethnic origin, religion or belief, disability and age should be ensured at every stage of AI development. AI applications should empower people and respect their fundamental rights; they should aim to empower citizens and enable access for people with disabilities as well. Thus, there is a need to develop ethical guidelines based on the existing regulatory framework and which should be applied by AI developers, providers, and users in the internal market, establishing an ethical level playing field in all countries. Therefore, a set of recommendations for a broader AI policy is appropriate. The basic requirements for reliable AI are such as to encourage its application, establishing the right environment of trust for effective AI development and use. In Fig. 1.11, the requirements are shown graphically.

Fig. 1.11 The basic requirements for reliable AI in Society 5.0.

4.1 *Human intervention and surveillance*

AI systems must help people make better and more informed choices in pursuit of their goals. They must promote the development of a thriving and equitable society by supporting human intervention and fundamental rights. Therefore, they must not reduce, limit, or mislead human autonomy. The general welfare of the user must always be central to the functionality of the system. Human surveillance helps ensure that AI systems do not endanger human autonomy or cause other adverse effects. Depending on the specific artificial intelligence-based system and its area of application, an appropriate level of control measures should be provided. These include the adaptability, accuracy, and explanation of such systems. Surveillance can be carried out through governance mechanisms that ensure that a human-intervention ("human-in-the-loop"), human-supervised ("human-on-the-loop") or human-controlled ("human-in-command") approach is adopted.

4.2 *Technical strength and safety*

For AI to be reliable, the algorithms must necessarily be secure, reliable, and robust enough to cope with errors or inconsistencies during all phases of the AI system lifecycle. In addition, algorithms must be adequately capable of handling wrong results. AI systems must also be resilient to both overt attacks and more devious attempts to manipulate data or algorithms and must ensure that a contingency plan is in place in case of problems. Their decisions must be accurate, or at least correctly reflect their level of accuracy, and their results must be reproducible. In addition, AI systems should contain safety mechanisms from the design stage to ensure that they are verifiably secure at every stage, especially considering the physical and mental safety of all people involved. This also includes minimizing and making reversible unintended effects or errors in system operation whenever possible. There should be processes in place to clarify and assess the potential risks associated with the use of AI systems in various application areas.

4.3 *Confidentiality and data governance*

Confidentiality and data protection must be ensured at all stages of the lifecycle of AI systems. Digital records of human behavior can enable AI systems to infer individuals' preferences, age and gender, sexual orientation, and religious or political beliefs. To enable people to have confidence in data processing, it is necessary to ensure that people have full control over their personal data and to ensure that data about them will not be used to harm or discriminate against them. In addition to safeguarding confidentiality and personal data, requirements must be met to ensure that AI systems are of high quality. The quality of the data sets used is critical to the performance of AI systems. With collecting data, these may reflect social conditioning or contain inaccuracies, errors, and material defects. This must be resolved before using any data set to train an AI system. In addition, the integrity of the data must be guaranteed. The processes and datasets used shall be tested and documented at all stages, such as planning, training, testing, and dissemination. This should also apply to AI systems that were not developed in-house but acquired elsewhere. Finally, access to data must be properly regulated and controlled.

4.4 Transparency

Traceability of AI systems should be ensured. It is important to record and document the decisions made by the systems and the whole process leading to the decisions, including a description of the data collection, and labeling and a description of the algorithm used. In this context, the explainability of the algorithms' decision-making process adapted to the people involved should be provided. The ongoing research to develop explainability mechanisms should be continued. Explanations on the extent to which AI system influences and defines the organizational decision-making, system design choices, and the logic behind its deployment should also be available. This ensures transparency not only of data and systems, but also of business models. Finally, it is important to communicate appropriately and in a manner pertinent to the case at hand, the capabilities, and limitations of the AI system to the various stakeholders involved. AI systems should also be identifiable as such, allowing users to know that they are interacting with an AI system and can identify the people responsible for it.

4.5 Diversity, non-discrimination, and fairness

The datasets used by AI systems for training and operation may be affected by inadvertent historical conditionings, incompleteness, and unsuitable governance models. If such conditionings are maintained, they could lead to direct and indirect discrimination. Damage can also result from the deliberate exploitation of consumer bias or unfair competition. The way AI systems are developed, such as the way an algorithm's programming code is written, can also be affected by conditioning. These concerns should be addressed at an early stage in the development of the system. The creation of diversified design groups and the establishment of mechanisms to ensure the participation, particularly, of citizens, in the development of AI may also be useful to overcome them. It is advisable to consult stakeholders who may be directly or indirectly affected by the system during its lifecycle. AI systems should consider and include the full range of human skills, competencies, and needs and ensure accessibility through a universal design approach that strives for equal access for all people.

4.6 Social and environmental welfare

To obtain a reliable AI, its impact on the environment should be considered. Ideally, all human (including future generations) beings should be able to benefit from biodiversity and a habitable environment. Therefore, the sustainability and ecological responsibility of AI systems should be encouraged. The same applies to AI solutions that address global challenges, such as the UN SDGs. The impact of AI systems should be considered not only from an individual perspective, but also from the perspective of whole society, including policy. AI systems can be used to improve social skills and in situations affecting the democratic process, including opinion-forming, political decision-making processes or electoral contexts.

4.7 Accountability

Accountability is the responsibility of administrators who employ public financial resources to report on their use of both the regularity of accounts and the effectiveness of management. Therefore, mechanisms ensuring the responsibility (in the sense of having to act) and accountability (that is to be accountable for the action done or caused to be done, to respond for the results obtained) of AI systems and their outcomes should be provided for both before and after their implementation. The verifiability of AI systems is critical in this context, as the evaluation of AI systems by internal and external auditors and the availability of such evaluation reports contribute greatly to the reliability of the technology. External verifiability should be ensured especially for applications that affect fundamental rights, including security-critical applications. Potential negative impacts of AI systems should be identified, evaluated, documented, and minimized. This process is facilitated using impact assessments, which should be proportionate to the magnitude of the risks posed by AI systems. Contrasts between requirements (which are often unavoidable) should be addressed in a rational and methodological way and considered. Finally, in case of unfair negative impact, accessible mechanisms should be provided to ensure adequate means of redress.

5. Conclusions

Although in human history many innovations have significantly marked the habits of citizens, the contemporary era has been characterized by an overwhelming digital revolution. As discussed in this chapter, the application areas of the most modern technologies have changed many ways of doing things in industry, as well as in the civil, energy, financial, and healthcare sectors. Technological development has brought about a degree of change that has led to a real new society, called Society 5.0, or Smart Society. The classification of innovation brought about by technological development is of a destructive nature. In fact, if in previous years the use of technology was conceived as a choice of the user who, depending on the context, could choose whether to use a technological support, today man no longer has this choice. Indeed, every human being is embedded in a society made up of digital information and is an integral part of it often without being a decision-maker in any process. Society 5.0 will be a society based on people's needs addressed and solved through technology, in line with the SDGs set by the UN in Agenda 2030. Despite the progressive aging of the population of industrialized countries, the European community seems well prepared to embrace this paradigm shift. Instead, a little less ready are Member Countries in terms of connectivity infrastructure, digital public services, human capital, and technical and technological skills to support this digital revolution. All policy choices implemented by the European Community and Member Countries in the policy plans that will affect the coming years include an extraordinary number of resources put in place to support the paradigm shift toward a super smart society, capable of simultaneously pursuing economic development, social and environmental protection goals. Soon, national governments will have to question the needs of the community and define their priority for investment and implementation. This will be followed by a period of social adaptation to new technological and digital applications. Moreover, as early as 2025, it will be possible to collect the most significant data and capture indicators to understand the results achieved by the implementation of digital development plans, as well as devise new lines of evolution to address the challenges of the coming decades.

References

Bibri, S.E. and Krogstie, J. (2017). Smart sustainable cities of the future: An extensive interdisciplinary literature review. Sustain. Cities Soc., 31: 183–212. https://doi.org/10.1016/j.scs.2017.02.016.

Cabinet Office (Council for Science, Technology, and Innovation). (2016a). The 5th Science and Technology Basic Plan (released on 22 January 2016). https://www8.cao.go.jp/cstp/english/basic/5thbasicplan.pdf. (Accessed 10 July 2022)

Cabinet Office (Council for Science, Technology, and Innovation). (2016b). Comprehensive Strategy on Science, Technology, and Innovation (STI) for 2016 (released on 24 May 2016). https://www8.cao.go.jp/cstp/sogosenryaku/2016.html. (Accessed 10 July 2022])https://www8.cao.go.jp/cstp/english/doc/2016stistrategy_summary.pdf (Summarized English version). (Accessed 10 July 2022)

Cabinet Office (Council for Science, Technology, and Innovation). (2017). Comprehensive Strategy on Science, Technology, and Innovation (STI) for 2017 (released on 2 June 2017), p. 2. https://www8.cao.go.jp/cstp/english/doc/2017stistrategy_main.pdf. (Accessed 10 July 2022)

Citaristi, I. (2022). International Telecommunication Union—ITU. In: The Europa Directory of International Organizations 2022. Routledge, pp. 365–369.

del Portillo, I., Eiskowitz, S., Crawley, E.F. and Cameron, B.G. (2021). Connecting the other half: Exploring options for the 50% of the population unconnected to the internet. Telecommunications Policy, 45(3): 102092. https://doi.org/10.1016/j.telpol.2020.102092.

Falanga, R., Verheij, J. and Bina, O. (2021). Green(er) Cities and Their Citizens: Insights from the Participatory Budget of Lisbon. Sustainability, 13: 8243. https://doi.org/10.3390/su13158243.

Gartner. (2021). Gartner identifies key emerging technologies spurring innovation through trust, growth, and change (23 August 2021). Available at https://www.gartner.com/en/

newsroom/press-releases/2021-08-23-gartner-identifies-key-emerging-technologies-spurring-innovation-through-trust-growth-and-change. (Accessed 10 July 2022)

GMS Association (GMSA). (2022). The Mobile Economy 2022. Available at https://www.gsma.com/mobileeconomy/wp-content/uploads/2022/02/280222-The-Mobile-Economy-2022.pdf. (Accessed 10 July 2022)

Gurjanov, A.V., Zakoldaev, D.A., Shukalov, A.V. and Zharinov, I.O. (2020). The smart city technology in the super-intellectual Society 5.0. J. Physics: Conf. Ser., 1679. https://doi.org/10.1088/1742-6596/1679/3/032029.

Haque, A.K.M.B., Bhushan, B. and Dhiman, G. (2021). Conceptualizing smart city applications: Requirements, architecture, security issues, and emerging trends. Expert Syst., 1: 23. https://doi.org/10.1111/exsy.12753.

Israilidis, J., Odusanya, K. and Mazhar, M.U. (2019). Exploring knowledge management perspectives in smart city research: A review and future research agenda. Int. J. Inf. Manag., 56: 101989. https://doi.org/10.1016/j.ijinfomgt.2019.07.015.

Masson-Delmotte, V., Zhai, P., Pirani, A., Connors, S.L., Péan, C., Berger, S. ... and Zhou, B. (2021). IPCC, 2021: Climate Change 2021: The Physical Science Basis. Contribution of Working Group I to the Sixth Assessment Report of the Intergovernmental Panel on Climate Change Cambridge University Press, Cambridge, United Kingdom and New York, NY, USA, https://doi.org/10.1017/9781009157896.

Nunes, S.A., Ferreira, F.A., Govindan, K. and Pereira, L.F. (2021). "Cities go smart!": A system dynamics-based approach to smart city conceptualization. J. Clean. Prod., 313: 127683. https://doi.org/10.1016/j.jclepro.2021.127683.

Palumbo, R., Manesh, M.F., Pellegrini, M.M., Caputo, A. and Flamini, G. (2021). Organizing a sustainable smart urban ecosystem: Perspectives and insights from a bibliometric analysis and literature review. J. Clean. Prod., 297: 126622. https://doi.org/10.1016/j.jclepro.2021.126622.

Reinsel, D. and Rydning, J. (2021). Worldwide Global StorageSphere Forecast, 2021–2025: To Save or Not to Save Data, That Is the Question. IDC Market Forecast US47509621 International Data Corporation (IDC), Framingham, Massachusetts, March 2021.

Siriwardhana, Y., Porambage, P., Liyanage, M. and Ylianttila, M. (2021). A survey on mobile augmented reality with 5G mobile edge computing: Architectures, applications, and technical aspects. IEEE Communications Surveys & Tutorials, 23(2): 1160–1192. https://doi.org/10.1109/COMST.2021.3061981.

United Nations. (2015). Draft outcome document of the United Nations summit for the adoption of the post-2015 development agenda. In: Draft Resolution submitted by the President of the General Assembly, Sixty-Ninth Session, Agenda Items 13(a) and 115, A/69/L.85, United Nations, New York, NY, USA.

Wang P. (2020). IoT Service Recommendation Scheme Based on Matter Diffusion, IEEE Access, 8: 51500–51509. https://doi.org/10.1109/ACCESS.2020.2979777.

Yigitcanlar, T. and Cugurullo, F. (2020). The Sustainability of Artificial Intelligence: An Urbanistic Viewpoint from the Lens of Smart and Sustainable Cities. Sustainability, 12: 8548. https://doi.org/10.3390/su12208548.

Yigitcanlar, T., Corchado, J., Mehmood, R., Li, R., Mossberger, K. and Desouza, K. (2021). Responsible Urban Innovation with Local Government Artificial Intelligence (AI): A Conceptual Framework and Research Agenda. J. Open Innov. Technol. Mark. Complex., 7: 1–16. https://doi.org/10.3390/joitmc7010071.

Yigitcanlar, T., DeSouza, K.C., Butler, L. and Roozkhosh, F. (2020). Contributions and Risks of Artificial Intelligence (AI) in Building Smarter Cities: Insights from a Systematic Review of the Literature. Energies, 13: 1473. https://doi.org/10.3390/en13061473.

Yigitcanlar, T., Foth, M. and Kamruzzaman, M. (2017). Towards Post-Anthropocentric Cities: Reconceptualizing Smart Cities to Evade Urban Ecocide. J. Urban Technol., 26: 147–152. https://doi.org/10.1080/10630732.2018.1524249.

Yigitcanlar, T., Kamruzzaman, Buys, L., Ioppolo, G., Sabatini-Marques, J., da Costa, E.M. and Yun, J.J. (2018). Understanding 'smart cities': Intertwining development drivers with desired outcomes in a multidimensional framework. Cities, 81: 145–160. https://doi.org/10.1016/j.cities.2018.04.003.

Yigitcanlar, T., Mehmood, R. and Corchado, J.M. (2021). Green Artificial Intelligence: Towards an Efficient, Sustainable and Equitable Technology for Smart Cities and Futures. Sustainability, 13: 8952. https://doi.org/10.3390/su13168952.

Žižek, S.Š., Mulej, M. and Potočnik, A. (2021). The sustainable socially responsible society: Well-being society 6.0. Sustainability, 13: 9186. https://doi.org/10.3390/su13169186.

2

Multi-Criteria Decision-Making Methods Applied to the Sustainability of Urban Transport

A Systematic Literature Review

David Ruiz Bargueño,[1] *Fernando Augusto Silva Marins,*[1]
José Antônio Perrella Balestieri,[1]
Pedro Ivan Palominos Belmar,[2] *Rubens Alves Dias*[1] and
Valerio Antonio Pamplona Salomon[1,*]

1. Introduction

The urbanization of the world, as we know it, has two main stages: The first stage began with the Industrial Revolution back in the 1760s, and the second stage after World War II in the 1940s. Economic and political processes mainly caused the growth and consolidation of cities. Despite their rapid growth, local or national governments were not so fast to respond to these

---

[1] Graduate Program in Engineering, School of Engineering and Sciences, Universidade Estadual Paulista, UNESP, Guaratinguetá, SP, 12516-410, Brazil.
[2] Department of Industrial Engineering, School of Engineering, Universidad de Santiago de Chile, Santiago, 27183001, Chile.
* Corresponding author: valerio.salomon@unesp.br

new realities. To make it worse, cities continued to grow, intensifying the mobility and transport needs of their citizens (Ruiz Bargueño et al., 2021).

The structure of the city had a diversified, poly-nuclear layout, with high rates of environmental degradation and internal insecurity. Consequently, all this resulted in the disorder of urban space, the deterioration of the architecture and public space, and the weakening of links between communities (Ascher, 2001). In the 21st century, urban transport has been challenged by the financial problems of most cities, the socialization of environmental problems, major traffic jams, and the humanization of urban management (Dangond Gibsone et al., 2011).

Urban mobility is the result of the specific characteristics of everyone and the strategies aimed at universal accessibility. This result demands coordinated actions and integration between transport and land use (Portugal, 2017). The lack of a culture of organizational learning in cities seems to be the most critical barrier for cities to learn from each other (May et al., 2017). There is a surprising scarcity of research on good practices in implementation, whether about public engagement, detailed design, or rapid response to problems (May, 2015).

Gradually, it is happening a shift from one approach in terms of transport to another in terms of people's mobility. Cities are dynamic bodies. They are continuing to grow, so is the demand for mobility of their citizens (Gebhardt et al., 2016). There may be similar places and people may face similar problems, but cities will never be the same, not even within the same country.

Nowadays, regarding the sustainability of the cities and their environment, urbanization affects more than half of the world's population, generating more than 85% of the global Gross Domestic Product (Ellen MacArthur Foundation 2019, United Nations 2019). Withal, cities exploit at least 75% of the natural resources (United Nations Environment Programme, 2017) and they produce 50% of global waste (Climate-KIC Nordic and C40 Cities, 2018). Urban areas are formed and inhabited by nearly 55% of the world's population but cover an area of just 2% of the total surface of the planet. The percentage of the population living in cities is expected to increase to 68% by 2050 (United Nations, 2014).

Citizens, policymakers, and planners perceive the negative repercussions of the linear and traditional urban approach (European Commission, 2020; Milios, 2018). In this regard, the United Nations has declared that the circularity of urban metabolism driven by digital technologies is a necessary tool to achieve the Sustainable Development Goals (SDGs) of the 2030 Agenda (Ciliberto et al., 2021; De Pascale et al., 2021; United Nations, 2015). Cities must strengthen the sustainable management of resources, facilitating the conservation, regeneration, restoration, and resilience of the ecosystem in the face of new and emerging challenges (United Nations, 2017). Urban metabolism constitutes a plurality of socio-technical-ecological processes and operations, which give rise to a multitude of flows of people, energy, byproducts, waste, emissions, raw materials, data, and information, among others that circulate within the city and the surrounding urban environment (Currie and Musango, 2016).

The design, installation, formatting, implementation, and remodeling of digital technologies must be conducted by considering the economic, social, and environmental dimensions of urban metabolism and at the same time proactively involving stakeholders in urban circularity initiatives (D'Amico et al., 2022). Then, it is necessary to support regulatory structures and greater multimodal integration for continuous and sustainable urban mobility (Venkat et al., 2019). The difficulties in achieving the expected results in urban mobility policies would not be so much due to the lack of technical tools, but to the inability to approve, legitimize, and implement policies by challenging the values of efficiency and individualism or prevailing supporters in many municipal governments (May, 2015).

According to its international society, multiple criteria decision-making (MCDM) can be defined as the study of methods and procedures by which concerns about multiple conflicting criteria can be formally incorporated into the management planning process (International Society of Multiple Criteria Decision-Making, 2022). In this sense, MCDM methods can be applied to the solution of common demands among municipalities, evaluating alternative plans for urban mobility (Ruiz Bargueño et al., 2021).

The main objectives of applying MCDM methods are the classification and selection of alternatives to support decision-making based on mathematical formulation and systematic methodology (Li et al., 2020; Lombardi Netto et al., 2021; Wallenius et al., 2008). The development of MCDM methods has been initially motivated by the variety of practical problems that requires consideration of various factors. MCDM continues to be developed providing modern decision-making techniques with advances in mathematical models and computational technology (Behzadian et al., 2010; Oliveira et al., 2022). There are dozens of methods for MCDM (Saaty and Ergu, 2015), including, alphabetically, to name a few:

- Analytic Hierarchy Process (AHP).
- Analytic Network Process (ANP).
- Data Envelopment Analysis (DEA).
- Fuzzy Sets Theory (FST).
- Multi-Attribute Utility Theory (MAUT).
- Potentially All Pairwise Rankings of All Possible Alternatives (PAPRIKA).
- Technique for Order of Preference by Similarity to Ideal Solution (TOPSIS).

This chapter presents a systematic literature review (SLR) of urban transport and MCDM. SLR is a research method to obtain a grouping and analysis of results on a given area of knowledge in a given time (Egger and Smith, 1998). The SLR better illustrates how the theme under study is been discussed and what are the main concepts, how they have been studied from different points of view, and how the field has evolved over time in order to synthesize it and identify theoretical gaps (Nakano and Muniz, 2018). The main research questions are: (Q1) How to characterize joint research in urban transport and MCDM? (Q2) What are the main trends and gaps for future research?

Figure 2.1 presents the research flow adopted in this chapter. The SLR, conducted in six phases including database searches, bibliometric study, and cluster analysis, is by the guideline proposed by Xiao and Watson, 2019.

This chapter is organized into four sections: Introduction, Bibliometrics, Results, and Conclusions. References are also provided after the Conclusions.

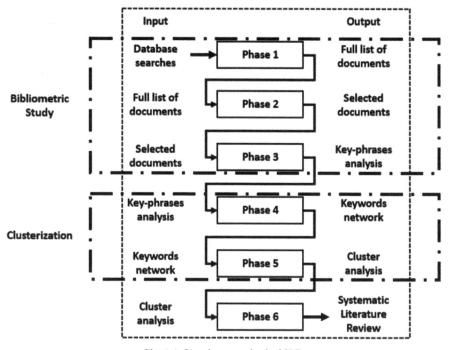

Fig. 2.1 Six-phase method of SLR.

2. Bibliometrics

The bibliometric analysis is an indispensable statistical tool to map the state of the art in each area of scientific knowledge (Oliveira et al., 2019). With bibliometrics, it is possible to identify essential information for various purposes, such as prospecting for research opportunities or support of scientific research.

Scopus is a source abstract and citation database curated by independent subject matter experts. Scopus delivers the most comprehensive overview of the world's research output in the fields of science, technology, medicine, social science, and arts and humanities (Elsevier B.V., 2022). Elsevier is also one of the leading publishers of international scientific journals.

The guidelines for searching the Scopus database were:

- Peer-reviewed articles.
- Articles in English, Portuguese, or Spanish languages.
- Full-text articles.
- Primary data extracted from the title, abstract, and keywords.
- Publication date between 2015 and 2022 or classic texts.
- Articles that meet the scope of the object searched.

The following query was applied: (TITLE-ABS-KEY (sustainab*) AND TITLE-ABS-KEY (city OR cities OR citize*) AND TITLE-ABS-KEY (urban AND mobility) OR TITLE-ABS-KEY (public AND transport*) AND TITLE-ABS-KEY (efficien*) AND TITLE-ABS-KEY (technolog*) OR TITLE-ABS-KEY (innovation) OR TITLE-ABS-KEY (optimization) OR TITLE-ABS-KEY (integrat*) OR TITLE-ABS-KEY (resource*) OR TITLE-ABS-KEY (environment*) OR TITLE-ABS-KEY (ecosystem*) OR TITLE-ABS-KEY (renewable*) AND TITLE-ABS-KEY (decision AND making OR decision-making OR mcdm) AND PUBYEAR > 2014 AND PUBYEAR < 2023 AND (LIMIT-TO (PUBSTAGE , "final")) AND (LIMIT-TO (LANGUAGE , "English")) AND (LIMIT-TO (SRCTYPE , "j")) AND (LIMIT-TO (DOCTYPE , "ar")) AND (LIMIT-TO (OA , "all")) AND (LIMIT-TO (SUBJAREA , "ENVI") OR LIMIT-TO (SUBJAREA , "SOCI") OR LIMIT-TO (SUBJAREA , "ENGI") OR LIMIT-TO (SUBJAREA , "ENER") OR LIMIT-TO (SUBJAREA , "MATH") OR LIMIT-TO (SUBJAREA , "ECON") OR LIMIT-TO (SUBJAREA, "COMP") OR LIMIT-TO (SUBJAREA , "BUSI")

Table 2.1 presents the keywords most found in the first search of the Scopus database.

Not all documents listed in Table 2.1 are aligned with the scope of this chapter. Only 43 documents were identified on MCDM and Urban Transport or Urban Mobility: 42 in the English language, 27 published in journals, and only 11 with open access (gold or green).

Figure 2.2 presents the distribution of the documents by subject area. Surprisingly, Social Sciences is the leading area, ahead of Decision Sciences, Engineering, and Environmental, all three tied for second place.

Table 2.1 Number of documents by keyword. (Source: Scopus database).

Keyword	Document
Sustainab*	570.397
City, cities, citizen*	51.721
Urban mobility	2.814
Public transport	4.989
Efficien*	502
Renew*	420
Ecosystem*	417
Environment*	413
Resource*	317
Integrat*	289
Optimization	221
Innovation	196
Technolog*	180

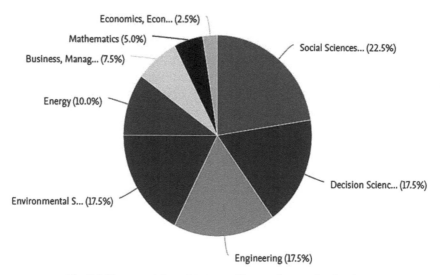

Fig. 2.2 Documents by subject area (Source: Scopus database).

The leading journals of publication and citations (with Web of Science's acronym) are:

- Ambiente e Sociedade.
- Decision Sciences Letters.
- Energies.
- Energy Policy (Energ. Policy).

- Entrepreneurship and Sustainability Issues.
- Gestao e Producao.
- International Journal of the Analytic Hierarchy Process (Int. J. AHP).
- Sustainability.
- Transport Policy (Transp. Policy).
- Urban Planning.

Figure 2.3 presents the Cite Score of the journals by year from 2011 to 2021. The journal`s Cite Score is a metric of the number of citations to recent articles published in that journal (Elsevier B.V., 2022). Energy Policy is the journal with the highest Cite Score, every year.

The next step is refining the list of documents accordingly to the scope of this chapter. It was done with SciVal, a platform for performance analysis of documents on the Scopus database (Universidade Estadual Paulista, 2022). On the others tools, SciVal performs a key-phrases analysis which consists of visual analysis of keywords' relevance. Figure 2.4 presents a cloud of keywords resulting from the SciVal application. The bigger the keyword's font, the higher its relevance, according to SciVal.

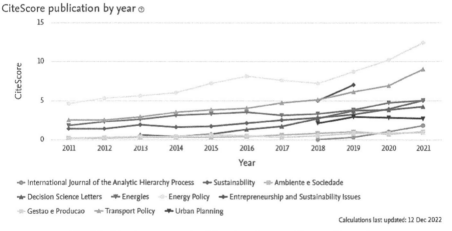

Fig. 2.3 Citation scores of publications by year (Source: Scopus).

VOSviewer is a software tool for constructing and visualizing bibliometric networks (Centrum voor Wetenschap en Technologische Studies, 2022). This software helps the identification of bibliometric networks from the collected data in the Scopus database. For this work, the settings for the VOSviewer application were:

- Map based on bibliographic data.
- Bibliographic database file: Scopus Elsevier (.csv).

A A A relevance of keyphrase | declining A A A growing (2012-2021)

Fig. 2.4 Cloud of keywords (Source: SciVal).

- Type of Analysis: Co-occurrence.
- Unit of Analysis: All keywords.
- Counting Method: Full counting.
- Minimum number of occurrences: 2.
- Number of keywords selected: 23 (as default).

Figure 2.5 presents the resulting five-cluster network of keywords.

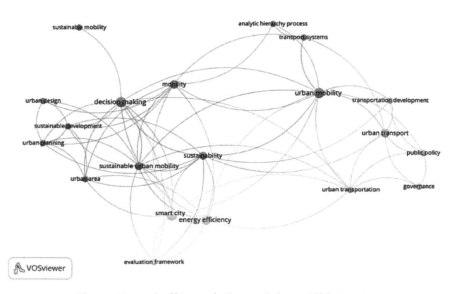

Fig. 2.5 Network of keywords (Source: Software VOSviewer).

The five clusters in the network generated by VOSviewer are identified by colors:

- Red Cluster with six keywords on Sustainability.
- Green Cluster with five keywords on Governance and Transport.
- Blue Cluster with three keywords on Urban Mobility but also including a keyword on AHP.
- Yellow Cluster with three keywords on Energy, Evaluation, and Smart City.
- Purple Cluster with two keywords on Decision-making and Sustainable Mobility.

The cluster of keywords are highly connected. There is no isolated cluster connecting with only another one or two. Each of the clusters connects itself with the other four. The only MCDM method associated with a keyword was the AHP.

After the cluster analysis and the key-phrases analysis, the list of 43 documents was reduced to the 28 documents presented in the next section.

3. Results

Table 2.2 presents the reduced set of documents resulting from key-phrase and cluster analyses. This table also includes the MCDM method. Matter of fact, Case Study (CS), Decision Support System (DSS), Delphi method (DM), Functional and Social Mix (FSM), Geographic Information System (GIS), Key Performance Indicators (KPI), Life Cycle Assessment (LCA), Literature Review (LR), Principal Components Analysis (PCA), Structural Equation Modeling (SEM), and Smartness and Sustainability Evaluation Tool (STILE) are not MCDM methods, but they are decision-making techniques or research methods adopted in some documents.

Next, some documents are commented on, starting with Alitaneh, 2019, who presented an attempt to solve various problems by the two factors of the mean and standard deviation of variables, introducing the coefficient of variation scales triangle of data as the best option for prioritization, scaling, pairwise comparison, and normalization of quantitative and qualitative variables. In this research, some 9-grade scales were proposed for scalar triangles with a new approach for the de-scalarization of quantitative variables (normalization).

Cestari et al. (2018) presented a public administration interoperability diagnosis method applying the AHP to calculate the capability of the entities to operate in collaborative or cooperative environments.

Dali et al. (2018) collected and analyzed the opinions of urban planners and the Maqasid al-Shariah scholars. Their aim was to develop an evaluation model identifying and ranking the Maqasid indicators and

Table 2.2 Key phrases and methods applied.

Document	Key Phrases	Methods
Alitaneh, 2019	Coefficient of variation, normalization, scales	AHP
Attard, Haklay and Capineri, 2016	Government, smart city, urban mobility	GIS
Bielińska-Dusza et al., 2021	Smart city, urban mobility	SEM
Bousdekis and Kardaras, 2020	Digital transformation, local government	AHP, FST, DM
Cestari, Loures and Santos, 2018	Capability, public administration	AHP
Colapinto et al., 2020	Environment, multifaceted development, sustainability	LR
Conejero and Silva César, 2017	Local productive arrangement	GIS
Dali et al., 2018	Cities, livability, prioritization	AHP
Degtiarev and Borisov, 2018	Expert assessment, prioritization	AHP, FST
Dhurkari and Swain, 2019	Innovation, partner selection, strategic alliances	AHP
García-Fuentes and De Torre, 2017	Smart city, urban mobility	STILE
García-Fuentes et al., 2021	Energy, smart city, urban mobility	STILE
Iannillo and Fasolino, 2021	Land-use, urban planning	FSM
Karasan, 2019	Investment, prioritization	AHP, FST
Keseru et al., 2016	Group decision-making, urban mobility	AHP
Lamata et al., 2018	Corporate social responsibility	AHP, TOPSIS
Mendoza et al., 2015	Solar energy, urban mobility, sustainability	LCA
Myrovali et al., 2020	Urban mobility, sustainability	DSS
Oswald Beiler and Philips, 2016	Pedestrian corridors, walkability	AHP
Pradhan et al., 2019	AHP, DEA, integration	LR
Qu 2016	Development, smart city	PCA
Quijano et al., 2022	Smart city, sustainability	KPI
Schiller and Xie, 2020	Cities, spatial planning, urban design	CS
Singh et al., 2019	Innovation, small enterprises, technology	AHP
Wang 2017	Smart city	AHP, FST
Xie and Wang, 2018	Bike sharing, urban mobility	CS
Yildiz et al., 2019	Environment design, sustainability, urban renew	AHP
Yilmaz and Ozkir, 2018	Consistency, pairwise comparisons	AHP

sub-indicators for livability and quality of life in cities of Malaysia. Maqasid al-Shariah is an Islamic legal doctrine; its specific aim was the preservation of five essentials of human well-being: religion, life, intellect, lineage, and property. The authors applied the AHP to create the framework and weights and, after that, they evaluated the calculation results with ATLAS. ti software (ATLAS.ti Scientific Software Development GmbH, 2022) and AHP's Super Decisions software (Creative Decisions Foundation, 2022).

Degtiarev and Borisov (2018) supported the premise that when an expert expresses judgements using natural language statements (e.g., words or phrases) inherent vagueness of language constructs can cause the interpretation to be imprecise. Their paper presents a model that gives weight to the constraints on domains of expert assessments as they are usually supplied with certain degrees of confidence. An empirical comparison of Fuzzy AHP using triangular fuzzy numbers and interval type-2 membership functions is also presented.

Dhurkari and Swain (2019) applied the AHP to solve partner selection problems during the formation of a strategic alliance for innovation. The research attempts to show the importance of partner selection as well as the traits that can lead to greater success of the prospective strategic alliance.

Karasan (2019) related that the intuitionistic fuzzy extensions are the most used type of fuzzy extension in the literature because they represent the decision makers' strength of commitment to the considered subject in a better and an effective way including membership and non-membership functions. In this study, hesitant intuitionistic fuzzy sets are used to extend the AHP in an investment prioritization problem based on relevant risk factors. Comparative analyses with intuitionistic fuzzy AHP and hesitant fuzzy AHP methods are conducted to validate the proposed method.

Keseru et al. (2016), referred to the importance of participatory decision-making, with multiple actors. This process can be complemented with an electronic system for a group decision support system to reach a mobility policy in the city center of Leuven, Belgium, based on the preferences of 34 stakeholders. The proposed method provides an opportunity for stakeholders to interactively weigh and evaluate scenarios in an online software tool.

Lamata et al. (2018), presented a Fuzzy AHP–TOPSIS approach allowing to rank firms based on how well they are doing good in the fields of corporate social responsibility.

Oswald Beiler and Phillips (2016), investigated how transportation planning agencies can improve the process of prioritizing projects to improve pedestrian infrastructure in Union County (Pennsylvania, USA). Identifying pedestrian infrastructure needs through the development of an index can resolve inconsistencies and inequities regarding funding sustainable transportation infrastructure. This research investigated how transportation planning agencies can improve the process of prioritizing pedestrian improvement projects. The study integrates pedestrian performance metrics with the AHP to assist in the prioritization of pedestrian pathway improvements. The index serves as a tool for agencies to enhance the decision-making process for planning pedestrian improvement projects.

Pradhan et al. (2019) combined the AHP with the DEA creating a DEAHP model. The study presented the flexibility and applicability of the integration model when it is used with other approaches and methods. The paper also presented a brief literature review on the integration of AHP with DEA.

Qu (2018) proposed an evaluation system of smart cities' construction potential which includes five aspects: information infrastructure, the level of economic development, science and technology supporting ability, the level of urban industrial development, and urban competitiveness. With a Kernel Principal Component Analysis (KPCA), the research made the sort and potential evaluation of 13 main cities of China and, by applying the clustering method, the researcher classified the cities based on the result of the evaluation.

Singh et al. (2019) explored the significance of various factors contributing to the manufacturing performance enhancement of small firms applying the AHP. Micro, SMSEs were defined by several factors and criteria, such as location, size, age, structure, organization, number of employees, sales volume, the worth of assets, ownership through development, and technology.

Wang (2018) decomposed the three Es of sustainability (economic, ecology, and equity) into eight quantitative indicators and their weights of importance are determined with the AHP for the construction of the overall metric. Once the indicators and weights are defined, the study applies Fuzzy comprehensive evaluation to grade specific data into three levels and eventually obtain a specific score of the overall metric in order to easily measure the success of smart growth of a city.

Yilmaz and Ozkir (2018) proposed an extended consistency analysis procedure to evaluate the consistency of incomplete pairwise comparison matrices. The authors support that if any judgement in a pairwise matrix is absent, the judgement acquisition phase yields an incomplete pairwise matrix. Since the current consistency analysis procedure is designed for complete matrices, the suitability for evaluating the inconsistency of incomplete matrices is questionable.

4. Conclusions

The research questions that this work aimed to answer were: (Q1) How to characterize joint research in urban transport and MCDM? (Q2) What are the main trends and gaps for future research? From what is presented in the last section, the answers to these questions are directly connected:

The AHP is the leading MCDM applied in the research of urban mobility.

Therefore, one main research gap is to study and apply different MCDM methods and compare the results with AHP. Indeed, this is alternatively done with hybridism of AHP, mainly with FST, and with the DEA, and the TOPSIS. However, there are a dozen of other MCDM methods that could be applied, starting with old European methods, such as the Élimination et Choix Traduisant la Realité (ELECTRE) and Preference Ranking Organization Method for Enrichment of Evaluations (PROMETTHE), but including the newer Full Consistency Method (FUCOM) and Step−wise Weights Assessment Ratio Analysis (SWARA).

Urban mobility and urban transportation are major issues worldwide. As they affect the quality of life of billions of people, not studying them is a lack of opportunity for researchers of MCDM methods, other than AHP. Those researchers may bring more relevance to their research by addressing this important subject of actual human life.

The delimitation of this work to a single database (Scopus) does not invalidate the results, since other databases, such as Clarivate, for instance, have similar contents. Still, the use of different databases and software and comparisons with our results deserve investigation.

References

Alitaneh, S. (2019). Theories on coefficient of variation scales triangle and normalization of different variables: A new model in development of multiple criteria decision analysis. *Int. J. AHP*, 11: 283–295.

Ascher, F. (2001). Les nouveaux principes de l'urbanisme. La Tour-d'Aigues: Éditions de l'Aube.

ATLAS.ti Scientific Software Development GmbH. (2022). The Qualitative Data Analysis & Research Software–ATLAS.ti. https://atlasti.com (Accessed 30 December 2022).

Attard, M., Haklay, M. and Capineri, C. (2016). The potential of volunteered geographic information (VGI) in future transport systems. *Urban Planning*, 1: 6–19.

Behzadian, M., Kazemzadeh, R.R., Albadvi, A. andAghdasi, M. (2010). PROMETHEE: A comprehensive literature review on methodologies and applications. *Eur. J. Oper. Res.*, 200: 198–215.

Bielińska-Dusza, E., Hamerska, M. and Żak, A. (2021).Sustainable Mobility and the Smart City: A Vision of the City of the Future: The Case Study of Cracow (Poland). *Energies*, 14: 7936.

Bousdekis, A. and Kardaras, D. (2020). Digital transformation of local government: A case study from Greece. *In*: W. Guédria et al. (Eds.) Proceedings of the CBI 20. *IEEE 22nd Conference on Business Informatics, Atwerp: IEEE*, pp. 131–140.

Centrum voor Wetenschap en Technologische Studies. (2022). VOSviewer: Visualizing Scientific Landscapes. http://www.vosviewer.com (Accessed 12 December 2022).

Cestari, J.M.A.P., Loures, E.F.R. and Santos, E.A.P. (2018). A method to diagnose public administration interoperability capability levels based on multi-criteria decision-making. *Int. J. Inf. Tech. Decis.*, 17: 209–245.

Ciliberto, C., Szopik-Depczyńska, K., Tarczyńska-Łuniewska, M., Ruggieri, A. and Ioppolo, G. (2021). Enabling the Circular Economy Transition: A sustainable lean manufacturing recipe for Industry 4.0. *Bus. Strateg. Environ.*, 30: 3255–3272.

Climate-KIC Nordic and C40 Cities. (2018). Municipality-led Circular Economy Case Studies. nordic.climate-kic.org/wp-content/uploads/sites/15/2018/05/Municipality-led-circular-economy-case-studies.pdf. Accessed 26 December 2022).

Colapinto, C., Jayaraman, R., Ben Abdelaziz, F. and La Torre, D. (2020). Environmental sustainability and multifaceted development: Multi-criteria decision models with applications. *Ann. Oper. Res.*, 293: 405–432.

Conejero, M.A. and Silva César, A. (2017). The governance of local productive arrangements (LPA) for the strategic management of geographical indications (GIS). *Ambient. Soc,*. 20: 293–314.

Creative Decisions Foundation. (2022). Super Decisions | Homepage. https://superdecisions. com/ (Accessed 31 December 2022).

Currie, P.K. and Musango, J.K. (2016). African urbanization: Assimilating urban metabolism into sustainability discourse and practice. *J. Ind. Ecol.*, 21: 1262–1276.

Dali, N.M., Abdullah, A. and Islam, R. (2018). Prioritization of the indicators and sub-indicators of Maqasid Al-Shariah in measuring liveability of cities. *Int. J. AHP*, 10: 348–371.

D'Amico, G., Arbolino, R., Shi, L., Yigitcanlar, T. and Ioppolo, G. (2022). Digitalisation driven urban metabolism circularity: A review and analysis of circular city initiatives. *Land Use Policy*, 105819.

Dangond Gibsone, C., Jolly, J.-F., Monteoliva Vilches, A. and Rojas Parra, A. (2011). Algunas reflexiones sobre la movilidad urbana en Colombia desde la perspectiva del desarrollo humano. *Papel Político*, 16: 485–514.

De Pascale, A., Arbolino, R., Szopik-Depczyńska, K., Limosani, M. and Ioppolo, G. (2021). A systematic review for measuring circular economy: The 61 indicators. *J. Clean. Prod.*, 281: 124942.

Degtiarev, K.Y. and Borisov, M.Y. (2018). Prioritization of alternatives based on analytic hierarchy process using interval type-2 fuzzy sets and probability-theoretical interval comparison. *Int. J. AHP*, 10: 447–468.

Dhurkari, R.K. and Swain, A.K. (2019). Application of AHP in partner selection for innovation in strategic alliances. *Int. J. Bus. Innov. Res.*, 19: 532–553.

Egger, M. and Smith, G.D. (1998). Meta-analysis bias in location and selection of studies. BMJ, 316: 61.

Ellen MacArthur Foundation. (2019). Circular Economy of Cities: Project Guide. *Isle of Wight: Ellen MacArthur Foundation.*

Elsevier, B.V. (2022). Scopus. www.scopus.com (Accessed 28 December 2022).

European Commission. (2020). Circular Economy Action Plan. Brussels: European Commission

García-Fuentes, M.Á. et al. (2021). Evaluation of Results of City Sustainable Transformation Projects in the Fields of Mobility and Energy Efficiency with Real Application in a District in Valladolid (Spain). *Sustainability*, 13: 9683.

García-Fuentes, M. and De Torre, C. (2017). Towards smarter and more sustainable cities: The Remourban Model. *Int. J. Entr. Suatain. Issues*, 4: 328–338.

Gebhardt, L. et al. (2016). Intermodal urban mobility: Users, uses, and use cases. *Transp. Res. Proc.*, 14: 1183–1192.

Iannillo, A. and Fasolino, I. (2021). Land-use mix and urban sustainability: Benefits and indicators analysis. *Sustainability*, 13: 13460.

International Society of Multiple Criteria Decision-Making. (2022). Mission of the Society. http://www.mcdmsociety.org/content/mission-society. (Accessed 26 December 2022).

Karasan, A. (2019). A novel hesitant intuitionistic fuzzy linguistic AHP method and its application to prioritization of investment alternatives. *Int. J. AHP*, 11: 127–142.

Keseru, I., Bulckaen, J. and Macharis, C. (2016). The multi-actor multi-criteria analysis in action for sustainable urban mobility decisions: The case of Leuven. *Int. J. Multicriteria Dec.*, 6: 211–236.

Lamata, M.T., Liern, V. and Pérez-Gladish, B. (2018). Doing good by doing well: An MCDM framework for evaluating corporate social responsibility attractiveness. *Ann. Oper. Res.*, 267: 249–266.

Li, T., Li, A. and Guo, X. (2020). The sustainable development-oriented development and utilization of renewable energy industry: A comprehensive analysis of MCDM methods. *Energy*, 212: 118694.

Lombardi Netto, A., Salomon, V.A.P., Ortiz-Barrios, M.A., Florek-Paszkowska, A.K., Petrillo, A. and De Oliveira, O.J. Multiple criteria assessment of sustainability programs in the textile industry. *Int. Trans. Oper. Res.*, 28: 1550–1572.

May, A.D. (2015). Encouraging good practice in the development of sustainable urban mobility plans. Case Stud. *Transp. Policy*, 3: 3–11.

May, A., Boehler-Baedeker, S., Delgado, L., Durlin, T., Enache, M., and Van Der Pas, J.W. (2017). Appropriate national policy frameworks for sustainable urban mobility plans. *Eur. Transp. Res. Rev.*, 9: 1–16.

Mendoza, J.M.F. et al. (2015). Development of urban solar infrastructure to support low-carbon mobility. *Energ. Policy*, 85: 102–114.

Milios, L. (2018). Advancing to a circular economy: Three essential ingredients for a comprehensive policy mix. *Sustain. Sci.*, 13: 861–878.

Myrovali, G., Morfoulaki, M., Vassilantonakis, B.-M., Mpoutovinas, A. and Kotoula, K.M. (2020). Travelers-led innovation in sustainable urban mobility plans. *Periodica Polytechnica Transportation Engineering*, 48: 126–132.

Nakano, D.N. and Jr. Muniz, J. (2018). Writing the literature review for an empirical paper. *Prod.*, 28: e20170086.

Oliveira, O.J., Silva, F.F., Juliani, F., Barbosa, L.C.F.M. and Nunhes, T.V. (2019). Bibliometric method for mapping the state-of-the-art and identifying research gaps and trends in literature: An essential instrument to support the development of scientific projects. *In*: S. Kunosic and E. Zerem (Eds.). Scientometrics Recent Advances, London: *InTech Open*, pp. 1–20.

Oliveira, V.A.R., Salomon, V.A.P., De Oliveira, G.C.R., Petrillo, A. and Neves, S.M. (2022). Systematic literature review of multi-criteria decision-making applied to energy management. *In*: M. Fathi, E. Zio and P.M. Pardalos (Eds.). *Handbook of Smart Energy Systems, Cham: Springer*, pp. 1–13..

Oswald Beiler, M.R. and Philips, B. (2016). Prioritizing pedestrian corridors using walkability performance metrics and decision analysis. *J. Urban Plann. Dev.*, 142: 04015009.

Portugal, L.S. (Ed.) (2017). Transporte, mobilidade e desenvolvimento urbano. Rio de Janeiro: Elsevier.

Pradhan, S., Olfati, M. and Patel, G. (2019). Integrations and applications of analytic hierarchy process with data envelopment analysis: A literature review. *Int. J. AHP*, 11: 228–268.

Qu, Y. (2016). Evaluation of smart city development potential in major cities of China based on the KPCA method. *In*: *International Conference on Management Science and Engineering (ICMSE)*, 855–862. Olten: IEEE.

Quijano, A. et al. (2022). Towards sustainable and smart cities: Replicable and KPI-driven evaluation framework. *Buildings*, 12: 233.

Ruiz Bargueño, D., Salomon, V.A.P., Marins, F.A.S., Palominos, P. and Marrone, L.A. (2021). State of the Art Review on the Analytic Hierarchy Process and Urban Mobility. *Mathematics*, 9: 3179.

Saaty, T.L. and Ergu, D. (2015). When is a decision-making method trustworthy? Criteria for evaluating multi-criteria decision-making methods. *Int. J. Inf. Technol.*, 14: 1171–1187.

Schiller, G., Gruhler, K. and Xie X. (2020). Assessing the efficiency of indoor and outdoor access-related infrastructure. *Buildings and Cities*, 1: 56–69.

Singh, D., Khamba, J.S. and Nanda, T. (2019). Justification of technology innovation implementation in Indian MSMEs using AHP. *Int. J. Serv. Oper. Man.*, 32: 522–538.

United Nations Environment Programme (2017). Resilience and Resource Efficiency in Cities. New York: United Nations.

United Nations (2017). New Urban Agenda. New York: United Nations.

United Nations (2015). Transforming OurWorld: The 2030 Agenda for Sustainable Development. New York: United Nations.

United Nations (2014) World Urbanization Prospects. New York: United Nations.

United Nations (2019).World Urbanization Prospects: The 2018 Revision. New York: United Nations.

Universidade Estadual Paulista (2022). SciVal–Projeto Unesp de internacionalização. https://www2.unesp.br/portal#!/propg/plano-de-internacionalizacao-da-unesp/inteligencia-print/scival---ferramenta-de-gestao/ (Accessed 29 December 2022).

Venkat, K., Maiti, S., Kanuri, C. and Mulukutla, P. (2019). Leveraging innovation for last-mile connectivity to mass transit. *Transp. Res. Proc.*, 41: 655–669.

Wallenius, J., Dyer, J.S., Fishburn, P.C., Steuer, R.E., Zionts, S. and Deb, K. (2008). Multiple criteria decision making, multiattribute utility theory: Recent accomplishments and what lies ahead. *Manage. Sci.*, 1336–1349.

Wang, J.X. (2017). Modeling for the measurement of smart city. *In: Proceedings of the International Conference on Management Science & Engineering*, 17–20. Nomi: IEEE.

Xiao, Lu, and Watson, Maria. (2019). Guidance on conducting a systematic literature review. *J. Plan. Educ. Res.*, 39: 93–112.

Xie, X.-F. and Wang, Z.J. (2018). Examining travel patterns and characteristics in a bikesharing network and implications for data-driven decision supports: Case study in the Washington DC area. *J. Transp. Geogr.*, 71: 84–102.

Yildiz, S., Kivrak, S. and Arslan, G. (2019). Contribution of built environment design elements to the sustainability of urban renewal projects: Model proposal. *J. Urban Plann. Dev.*, 145: 04018045.

Yılmaz, S.K. and Ozkir, V. (2018). Extended consistency analysis for pairwise comparison method. *Int. J. AHP*, 117–134.

3

Green Materials
Sustainable Materials, Green Nanomaterials

Narinder Singh,[1,]* *Francesco Colangelo*[2] and *Ilenia Farina*[2]

1. Introduction

In the industrialized environment, the concept of green buildings for both social and environment sustainability has been acknowledged. A smart green building offers substantial quality management and assurance, efficiency, convenience, accessibility, and systematic monitoring of all activities. Smart green buildings can be constructed using three support systems: the context of the building area, the components employed throughout construction, and the execution of the suggested task. Multiple varieties of sustainable, intelligent, indigenous, and recycled resources, etc., enhance the environmental reliability and integrity of the proposed structure (Roy et al., 2016). There are various green materials and processing technologies that exist to date that serve to decrease time and costs, improve the quality of work waste management, budget-friendly housing, temperature resistance system, etc. (Huang et al., 2019). Cost is one of the

[1] Department of civil engineering, University of Salerno, Fisciano, Italy.
[2] Department of Engineering, University of Naples "Parthenope", Centro Direzionale, Naples, Italy.
  Emails: ilenia.farina@uniparthenope.it; francesco.colangelo@uniparthenope.it
* Corresponding author: narinder.singh@assegnista.uniparthenope.it

project's most essential considerations. In most circumstances, the initial cost of constructing adaptive green buildings is moderately expensive, therefore we must prioritize low maintenance, minimum material waste, cost overruns, and government assistance. To minimize carbon dioxide (CO_2) emissions and have a good effect on the climate, it is necessary to convert to renewable power (Rahman et al., 2017). A viable solution for residual energy structures is photovoltaic modules mounted on the façade for cooling (and heating) as well as roof implementations of the remaining energy applications. Due to the optimal utilization of the sun's atmosphere in the morning and evening, the main benefits of facade-integrated photovoltaics include climate independence, cooling, and equilibrium production. Green building (GB) is the most efficient approach to use green sustainability and architecture in the contemporary, technologically advanced world, according to the consensus of researchers (Li et al., 2021). For modern applications in the aviation, industrial, and social industries, as well as materials engineering, optic systems, artificial intelligence (AI), nanotechnology, and biotech, smart or intelligent materials are required. In the 1980s, intelligent objects began to appear. As intelligent beings natural environments evolve, so do their structures (Vogl et al., 2020).

Conventional concrete uses cement, gravel (fine aggregate), coarse aggregate (CA), and liquid in terms of the use of green materials in building and construction engineering. The building sector is the central pillar of global economic and social growth. It accounts for approximately 6% of the global gross domestic product (Ahmad and Zhang, 2020). Decades of significant industrialization and urbanization on a global scale have contributed to the sector's expansion. This expansion places an enormous strain on natural resources and has a significant impact on the environmental system. According to Levin (Duan et al., 2022), the building industry contributes between 15–45% to the total degradation of the environment in eight major lifecycle assessment (LCA) categories, including depletion of raw materials, energy use, freshwater consumption, and air emissions and research indicates that the situation has not improved. Recently, there has been a greater emphasis on investigating strategies to reduce environmental concerns caused by humans in addition to identify alternatives that contribute to sustainable growth. To achieve sustainability, various waste substances can be substituted for cement, sand, and CA (Khunt et al., 2022). This sort of research aims to develop mixture designs for sustainable concrete and evaluate the fresh state concrete features and hardened concrete characteristics of sustainable concrete that has been developed. Typically, CA covers between 65–75% of concrete. Global supply of CA is finite; hence a replacement is necessary (Fletcher, 2011). In addition, India produces a substantial amount of construction and demolition (C&D) debris annually. The disposal and usage of C&D waste is another obstacle. Sustainable concrete incorporates waste resources such as ground-granulated blast-furnace slag (GGBS), fly ash (FA), and other sustainable construction materials (SCMs) in place of cement, recycled coarse aggregate (RCA) in place of CA, and manufactured sand (m-sand) in place of FA (Pellegrino et al., 2019). Numerous studies used various SCM materials, including FA, GGBS, Silica Fume, and Lime as replacements for cement, RCA as replacements for CA, and m-sand as replacements for fine aggregate (Rajesh et al., 2021).

Furthermore, the manufacture of construction products has a significant impact on the environment. The demand for cement-based items has increased in parallel with the building industry's robust infrastructure expansion (Environment et al., 2018). Cement is a widely suggested building material throughout the world. In the previous decade, cement production reached 4 billion tons, a significant rise over the 2000s. Cement production and utilization account for 8% of global CO_2 emissions (Favier et al., 2018), with 50% attributable to the formation of clinkers, 40% to fuel combustion, and 10% to transportation, raw material processing, drying, crushing, and blending. China has supplied more than 50% of the world's cement production over the past three decades (Zou et al., 2017). Cement consumption is unlikely to decrease soon due to the continuation of global building programs and the availability of abundant cement deposits. Over 90% of the CO_2 was produced during the calcination and combustion processes, which are responsible for the emission of around 0.9 kg CO_2 every kilogram of cement formed. In fact, the requirement for non-renewable sources in cement manufacture have generated a new environmental risk (Guezuraga (Guezuraga et al., 2012). As a result, industry leaders and researchers have found novel solutions to these rising issues. Numerous attempts to find substitutes for cement have been described in numerous scholarly literatures (Demaria et al., 2018; Behera et al., 2014; Plevin et al., 2014). By converting waste items into cement substitutes, cement consumption and the environmental damage caused by open trash disposal may be reduced. In addition, in the field of sustainable plastic, polyethylene terephthalate (PET) is widely utilized in the production of water bottles, food packaging, and several other plastic items. However, improper disposal of huge quantities of PET trash may result in significant environmental issues. Utilizing PET-derived products as a performance-enhancing asphalt additive is one option for recycling and reusing this waste material. Likewise, the disposal of used tires is a significant environmental issue. Utilizing crumb rubber (CR) from used tires improves asphalt's rheological qualities, but is frequently hampered by other issues, such as poor storage stability, according to research (Demaria et al., 2013).

In the subject of techniques for processing green building materials, numerous approaches have been established and investigated to date and are discussed in this chapter such as green building techniques for cost effective housing, Application of building information modeling (BIM) and ontology to produce smart sustainability initiatives, wall technology for the hot summer and cold winter weather in a green building, Performance Analysis of a Facade Integrated Photovoltaic Powered Cooling System, and the Smart homes method. The smart home is retrofitted with smart materials, and its activities can be managed through remote. In addition, the 3D plastering technology reduces CO_2 emissions by up to 99% compared to conventional methods (Son et al., 2021).

Finally, this chapter describes green materials in terms of their sustainability and construction engineering. Detailed discussion of the materials that are paving the way for a better environment and building. Numerous green materials have been identified, including geopolymers, construction waste, plastic materials, and sustainable green concrete. In addition, the processing technologies for eco-friendly materials are elaborated upon in the following sections. In the following parts, we will discuss the green and sustainable aspects of materials and construction.

2. Sustainable Approach towards Sustainable Materials

A green structure necessitates superior materials and structures for adaptable supportability compared to a conventional construction. In India, the market for green building supplies and services is expanding alongside with the rising trend of green building development. Figure 3.1 depicts

Fig. 3.1 Rules for green building construction (Patel and Patel, 2021).

the green building construction rules (Patel and Patel, 2021). During the planning and execution of a construction project, the practical development combines many methodologies. Utilizing eco-friendly materials is one of the feasible plan activities and growth systems. The green materials are environmentally responsible because they contribute to the decline of natural effects (Greenomics) (Patel and Patel, 2021). Asset efficiency, indoor air quality, energy efficiency, and cost-effectiveness should be the defining characteristics of the practical building materials. Green building grade for coordinated environmental assessment. A green building material that has a negligible or positive influence on the environment (Lu et al., 2019). Most green materials are comprised of recyclable materials that aid the ecosystem and make efficient use of discarded energy; their production also requires less energy.

In the research by Nematzadeh et al. the sustainable approach has been used by using the recycled polymers in aggregates and fibers (Nematzadeh et al., 2020).

Aggregates and fibers are made from recycled polymer. The flexural strength, compressive strength, workability, density, and split tensile strength of these materials are all measured and compared to virgin aggregates, Electric arc furnace steel slags, industrial waste, are replaced in concrete by aggregates to maintain sustainability by researchers (Saha and Sarker, 2017). Numerous cubes are cast and tested for strength using furnace slag. During a year of data collection it was discovered that the blocks' usability is substantially comparable to that of normal blocks (Patil et al., 2022). Blast furnace slag and olive stone biomass ash (OBA) are used in this innovative alkali activator to create sustainable blocks. The usage of OBA cement reduces CO_2 concentration by up to 96% when compared to ordinary Portland cement. Compared to ordinary Portland cement (OPC), these blocks only have one drawback: they absorb more water. For mycelium, the best sentence is "waste to best". It is a highly productive, cost-effective, and a promising green building material. As a result, it consumes less energy and is an excellent insulation. Rice husk is a byproduct of the rice industry. Rice husk ash is a renewable and environmentally friendly material in the construction sector since it substitutes a particular amount of cement in the concrete mix. In India, this drug is inexpensive and frequently available. Using sugarcane bagasse ash in place of up to 20% of the cement in self-compacting concrete improves rheological test results (Le et al., 2018). Timber-framed construction, sun-dried walling, and rammed earth are all examples of vernacular materials and processes that are more efficient in green building (Agyekum et al., 2020).

The rapid rise in pollution levels in soils in recent years can mostly be attributed to the construction industry. It is possible to employ a variety of home trash as sustainable materials in construction and building. The use of agricultural leftovers as a substitute for cement has been studied. Partial cement alternatives have been tested with rice and sugar cane bagasse ash as well as sawdust. Even though industrial outputs showed great structural performance, there are many advantages to using agro-based wastes, including their accessibility, affordability, light weight, recyclable nature, energy efficiency, and environmental acceptability (Nandhini and Karthikeyan, 2022). In the following section, features and green applications of eggshells (ES) are explained in detail.

2.1 Eggshells as green materials

Eggshells, the outer layer of an egg's shell, are an example of a waste product from agriculture. About 10% of the egg's weight is made up of the shell chambers. The world's egg production is on the rise, resulting in an estimated 8 million metric tons of eggshell waste every year. A vast number of eggshells are thrown away, even though they can be utilized to manufacture ceramics with calcium phosphates, to make biofuels, and even as an extractor for cleaning contaminants and fertilizer. ES include calcium carbonate, which is a key ingredient in cement binder paste (Calcium Silicate Hydrate (CSH)) (Shiferaw et al., 2019). As a result, it can be used as a dry powder alternative for cement in building materials. To assist mitigate environmental issues, ES powder can be used in the production of construction materials. This is because chicken litter is clearly recycled, and cement is widely used (Mieldažys et a., 2019). Cement pastes, cement mortars, concrete, soil and brick stabilizers, earth cement and sand Crete blocks have all undergone several experiments to determine whether ES powder can be successfully used in any of these applications. ES powder as a building material has not been evaluated in a comprehensive, state-of-the-art way. Eggshell powder's use in building materials will be the primary focus of the report, which will include an in-depth analysis of the most recent research findings (Hut, 2014). Eggshells are generated in large quantities because of commercial egg production. Open dumping is the most important environmental issue that needs to be addressed (Waheed et al., 2019). A long-term solution may be attainable if ES uses can be found. Much effort has been made in recent years to convert ES waste into useful products. Es have been shown to be useful in a wide range of ways, from simple food additives to complex processes like metal immobilization and implanting. As far back as the early 2000s, ES was used as a source of calcium to make calcium phosphates (Mardziah et al., 2021).

Eggshell is a food processing agro-waste item that has led to negative environmental repercussions in our surrounds, in homes, and in industries. The production of ES waste is expected to rise in tandem with global population growth, necessitating higher disposal costs for the huge volumes of trash generated by the food processing industry. According to the Food and Agriculture Organization's latest data, 6.4 million metric tons of Es trash will end up in landfills around the world (FAO) (Nandhini and Karthikeyan, 2022). When not properly disposed of, ES, a common agricultural waste, can pollute the ecosystem. Even though its waste has been used in soil remediation to correct pH, a substantial amount of material is still thrown away in landfills without any further use (Ngayakamo and Onwualu, 2022). The spread of urban pests and the resulting foul stench have made improper ES waste disposal a source of worry. This has resulted in a significant amount of manufacture of ES because egg breaker firms generate more trash than families and restaurants do. More than 82.17 million tons of eggs were produced worldwide in 2019, up from 73.9 million tons in 2016 (Ngayakamo and Onwualu, 2022).

Eco-friendly construction can benefit from the use of Es, a bio-waste product. For every year that goes by, the food business produces enormous amounts of ES, which are then thrown out in landfills without proper treatment. Eggshells should be disposed of in sanitary landfills rather than in the open rubbish (Waheed et al., 2019). Because dangerous gases are created by these pollutants, they could constitute a threat to human life and health. In terms of physical and chemical properties, powdered eggshell (PES) is a little different from cement particles. Cement with large volume FA was partially substituted with ES to ameliorate the poor performance of concrete containing FA (HVFA) (Rashad, 2015). FA cement was hydrated faster by using calcium carbonate ($CaCO_3$) derived from ES. When CaCO3 is introduced to the HVFA cementitious matrix, the carbonate ion from CC and the aluminate hydrate from Portland cement hydration interact to produce carbo-aluminates.

2.2 Ferrochrome Slag (FCS)

FCS, a by-product material, has recently been researched to see if it may be used in civil engineering projects. Due to the better chemical, physical, and mechanical features, FCS has recently gained the attention of more researchers in the field of engineering, including concrete, bricks, hot bitumen mix (HBM), and base layers. This material can serve as a binder or an aggregate replacement in concrete based on the results of a variety of studies (Fares et al., 2021).

Aggregate is employed in a wide range of civil engineering projects on a regular basis. In 2017, aggregate consumption was expected to reach 28 billion tons per year. Aggregate is a crucial component in concrete construction. It typically accounts for between 60–75% of the volume of the concrete. A considerable possibility is seen in the use of concrete aggregate made from waste and byproduct resources. Research into the use of a wide range of wastes and byproducts has been extensive, including steel slag and copper slag as well as plastic and rubber waste and glass and ceramic waste. FCS has shown a strong ability to improve concrete properties when used as aggregate in concrete. FCS properties and its use as concrete aggregate have received only sporadic attention from researchers. Sahu et al. (Sahu et a., 2016) examined FCS's mineralogical phases and environmental compatibility. FCS is used in a wide range of applications, including civil engineering projects and ceramics, and the hazardous hexavalent-chromium treatment was also covered in a study by Al-Jabri (Al-Jabri, 2018) gave a brief overview of the use of FCS in concrete and road construction. An evaluation of the mechanical characteristics of concrete with FCS as an aggregate was conducted by Lakshmi and Anu (PS and Anu, 2018) using very limited research data. But no systematic analysis of current research efforts on FCS characteristics and its use as concrete aggregate has been conducted.

Ferrochrome (FC) alloy, which is primarily composed of chromium and iron and contains 50–70% by weight of chromium as well as varying proportions of iron, carbon, and other components, generates FCS as a byproduct. In 2017, an estimated 11.7 million metric tons of FC were produced around the world. A 3.6% yearly growth rate has also been predicted to keep up with the rising demand. Chrome ore and/or fine concentrates are utilized in the manufacture of FC from chromite, which is the primary raw material. FC is mostly produced in China and South Africa, the two countries with the most chromite resource. They make up more than two-thirds of the global FC output. In the stainless-steel sector, more over 80% of the FC that is produced is utilized. When it comes to making this, the most common method is to use submerged electric arc furnaces. Ferrochrome slag is the name given to the slag that is produced during the manufacturing of FC (FCS). Each ton of FC produces between 1.1 and 1.6 tons of FCS. According to the type of food being fed, the number of FCS can vary greatly (Fares et al., 2021).

2.3 Bamboo as Key 'Green' Material

Bamboo is a perennial, evergreen plant that belongs to the family of grasses. The culm, or jointed stem, that distinguishes bamboo from other grasses is characteristic of both plants. Blanketed culms are the norm, although there are varieties of bamboo that have solid culms instead (DeBoer and Bareis, 2000). Although the bamboo internode has a low splitting force, interlayer propagation is made more difficult by the presence of nodes (Borowski et al., 2022). One of bamboo's greatest assets is the wide variety of "end products" and uses it can serve. These include a wide range of "end products," such as furniture and other "end products," as well as "end uses," such as construction materials. Nodes, the reliable joints at the beginning and end of each culm segment, are the foundation of the culm. Increased strength is achieved with the use of bamboo nodes. A bulge around the culm segment ends identifies a node. In comparison to the 2–5% annual rise in tree biomass, bamboo biomass grows at a rate of 10%–30% per year, indicating a superior raw material performance (Lobovikov et al., 2005). Rhizomes, the underground stems of nearly every species of bamboo, spread rapidly and aggressively. Many bamboo variants are cold-resistant and can thrive in temperatures as low as 15–20°F; as a result, more than 50 million hectares of land are occupied by bamboo species (DeBoer and Bareis, 2000). Because of its rapid growth, bamboo farming is seen to be a promising technique for mitigating climate change. Because of its rapid longitudinal growth during its second and third weeks, the dynamics of bamboo's growth are quite intriguing. Growth dynamics begin to accelerate significantly after the first 12 days of bamboo's growth (i.e., it grows 6 cm each day for the next four days). It is possible for bamboo to grow up to 80 cm per day in the third week. Bamboo is a powerful CO_2 absorber because of its unique features. If an additional 10 million hectares of bamboo were planted over the course of 30 years, the plants and their products would avoid almost 7 gigatons of CO_2 emissions (Dwivedi et al., 2019).

2.4 Plastics waste for sustainable future

In the packaging industry, plastic is a crucial material because of its low cost, flexibility, and long-term useability. Around six decades ago, a mass manufacturing of plastics began, but it has since been accelerated so much that 8.3 billion metric tons have been produced—mostly in disposable products that end up in the trash bin (Moore, 2008). In its most basic form, plastic is nothing more than a long chain polymer compound made by a polymerization reaction from repeating structural units of specified monomers. At high temperatures and pressures, as many as 20,000 ethene molecules can be linked together to form a single polyethene chain. Due to the large number of polymeric chains in this plastic, it is extremely difficult to degrade it naturally. Most of the plastics we use in packaging and everyday consumer goods are made from fossil-fuel-based feedstocks. There are a wide range of regularly used polymers in the plastics manufacturing industry, including polyethylene (PE), polytetrafluoroethylene (PTFE), polypropene (PP), polystyrene (PS), polyvinylchloride (PVC), and nylon (Mirji and Lobo, 2017). Since most traditional plastic goods are made of non-biodegradable polymers that take more than 100 years to decompose in moist soil, it is no wonder that these materials have a profound effect on our daily lives and that it is difficult to dispose of these wastes in a sustainable manner. Since plastic waste is recycled only 16%, more than 380 million tons of plastics are created each year (Rafey and Siddiqui, 2021; Wang, 2019). Plastics recycling codes have been established by the Society of the Plastics Industry (SPI), primarily to help in identifying the polymers used in manufacturing and to speed up the recycling process. It is critical to devise a method for properly managing plastic trash that has a positive effect on the environment through the recycling, repurposing, or disposal of the material. In terms of the quality and complexity of recycled materials compared to the original items, the products are categorized as primary, secondary, and quaternary, respectively. After recycling, primary products are typically used since they retain their original quality. Recycled plastic pellets are made from plastic waste that has been separated and collected from the rest of the mixed garbage after a mechanical recycling process is completed. This means that secondary and tertiary products can't be used for the same purpose again. Chemical processes are primarily responsible for the development of these recycled plastics' unique properties. When plastic garbage is incinerated, the combustion products can be utilized to generate electricity. Conventional plastic recycling efficiency, on the other hand, remains an open question. In comparison to landfills, just a small percentage of plastics have been recycled to date (Singh and Singh, 2021; Singh, 2021).

2.5 *Green engineered cementitious composites*

Using a polyvinyl alcohol fiber and a polyethylene volume fraction of around 2%, Kaoleian et al. (Keoleian et al., 2005) have constructed engineering cementitious composites (ECCs) that mimic the proportion of concrete fiber (i.e., water, cement, fine aggregate, without coarse aggregate, fiber, and chemical additives). An innovative cement-based material with excellent mechanical properties, high ductility, and long-term durability was developed by Kaoleian et al. (Keoleian et al., 2005) using analytical methods from engineering micromechanics. It is quite difficult for traditional concrete to meet the 1.4% ductility standard for cement-based materials that are robust to temperature changes and dynamic live loads, but the ECC can readily do so (Suhendro, 2014).

Recycling carpet as a fiber and using more waste FA in the concrete mix are examples of ECC materials that have been adapted to meet environmental protection requirements while still maintaining good mechanical performance, such as: It was dubbed "Green ECC" for the ECC material produced. 318 kg/m^3 cement, 701 kg/m^3 class F flay ash, 701 kg/m^3 sand, 289 kg/m^3 water, and 26 kg/m^3 PVA fiber make up the rest. Compressive strengths of the traditional ECC mix, current ECC mix, high fly ash mix, and ECC with carpet fiber were 35; 65; 33; and 55 MPa correspondingly. Kaoleian et al. (Keoleian et al., 2005) claim that tailored micro-structures can compensate for a decline in performance due to the substitution of cement with other materials, or the influence of waste utilization in the concrete mix.

2.6 Composite materials

This century has seen a revolution in material science and engineering thanks to advances in nanotechnology and nanoscience. Two or more distinct physicochemical components make up intelligent molecules. There are two parts to the compounds: a matrix and a filling. The matrix acts as a building block for the rest of the system, acting as a template for how the various elements should be connected and interacted. A filter's mechanical strength and other properties can be improved by strengthening it within a matrix rather than outside of it (Yang et al., 2004). Civil engineers and the construction industry both consider combinations as critical variables in the construction of new infrastructure and the repair of deteriorated existing infrastructure. Fiber-reinforced polymer composites (FRPCs) have advanced in recent decades due to their exceptional technical characteristics. The low cost of FRPCs also makes them a viable alternative to aged steel used as an additive in concrete structures. The use of FRPC sheets to preserve, restore, and reinforce existing concrete structures has also aroused public attention (Bernatas et al., 2001). When it comes to building maintenance, infrastructure integration, and fixing problematic piping systems, CF reinforced epoxy matrices are frequently used. The process yields high-performance materials with a wide range of desirable properties, making them ideal for use in civil engineering (Fig. 3.2) (Patil et al., 2022).

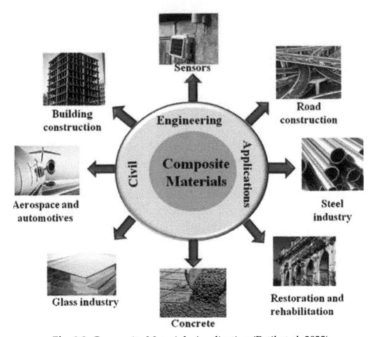

Fig. 3.2 Composite Materials Application (Patil et al. 2022).

2.7 Sustainable blocks

In the preparation of earth-based construction blocks, cementing ingredients are commonly used to stabilize the soil. Binding material innovation tends to focus on low carbon systems, many of which use alkaline activation (Campos Teixeira et al., 2020). As a mineral precursor, blast furnace slag (BFS) and olive stone biomass ash (OBA) were employed.. Alkali activator, a major source of carbon emissions, has been replaced with OBA, a less polluting alternative. Compacted dolomitic soil blocks were created by using a mixture of OBA/BFS and water. Making BFS-stabilized soil-compacted blocks with OBA as an alkali activator alternative has been found to emit just 3.3 $kgCO_2$/ton stabilized soil, 96% less CO_2 than standard Portland cement stabilization (see Fig. 3.3) (Patil et al., 2022).

Fig. 3.3 Sustainable blocks (Patil et al., 2022).

2.8 Electric arc furnace (EAF) steel slag

EAF is a steelmaking process that makes use of waste metal. Huge amount of all steel products worldwide is made from scrap steel. When three graphite electrodes form an electric arc, a great amount of heat is generated, causing the object to become liquid. It is easy to remove slags since they are lighter than liquid metal. After rapidly cooling the oxidized and excess liquid to room temperature from roughly 1600°C, a secondary crushing and filtering process produces EAF slag for use in the production of steel (see Fig. 3.4) (Patil et al., 2022).

2.9 Green plastic vs nonbiodegradable plastics

Green polymers, also known as biodegradable bio-based polymers, are promising solutions to a circular economy that can help solve the

Fig. 3.4 EAF slag (Patil et al. 2022).

problem of waste management (Shevchenko et al., 2022). Biomass or microorganisms, or synthetic polymers, are the most common sources of green polymers. Biomass-based green polymers can be made from a variety of natural resources, including animal or plant proteins, cellulose, carbohydrates, biogenic byproducts, and lactic acid. Forty-six percent of the market for biomass-based green polymers production is made up of biogenic residues in addition to the waste that is generated in the forestry and industrial sectors. Except for biogenic residues, carbohydrates, sugars, cellulose, and oils from edible and non-edible plants provide 20%, 17%, 8%, and 9%, respectively, of the biomass needed for green plastics (Nandy el al., 2022). Biomass feedstock for green polymers is still a long way off from fossil-based polymers in terms of volume, though. Only 3.8 million tons of biomass-based polymer were produced in 2019, which is less than 1% of the total production of conventional fossil-based plastics. PHAs, TPS, and PLA are the most promising biodegradable polymers on the market, with growth rates relative to each other of more than thrice. PLA is a biodegradable polyester made from lactic acid by fermenting carbohydrate sources such as corn starch or sugarcane under regulated conditions. A lactide monomer or lactic acid monomer can serve as the building components for this protein. It has been observed that the PLA bottle degrades completely in 30 days (Kale et al., 2007). There are a few uses for unaltered PLA because of its brittleness and low resistance to oxygen. Consequently, physical and chemical changes and blends, with other environmentally friendly polymers, are being used to increase PLA's processibility. TPS plastics, like PLA, have low stability and become brittle when water molecules are removed from them. Natural or synthetic polymers, fillers, and fibers can also be used to enhance the qualities of TPS (Nazrin et al., 2020). TPS are typically made from starch generated from sweet potatoes etc.) that is subjected to regulated conditions of shear, temperature, pressure, and water content to form semi-crystalline or amorphous polymers. TPS plastics, such as wraps, containers, disposable dining utensils, antistatic, loose fill, molded protective packaging, waste bags, etc., have recently been commercialized (Nazrin el al., 2020).

PHA, a microbial fermentation-derived polymer with great potential, is yet another intriguing biomass-based material. For example, PHA can be

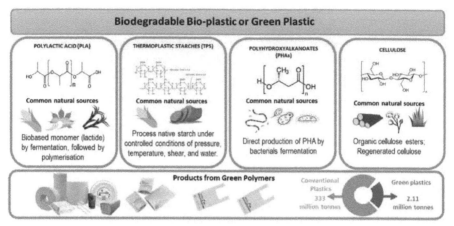

Fig. 3.5 Recent advancements in green technology (Nandy et al., 2022).

made into hard or soft plastics, as well as crystalline or amorphous materials. The UV resistance of these polymers is exceptional. The drawbacks of PHA are like those of other biomass-based polymers, including poor mechanical qualities, incompatibility with typical thermal processing procedures, and heat deterioration. It's still possible to improve the efficiency of biomass-based polymers through further chemical processing, despite these limitations (Vink et al., 2003). As per a 2019 global data assessment by European Bioplastics and the Nova-Institute, more than half of the 2.11 million tons of bioplastics generated globally were disposable biopolymers, comprising PHA, PLA, starch blends etc. Biodegradable plastic output is predicted to reach 1.33 million kilograms by 2024, mainly due to PHA's rapid expansion (Oksman et al., 2014).

3. Green Nanomaterials

Recently, nanomaterials have proven to be more effective in a variety of fields, such as medicine, energy, and advanced manufacturing. There are now a lot of products on the market that use nonrenewable resources and generate hazardous waste as a side effect. Sustainable society in the 21st century may be made possible by green nanotechnology, a mix of nanotechnology and the concepts and practices of green chemistry (Vergragt, 2006). To decrease or eliminate the use and generation of hazardous compounds, a chemical philosophy known as "green chemistry" is based on a set of principles (Krishnaswamy and Orsat, 2017). Environmentally friendly approaches to nanomaterial synthesis are becoming increasingly common in the field of green nanotechnology. As reducing agents for nanoparticle syntheses, nature supplies us with several chemical compounds including plant extracts and biopolymers, vitamins and proteins as well as glutathione and sugars (e.g., glucose) (Nikolaidis, 2020). Plant extracts have received the greatest attention so far. A natural reducing agent, plant extracts are seen as one of the most promising because of their abundance (Duan et al., 2015). For example, metal nanoparticles that can be used in electronics and medicinal applications can be synthesized utilizing plant extracts as reducing agents. The use of gold and silver nanoparticles in biomedical applications, such as medication and gene delivery, has recently become a hot study topic. Nontoxic green reduction agents, such as plants, algae, bacteria, and fungus, are employed to enhance biocompatibility. As a reducing and stabilizing agent, Elia et al. used Salvia officinalis, Lippia citriodora, Pelargonium graveolens, and Punica granatum extracts to produce gold nanoparticles (Elia et al., 2014). In the manufacture of metal nanoparticles, reducing and stabilizing agents derived from biopolymers are also common. For the large-scale synthesis of nanoparticles, polymeric carbohydrate molecules have already been used in numerous industries. Biopolymers such as cellulose, chitosan, and dextran, which are derived from plants, crustacean exoskeletons, and sugarcane, can be used in the manufacture of nanoparticles (Elia et al., 2014).

3.1 Intelligent composite materials

Using intelligent composite materials to construct buildings reduces the likelihood of costly and time-consuming mistakes. In concrete and road construction, self-sensors, carbon fibers, and other materials are used (Monteiro et al., 2017; Farina et al., 2022). For smart gypsum boards, PCM (phase change material) nanoencapsulation is employed in amounts varying from 1% to 30% of the board's weight. Its substantial thermal energy content made this composite material effective at lowering the temperature in the room (Mohseni et al., 2019). Smart houses are equipped with a variety of features, such as sensors and actuators, networking, and energy efficiency. Smart houses are designed to save money and energy when they are built. Fiber optics, piezoelectric materials, magnetic-rheological (MR) fluids, and electro-rheological fluids (ER) are all covered in great depth. Studies believe that MRs are much more beneficial than ERs due to the lower yield stress and higher voltage energy delivery of ERs. SMAs are vital in aseismic design since they act as an immensely efficient means of isolation. Smart structures use IoT technologies to collect information.

3.2 Nanoengineering

Nanotubes of carbon Because of their hexagonal structure and the fact that they are carbon nanotubes, it is possible to imagine them as rolled-up graphite sheets. Dome-shaped half fullerene molecules cap the ends. The mechanical characteristics of nanotubes are strongly influenced by the atomic configuration in the nanostructure. Carbon nanotubes (CNTs) are produced in both single- and multi-walled varieties (MW CNTs) (Heise et al., 2009). It has greatly improved the mechanical, physical, and chemical properties of the materials, making it possible to overcome many of the constraints of conventional materials. Construction materials based on cementitious binders have a wide range of properties that are influenced by micro and nanoscale formations (Mukhopadhyay, 2011). CNsT exhibit a prospective role to modify/improve the features of conventional construction materials such as concrete and steel to generate high performance construction material. If certain difficulties, such as the uniform distribution of CNTs in the composite and the bond behavior of CNT-modified concrete, can be solved, a crack-free, durable concrete can be achieved. When it comes to a nanotube structure, the grapheme's symmetry and unique electronic structures substantially influence its electrical properties. Compared to copper, CNTs have a conductivity eight times greater. There are potential possibilities for nanoscale wires and efficient sensing purposes since they can carry current densities comparable to those of any regular metallic wire. Pham et al. (Pham et al., 2008) have developed carbon nanotubes polymer composite sheets that can be employed as a strain sensor. Li, Thostenson and Chou (Li et al., 2008) and others have reported on the use of carbon nanotubes as strain sensors for civil structures' structural health monitoring systems. Concrete nanotubes can, in other words, be an important aspect of civil engineering's ability to reimagine its position in society. x Compressive strength formula for carbon nanotube cement composite Compressive strength of a CNT/cement composite can be computed using an analytical method described by Ghasemzadeh et al. (Ghasemzadeh et al., 2011). Using Von Mises'criterion, elasticity relationships, and Representative Elementary Volume (REV) as a component with an indicator, the composite was examined.

3.3 *Nano concrete with carbon nanotube*

A new type of concrete, known as nano concrete, is being developed by researchers at the University of Minnesota. As a new technology, qualities of CNTs are continuously being researched. In addition, this is an important milestone for nanotechnology. A wide array of Carbon Nanotubes and specific Carbon Nanotube compounds are available for research and development of novel applications from Nanocyl (Fig. 3.6) (Bi et al., 2012).

When stress is applied to CNTs, the electrical potential of the material changes. For sensor applications, the relationship between stress and changes in electric potential is reliable and extremely linear. Concrete could be transformed into a smarter, more environmentally friendly material by incorporating it into civil engineering projects. Experimental studies have examined the mechanical characteristics of cementitious composites

Fig. 3.6 Nanocyl NC 7000 (Bi et al., 2012).

containing CNTs with diameters of 10 and 20 microns (Yazdanbakhsh et al., 2007). Compressive and flexural strength tests at 7 and 28 days showed that the cementitious composites containing CNTs had a higher compressive and flexural strength than those with no CNTs. SEM analysis of the microstructures of the composites containing and excluding CNTs revealed that the CNTs filled the pore space in the composites under study.

3.4 Self-sensing carbon nanotubes

CNT/cement composites with self-sensing properties were developed by Yu and Kwon (Kwon, E. and Yu, 2012). CNT's piezoresistive characteristic enables the composite to detect pavement stress/stain. Concrete pavements can be strengthened and made more durable by using CNTs as reinforcing elements. Researchers have designed and tested CNT/cement composites that are piezoresistive. When compressive stress levels increase, so does the electrical resistance in the composite. Composites made using various fabrication methods are also examined for their piezoresistive responses. Dispersing CNTs into the cement matrix is made easier with the use of surfactant coated CNTs, which exhibit promising piezoresistive capabilities. The self-sensing concrete was put to the test by applying dynamic loads in a controlled environment in both the lab and on the road. Researchers found that CNT/cement composites are effective stress and strain sensors in tests [79].

3.5 *Green chemistry and industrial ecology*

Durability, maintenance, and environmental problems can all be addressed by improving the cement used in concrete. The use of cements with alternative compositions, binding phases, and green chemistry holds great promise. For example, mineral wastes and recycled resources can be used to produce cement that requires less energy to produce. Introducing industrial ecology and green chemistry concepts as a driving force for the study, development, and commercial appeal of alternative and sustainable cements was introduced by researchers (Demaria et al., 2013). According to Phair [80], Alkali-activated cements, magnesium, and sulfoaluminate-based cements, as well as typical Portland cements, were analyzed and evaluated in terms of their chemistry and characteristics. An emphasis is placed on the more environmentally friendly alternative cements' physical and chemical qualities, as well as their performance in various construction settings. In terms of environmental, engineering, and economic properties, alternative cements offer a lot of potential. Inorganic polymers (geopolymers) have been reported to be used in sustainable concrete by Duxson et al. (Duxson et al., 2007), for example. Alkali-activated aluminosilicates have lower CO_2 emissions than Portland cement, which is used to make it. The produced concrete is said to be strong enough and chemically resistant enough as well. To obtain the same level of quality as high-performance concrete, geopolymer concrete must also be manufactured. However, successful application of this method necessitates a solid grasp of geo-polymerization chemistry. Microstructure and mechanical characteristics of aggregates (compressive strength and tensile strength) Torgal et al. (Pacheco-Torgal et al., 2007) propose a GMWM binder made from geopolymer mine waste (2007). The ratio of aggregate to binder, aggregate size, and aggregate type are just a few of the aggregate considerations (schist, granite, and limestone). However, the data obtained show that GMWM has strong power (press and drag) and grain size effects on tensile strength at a young age Inorganic polymer (geopolymer) technology has been studied by Duxson et al. (Duxson et al., 2007) in the construction of green concrete. Alkali-activated aluminosilicates are a new kind of cement that emits a fraction of the greenhouse gas emissions of standard Portland cements while exhibiting superior strength and chemical resistance, as well as a host of other intriguing qualities. One of the biggest obstacles to widespread use of geopolymer technology is a lack of long-term durability data (more than 20 years). Additionally, various regulations in Europe and North America, such as minimum clinker concentration levels and chemical compositions in cements, are difficult to meet. With accelerated durability testing revealing encouraging improvements for salt scaling and freeze–thaw cycles, these challenges are still being worked on. As with most high strength concrete, geopolymer concrete complies with performance-based requirements (Noushini and Castel, 2018).

4. Sustainability Impacts: Waste management and Green Economy

Materials research and engineering must develop a new way to modernize ecotechnology systems while simultaneously reducing waste products and the continued reliance on raw materials and recycling disposals at a high level, given the current worrying scenario. Circular economy and the use of as many biocompatible materials as feasible will assist us in resolving this issue (Brennan et al., 2015). Using natural resources and disposing of garbage is not a priority in a traditional economic system, which is based on a scarcity mentality. The three pillars of sustainability, community, environment, and economics are ignored in this traditional economy structure (Lehtonen, 2004). The concept of a "circular economy" has been introduced to make a significant impact on the future of the planet's sustainability. The following industries are the primary focus of this concept: By minimizing the use of raw/natural resources, increasing product life and reusability, as well as recycling waste, we aim to reduce our environmental impact. Together with the UN Environment Program (UNEP), UNIDO has been running a program called Resource Efficient and Cleaner Production (RECP) since 1994 under a shared flagship. research focuses on more cost-effective and environmentally friendly production methods (RECP). Reducing the impact of industrial activity on the environment, increasing economic output, and enhancing the well-being of workers and local communities are all part of the RECP's sustainable strategy. There are already various eco-industrial parks promoted by UNIDO around the world (UNIDO's Environment Solution, Resource Efficient and Cleaner Production (RECP)) [86]. The goal is to create an impact on the circular economy and a long-term sustainable future by extending resource efficiency beyond the bounds of a single organization. UNIDO set a 2017 deadline for the completion of 33 eco-industrial parks in 12 developing countries, including Vietnam, China, Colombia, Ethiopia, India, Indonesia, Morocco, Peru, Senegal, and South Africa, to demonstrate the immense potential of the eco-industrial park. By this time, between 2016 and 2018, participating firms in Viet Nam invested approximately $11 million USD in RECP adoption, and the total annual return was $9.6 million. Cement co-processing industries in Vietnam's Can Tho province save 7.5 million tons of coal and cut CO_2 emissions by 14.5 million tons annually (Report, 2014–2019). Eco-towns (or EIPs) were built by the Japanese Ministry of Environment and the Ministry of Economy, Trade, and Industry (METI) with a total investment of more than 1.5 billion USD, where 47 recycling plants were constructed, and 732,000 tons of garbage was reused or recycled (METI (Ministry of Economy, Trade and Industry) (2006)). Using recycled materials and saving 272,000 tons of raw material, a cement firm in Kawasaki Eco-town cut CO_2 emissions by 43,000 tons annually by 2009, according to the city's waste management report (Hashimoto et al., 2010). Japan is making steady progress toward a circular economy, with a current recycling rate of 70% and a target of 80% by the end of the decade. For the circular economy and EIPs, China and South Korea have national strategies like Japan's. EIPs in China recycled plastics to save 14 million metric tons of greenhouse gases in 2016 (Liu et al., 2018). Around CNY 70 trillion (11.2 trillion USD) in spending on high-quality items and services might be saved through China's circular economy by the year 2040, comparable to 16% of China's predicted GDP (Morlet, 2018). In Europe, the concept of a circular economy first evolved in the 1980s and 1990s but has only recently been put into practice. Several circular economy action plans have been adopted with the active involvement of the European Commission (EC) and all EU countries, with priority given to specific sectors such as electronics, information and communication technology, batteries, construction and buildings, textiles, packaging, plastics, and chemical industries (Pearce et al., 1990). Circular economy principles in the EU can create 700,000 new jobs by 2030, according to a recent study (Econometrics, 2018), with a 0.5% increase in EU GDP. For our planet's sake, it is imperative that we embrace the principles of the circular economy and green technologies.

5. Emerging Green Processing Technologies

Emerging technologies are the method for the evolution of green buildings under varied circumstances. For example, in a wall technology the temperatures inside and out of a structure were kept stable during periods of extreme heat or cold. Using a computer model, researchers created a prefabricated wall that can withstand extreme heat and cold without compromising the comfort of its occupants. The results (Wu and Ge, 2018) shows that the new technology is more technically and economically feasible than the traditional approach. In China, each residential building has a retrofit system installed to treat the wastewater created by that building. An anaerobic-anoxic-oxic membrane bioreactor (A2O-MBR) system is used to recycle the wastewater created in the green building. Analytical methods are used to gather and examine samples of waste materials. For nearly a year, samples have been analyzed in a variety of ways (Wang et al., 2019; Ge et al., 2020). Thirty-seven different green building techniques for low- and moderate-income dwellings were examined in this study. In this study, EQT and ELT models with grading indicators for maturity, economics, environmental burden, and quality were developed. ELT seeks to reduce the number of resources used to operate a building, whereas EQT seeks to improve the structure's interior spaces. Therefore, 11 technologies have been selected to suit the requirements such as the interior environment, material classification, energy consumption, water usage and resource implementations, etc. BIM and ontology models have been promoted by researchers as a foundation for the development of smart buildings. This kind of technology is useful for a wide range of multidisciplinary tasks. The jess rule engine is used to combine SWRL and OWL models to produce the most accurate outcomes during construction and operation. Ontology models, BIM, and other tools are used to identify any potential problems before they occur on the construction site. An ontology model-based BIM system organizes all processes in a methodical manner, reducing construction time while maintaining quality (Khattra et al., 2021). Each device's energy consumption is assessed to adapt existing buildings into green buildings. Green Retrofit Technologies and Green Retrofit Policies are then released to the public for education and awareness raising purposes. A Technology Acceptance Model (TAM) has been presented in Iran, which considers a variety of elements, including public awareness, education, environmental concerns, and an estimate of the project's cost. This methodology helps governments adopt and execute green construction practices (Rajaee et al., 2019). The effectiveness of photovoltaic cooling systems for facades is determined for a variety of environmental circumstances. The cooling demand is met by a solar system with an overall efficiency of up to 73.8% (Bröthaler et al., 2021). When it comes to the building process, Singapore has incorporated 5G technology to develop environmentally friendly structures. AI, (IoT), Machine Learning (ML) Blockchain, traffic monitoring, and other technologies are employed in these low-cost constructions. Aside from being more expensive, traditional plywood formwork had a higher energy footprint. Combining free-form panel (FCP) and 3D plastering techniques solves this issue. It minimizes both the project's cost and the time it takes to complete. A case study of the "Dongdaemun Design Plaza" skyscraper demonstrates that 3DPT saved 1,196 tons of CO_2 in the study's conclusion (Son et al., 2021).

In Table 3.1, various emerging technologies have been discussed based on the green materials processing. It has been seen that the processing technologies also plays a vital role on the environment. As the technology to be used also leave immense amounts of carbon footprints hence affect the environment. The green material along with the better technologies would pave the way for the cleaner and greener environment.

Table 3.1 The technology processes in green context (Patil et al., 2022).

	Materials and Methods	Features	Conclusion
1	A2O-MBR treatment of wastewater.	According to the results of a study, the effluent removal ratio is higher, and the cost-benefit ratio is reached.	An estimated 61% of the water can be regenerated using the suggested technology, which shows that it is both technically and economically viable.
2	The application of GBT to affordable housing.	GBTs are validated using case studies, and then low-cost techniques are proposed, EQT and ELT models are provided, and a framework for construction and cost management is proposed.	Any person who wants to lessen their carbon impact on the environment is given a wide range of low-cost options.
3	BIM and ontology implementation to produce intelligent green buildings.	Ontology and BIM assessment methods are used to check the design prior to the construction phase. Time and money can be saved by using SWRL and Jess rule ontology models. After a case study was completed, the results were analyzed.	Engineers, contractors, and others benefit from the method's advantages because most activities are shown before spending money. In addition, complications and mistakes are kept at bay.
4	New wall technology for climate zones with hot summers and cold winters in green buildings.	Energy-efficient prefabricated wall. Inside, the temperature has dropped significantly. To evaluate the wall's environmental impact, the Flotherm software is employed.	This technology's properties are examined at a specific temperature. Efforts to improve energy efficiency are made, and results are better.
5	Green building retrofitting technology and policies.	Based on the rankings, methods for selecting energy-efficient equipment and appliances, for reducing the energy consumption of air conditioners, for reducing tax policies, for developing building standards, and so on, are offered.	Rather than developing new structures, a green retrofit of an existing building can provide an immediate solution. People are more likely to support green building policies if they see them in action.
6	An Iranian socio-psychological template for green construction.	Introducing the Technology Acceptance Model (TAM). There are five primary components to this paradigm, and they are: subjective understanding, social influence, social trust, environmental attitudes, and perceived costs.	The TAM Model is a useful tool for educating people in underdeveloped nations about green building practices. In GBTs, the Iranian government has taken the lead in encouraging stakeholders to use green technologies.

Table 3.1 contd. ...

Table 3.1 contd....

	Materials and Methods	Features	Conclusion
7	A Façade Integrated Photovoltaic-Powered Cooling System's Performance Evaluation.	Power for the COOLSKIN system comes from solar panels. Outside the building, a one-year case study evaluates the system's effectiveness.	After a particular amount of time, a system begins to behave in a consistent fashion. Behavior in severe weather is crucial due to voltage drop difficulties that have been encountered.
8	5G technology in Singapore enabling smart energy governance and smart buildings.	During the phases of building, operation, and maintenance, 5G technology analyses the data. Smart sensors, energy management, blockchain, artificial intelligence, machine learning, and DRL techniques, among others, are being discussed.	Using 5G technology, smart buildings can be built at low cost with great performance. Singapore's carbon footprint has decreased by up to 36% in the past 25 years.
9	Smart house innovation.	It enhances automation, remote access, and monitoring of daily tasks. Many components in smart homes contribute to listening and measuring, communication, control, and processing data.	The goal of a smart home is to use less energy, save money on utility bills, and make it easier to manage all the linked applications, such as electrical appliances, HVAC systems, and lighting.
10	3D Plastering technique.	It is reduced because of this research, embodied CO_2, and free-form concrete panel production (FCP) have been examined. When compared to the conventional approach, this technology can minimize CO_2 emissions by up to 99%.	This method allows for the fabrication of free-form concrete panels, which decreases both time and cost. Suitable for small and medium-sized homes.

6. Conclusions

The development of green technology has made it increasingly viable to utilize ecologically friendly materials in construction. It is possible to reduce the embodied energy and maximize the usage of renewable resources by using recycled industrial waste. Green materials and smart material concept are getting more and more attention these days the cause of diverse range of benefits and properties. green materials are the alternative to those currently available materials. current materials are abundantly available but are harmful and contribute towards CO_2 emission hence greenhouse effect. In this chapter the light has been shed on various kinds of green/smart and green nanomaterials with the green processing technologies. green technologies are also very important in the generation of green materials and combination of green technologies and green materials could be the solution for the next generation materials. It has been seen that there are many green materials which are being used nowadays such as eggshells, FCS, Bamboo fibre, plastic waste, ECC, composite materials, sustainable blocks, and EAF steel slag, etc. Plastic turned out to be the biggest contributors towards being green materials. Since the plastic waste is abundantly available. In addition, composite materials have also been contributing immensely to the green environment. Impact of sustainability on green economy has also been discussed in the chapter along with the emerging technologies and the process of green materials. Green nanomaterials such as intelligent composite material, nano engineering, nano concrete with carbon nano tubes, self-sensing carbon nanotubes, nano silica have also marked their presence in green materials. This chapter covers the wide range of green concepts in terms of materials, and technology and would provide the concrete knowledge on such topics.

References

Agyekum, K., Kissi, E. and Danku, J.C. (2020). Professionals' views of vernacular building materials and techniques for green building delivery in Ghana. *Scientific African*, 8: e00424.

Ahmad, T. and Zhang, D. (2020). A critical review of comparative global historical energy consumption and future demand: The story told so far. *Energy Reports*, 6: 1973–1991.

Al-Jabri, K.S. (2018). Research on the use of Ferro-Chrome slag in civil engineering applications. *MATEC Web of Conferences, EDP Sciences*, p. 01017.

Banik, R.L. (2015). *Bamboo Silviculture*. Springer, pp. 113–174.

Behera, M., Bhattacharyya, S., Minocha, A., Deoliya, R. and Maiti, S. (2014). Recycled aggregate from C&D waste & its use in concrete–A breakthrough towards sustainability in construction sector: A review. *Construction and Building Materials*, 68: 501–516.

Bernatas, R., Dagreou, S., Despax-Ferreres, A., and Barasinski, A. (2021). Recycling of fiber reinforced composites with a focus on thermoplastic composites. *Cleaner Engineering and Technology*, 5: 100272.

Bi, J., Hariandja, B., Imran, I. and Pane, I. (2012). Material Development of Nano Silica Indonesia for Concrete Mix. *Advanced Materials Research*, Bach, Switzerland: Trans. Tech. Publ., pp. 277–280.

Borowski, P.F., Patuk, I. and Bandala, E.R. (2022). Innovative Industrial Use of Bamboo as Key 'Green' Material. *Sustainability*, 14(4): 1955.

Brennan, G., Tennant, M. and Blomsma, F. (2015). Business and production solutions: Closing loops and the circular economy. *In*: H. Kopnina and E. Shoremn-Ouimet (Eds.). *Sustainabilty*. Eartscan/Routledge, pp. 219–239.

Bröthaler, T., Rennhofer, M., Brandl, D., Mach, T., Heinz, A., Újvári, G., Lichtenegger, H.C. and Rennhofer, H. (2021). Performance Analysis of a Facade-Integrated Photovoltaic Powered Cooling System. *Sustainability*, 13(8): 4374.

Campos Teixeira, A.H., Soares Junior, P.R.R., Silva, T.H., Barreto, R.R. and De Silva Bezerra, A.C. (2020). Low-Carbon Concrete Based on Binary Biomass Ash–Silica Fume Binder to Produce Eco-Friendly Paving Blocks. *Materials*, 13(7): 1534.

DeBoer, D. and Bareis, K. (2000). Bamboo building and culture Deboer Architects. Recuperado de http://jubilee101. com/subscription/pdf/Bamboo-Construction/Bamboo-Building-and-Culture--27pages. pdf.

Del Moral, B., Baeza, F.J., Navarro, R., Galao, O., E. Zornoza, E.,Vera, J., Farcas, C. and Garcés, P. (2021). Temperature and humidity influence on the strain sensing performance of hybrid carbon nanotubes and graphite cement composites. *Construction and Building Materials*, 284: 122786.

Demaria, F., Schneider, F., Sekulova, F. and Martinez-Alier, J. (2013). What is degrowth? From an activist slogan to a social movement. *Environmental Values*, 22(2): 191–215.

Duan, H., Wang, D. and Li, Y. (2015). Green chemistry for nanoparticle synthesis. *Chemical Society Reviews*, 44(16): 5778–5792.

Duan, Z., Huang, Q. and Zhang, Q. (2022). Life cycle assessment of mass timber construction: A review. *Building and Environment*, 109320.

Duxson, P., Provis, J.L., Lukey, G.C. and Van Deventer, J.S. (2007). The role of inorganic polymer technology in the development of 'green concrete'. *Cement and Concrete Research*, 37(12): 1590–1597.

Dwivedi, A.J., Kumar, A., Baredar, P. and Prakash, O. (2019). Bamboo as a complementary crop to address climate change and livelihoods: Insights from India. *Forest Policy and Economics*, 102: 66–74.

Econometrics, C. (2018). *Impacts of Circular Economy Policies on the Labour Market*. Final report and annexes,

Elia, P., Zach, R., Hazan, S., Kolusheva, S., Porat, Z.E. and Zeiri, Y. (2014). Green synthesis of gold nanoparticles using plant extracts as reducing agents. *International Journal of Nanomedicine*, 9: 4007.

Environment, U., Scrivener, K.L., John, V.M. and Gartner, E.M. (2018). Eco-efficient cements: Potential economically viable solutions for a low-CO2 cement-based materials industry. *Cement and Concrete Research*, 114: 2–26.

Fares, A.I., Sohel, K., Al-Jabri, K. and Al-Mamun, A. (2021). Characteristics of ferrochrome slag aggregate and its uses as a green material in concrete: A review. *Construction and Building Materials*, 294: 123552.

Fares, A.I., Sohel, K.M.A., Al-Jabri, K. and Al-Mamun, A. (2021). Characteristics of ferrochrome slag aggregate and its uses as a green material in concrete: A review. *Construction and Building Materials*, 294.

Farina, I., Moccia, I., Salzano, C., Singh, N., Sadrolodabaee, P. and Colangelo, F. (2022). Compressive and Thermal Properties of Non-Structural Lightweight Concrete Containing Industrial Byproduct Aggregates. *Materials*, 15(11): 4029.

Favier, A., De Wolf, C., Scrivener, K. and Habert, G. (2018). *A sustainable future for the European Cement and Concrete Industry: Technology assessment for full decarbonisation of the industry by 2050*. ETH Zurich.,

Fletcher, R. (2011). Sustaining tourism, sustaining capitalism? The tourism industry's role in global capitalist expansion. *Tourism Geographies*: 13(3) 443–461.

Ge, J., Zhao, Y.,Luo, X., and Lin, M. (2020). Study on the suitability of green building technology for affordable housing: A case study on Zhejiang Province, China. *Journal of Cleaner Production*, 275: 122685.

Ghasemzadeh, H. and Akbari Jalalabad, E. (2011). Computing the compressive strength of carbon nanotube/cement composite. *International Journal of Civil Engineering*, 9(3) 223–229.

Guezuraga, B. Zauner, R. and Pölz, W. (2012). Life cycle assessment of two different 2 MW class wind turbines. *Renewable Energy*, 37(1): 37–44.

Hashimoto, S. Fujita, T., Geng, Y. and Nagasawa, E. (2010). Realizing CO2 emission reduction through industrial symbiosis: A cement production case study for Kawasaki. *Resources, Conservation and Recycling*, 54(10): 704–710.

Heise, H., Kuckuk, R., Ojha, A., Srivastava, A., Srivastava, V. and Asthana, B. (2009). Characterisation of carbonaceous materials using Raman spectroscopy: A comparison of carbon nanotube filters, single-and multi-walled nanotubes, graphitised porous carbon and graphite. *Journal of Raman Spectroscopy: An International Journal for Original Work in all Aspects of Raman Spectroscopy, Including Higher Order Processes, and also Brillouin and Rayleigh Scattering*, 40(3): 344–353.

Huang, Z., Lu, Y., Wong, N.H. and Poh, C.H. (2019). The true cost of 'greening' a building: Life cycle cost analysis of vertical greenery systems (VGS) in tropical climate. *Journal of Cleaner Production*, 228: 437–454.

Hut, M.N.S. (2014). *The Performance of Eggshell Powder as an Additive Concrete Mixed.*, UMP.

Kale, G., Auras, R. Singh, S.P. and Narayan, R. (2007). Biodegradability of polylactide bottles in real and simulated composting conditions. *Polymer Testing*, 26(8): 1049–1061.

Keoleian, G., Kendall, A.M., Lepech, M.D. and Li, V.C. (2005). Guiding the design and application of new materials for enhancing sustainability performance: Framework and infrastructure application. *MRS Online Proceedings Library* (OPL), 895.

Khattra, S.K., Singh, J. and Rai, H.S. (2021). A Statistical Review to Study the Structural Stability of Buildings Using Building Information Modelling. *Archives of Computational Methods in Engineering*, 1–18.

Khunt, Y., Khamar, N., Nathwani, V., Joshi, T. and Dave, U. (2022). Investigation on mechanical parameters of concrete having sustainable materials. *Materials Today: Proceedings*

Krishnaswamy, K. and Orsat, V. (2017). Sustainable delivery systems through green nanotechnology. *In: Nano-and Microscale Drug Delivery Systems*. Elsevier, pp. 17–32.

Kwon, E. and Yu, X. (2012). Carbon Nanotube Based Self-Sensing Concrete for Pavement Structural Health Monitoring. University of Minnesota, Duluth. URL: https://rosap.ntl. bts.gov/view/dot/40249.

Le, D.-H., Sheen, Y.-N. and Lam, M.N.-T. (2018). Fresh and hardened properties of self-compacting concrete with sugarcane bagasse ash–slag blended cement. *Construction and Building Materials*, 185: 138–147.

Lehtonen, M. (2004). The environmental–social interface of sustainable development: Capabilities, social capital, institutions. *Ecological Economics*, 49(2): 199–214.

Li, C., Thostenson, E.T. and Chou, T.-W. (2008). Sensors and actuators based on carbon nanotubes and their composites: A review. *Composites Science and Technology*, 68(6): 1227–1249.

Li, Y., Rong, Y., Ahmad, U.M., Wang, X., Zuo, J. and Mao, G. (2021). A comprehensive review on green buildings research: Bibliometric analysis during 1998–2018. *Environmental Science and Pollution Research*, 28(34): 46196–46214.

Liu, Z., Adams, M., Cote, R.P., Chen, Q., Wu, R., Wen, Z., Liu, W. and Dong, L. (2018). How does circular economy respond to greenhouse gas emissions reduction: An analysis of Chinese plastic recycling industries. *Renewable and Sustainable Energy Reviews*, 91: 1162–1169.

Lobovikov, M., Paudel, S., Ball, L. Piazza, M., Guardia, M., Wu, J. and Ren, H. (2005). World bamboo resources: A thematic study prepared in the framework of the global forest resources assessment. Food & Agriculture Org., 2007.

Lu, W., Chi, B., Bao, Z. and Zetkulic, A. (2019). Evaluating the effects of green building on construction waste management: A comparative study of three green building rating systems. *Building and Environment*, 155: 247–256.

Mardziah, C., Ramesh, S., Tan, C., Chandran, H., Sidhu, A., Krishnasamy, S. and Purbolaksono, J. (2021). Zinc-substituted hydroxyapatite produced from calcium precursor derived from eggshells. *Ceramics International*, 47(23): 33010–33019.

METI (Ministry of Economy, Trade and Industry). (2006). Ex-post Evaluation Report on the Construction and Improvement of Recycling Plants and Facilities, Ministry of Economy, Trade and Industry, [in Japanese].

Mieldažys, R., Jotautienė, E. and Jasinskas, A. (2019). The opportunities of sustainable biomass ashes and poultry manure recycling for granulated fertilizers. *Sustainability*, 11(16): 4466.

Mirji, R. and Lobo, B. (2017). Computation of the mass attenuation coefficient of polymeric materials at specific gamma photon energies. *Radiation Physics and Chemistry*, 135: 32–44.

Mohseni, E., Tang, W. and Wang, S. (2019). Development of thermal energy storage lightweight structural cementitious composites by means of macro-encapsulated PCM., *Construction and Building Materials*, 225: 182–195.

Monteiro, A.O., Cachim, P.B. and Costa, P.M. (2017). Self-sensing piezoresistive cement composite loaded with carbon black particles. *Cement and Concrete Composites*, 81: 59–65.

Moore, C.J. (2008). Synthetic polymers in the marine environment: A rapidly increasing, long-term threat. *Environmental Research*, 108(2): 131–139.

Morlet, A. (2018). *The Circular Economy Opportunity for Urban and Industrial Innovation in China*. Ellen MacArthur Foundation.

Mukhopadhyay, A.K. (2011). Next-generation nano-based concrete construction products: A review. *Nanotechnology in Civil Infrastructure*, 207–223.

Nandhini, K. and Karthikeyan, J. (2022). Sustainable and greener concrete production by utilizing waste eggshell powder as cementitious material: A review. *Construction and Building Materials*, 335: 127482.

Nandy, S., Fortunato, E. and Martins, R. (2022). Green economy and waste management: An inevitable plan for materials science. *Progress in Natural Science: Materials International*.

Nandy, S., Fortunato, E. and Martins, R. (2022). Green economy and waste management: An inevitable plan for materials science. *Progress in Natural Science: Materials International*, 32(1): 1–9.

Nazrin, A, Sapuan, S., Zuhri, M., Ilyas, R., Syafiq, R. and Sherwani, S. (2020). Nanocellulose reinforced thermoplastic starch (TPS), polylactic acid (PLA), and polybutylene succinate (PBS) for food packaging applications. *Frontiers in Chemistry*, 8: 213.

Nematzadeh, M., Shahmansouri, A.A. and Fakoor, M. (2020). Post-fire compressive strength of recycled PET aggregate concrete reinforced with steel fibers: Optimization and prediction via RSM and GEP. *Construction and Building Materials*, 252: 119057.

Ngayakamo, B. and Onwualu, A.P. (2022). Recent Advances in Green Processing Technologies for Valorisation of Eggshell Waste for Sustainable Construction Materials. *Heliyon*, e09649.

Nikolaidis, P. (2020). Analysis of green methods to synthesize nanomaterials *In: Green Synthesis of Nanomaterials for Bioenergy Applications*. Cyprus University of Technology, pp. 125–144.

Noushini, A. and Castel, A. (2018). Performance-based criteria to assess the suitability of geopolymer concrete in marine environments using modified ASTM C1202 and ASTM C1556 methods. *Materials and Structures*, 51(6): 1–16.

Oksman, K., Mathew, A.P., Bismarck, A., Rojas, O. and Sain, M. (2014). *Handbook of Green Materials: Processing Technologies, Properties and Applications* (in 4 volumes). World Scientific.

Pacheco-Torgal, F., Castro-Gomes, J. and Jalali, S. (2007). Investigations about the effect of aggregates on strength and microstructure of geopolymeric mine waste mud binders. *Cement and Concrete Research*, 37(6): 933–941.

Patel, P. and Patel, A. (2021). Use of sustainable green materials in construction of green buildings for sustainable development. *In: IOP Conference Series: Earth and Environmental Science*. IOP Publishing, p. 012009.

Patil, M., Boraste, S. and Minde, P. (2022). A comprehensive review on emerging trends in smart green building technologies and sustainable materials., *Materials Today: Proceedings*.

Pearce, D.W. Turner, R.K. and Turner, R.K. (1990). *Economics of Natural Resources and the Environment*. Johns Hopkins University Press.

Pellegrino, C., Faleschini, F. and Meyer, C. (2019). Recycled materials in concrete. *Developments in the Formulation and Reinforcement of Concrete*, 19–54.

Phair, J.W. (2006). Green chemistry for sustainable cement production and use. *Green Chemistry*. 8(9): 763–780.

Pham, G.T., Y.-B. Park, Liang, Z., Zhang, C. and Wang, B. (2008). Processing and modeling of conductive thermoplastic/carbon nanotube films for strain sensing. *Composites Part B: Engineering*, 39(1): 209–216.

Plevin, R.J., Delucchi, M.A. and Creutzig, F. (2014). Using attributional life cycle assessment to estimate climate-change mitigation benefits misleads policymakers. *Journal of Industrial Ecology*, 18(1): 73–83.

PS, L.P. and Anu, V. (2018). Use of Ferrochrome slag as aggregate in concrete: A review. *IRJET*, 5(11) Nov.

Rafey, A. and Siddiqui, F.Z. (2021). A review of plastic waste management in India: Challenges and opportunities. *International Journal of Environmental Analytical Chemistry*, 1–17.

Rahman, F.A., Aziz, M.M.A., Saidur, R., Bakar, W.A.W.A., Hainin, M., Putrajaya, R. and Hassan, N.A. (2017). Pollution to solution: Capture and sequestration of carbon dioxide (CO_2) and its utilization as a renewable energy source for a sustainable future. *Renewable and Sustainable Energy Reviews*, 71: 112–126.

Rajaee, M., Hoseini, S.M. and Malekmohammadi, I. (2019). Proposing a socio-psychological model for adopting green building technologies: A case study from Iran. *Sustainable Cities and Society*, 45: 657–668.

Rajesh, K.N., Raju, P.M., Mishra, K. and Madisetti, P.K. (2021). A review on sustainable concrete mix proportions. *In: IOP Conference Series: Materials Science and Engineering*. IOP Publishing, p. 012019.

Rashad, A.M. (2015). An investigation of high-volume fly ash concrete blended with slag subjected to elevated temperatures. *Journal of Cleaner Production*, 93: 47–55.

Report, P. (2014–2019). Eco-Industrial Park Initiative for Sustainable Industrial Zones in Viet Nam, Ministry of Planning and Investment of Vietnam.

Roy, S. Mishra, H. and Mohapatro, B. (2016). Creating Sustainable environment using smart materials in smart structures. *Indian Journal of Science and Technology*, 9(30).

Saha, A.K. and Sarker, P.K. (2017). Sustainable use of ferronickel slag fine aggregate and fly ash in structural concrete: Mechanical properties and leaching study. *Journal of Cleaner Production*, 162: 438–448.

Sahu, N., Biswas, A. and Kapure, G.U. (2016). A short review on utilization of ferrochromium slag. *Mineral Processing and Extractive Metallurgy Review*, 37(4): 211–219.

Shevchenko, T., Ranjbari, M., Shams Esfandabadi, Z., Danko, Y. and Bliumska-Danko, K. (2022). Promising Developments in Bio-Based Products as Alternatives to Conventional Plastics to Enable Circular Economy in Ukraine. *Recycling*, 7(2): 20.

Shiferaw, N., Habte, L., Thenepalli, T. and Ahn, J.W. (2019). Effect of eggshell powder on the hydration of cement paste. *Materials*, 12(15): 2483.

Singh, N. (2021). *Additive Manufacturing with Functionalized Nanomaterials*. Elsevier, pp. 1–34.

Singh, N. and Singh, G. (2021). Advances in polymers for bio-additive manufacturing: A state of art review. *Journal of Manufacturing Processes*, 72: 439–457.

Son, S., Lee, D., Oh, J. and Kim, S. (2021). Embodied CO_2 Reduction Effects of Free-Form Concrete Panel Production Using Rod-Type Molds with 3D Plastering Technique. *Sustainability*, 13(18): 10280.

Suhendro, B. (2014). Toward green concrete for better sustainable environment., *Procedia Engineering*, 95: 305–320.

UNIDO's Environment Solution, Resource Efficient and Cleaner Production (RECP) Programme. https://www.unido.org.

Vergragt, P.J. (2006). How technology could contribute to a sustainable world. *GTI Paper Series*, 28.

Vink, E.T., Rabago, K.R., Glassner, D.A. and Gruber, P.R. (2003). Applications of life cycle assessment to NatureWorks™ polylactide (PLA) production. *Polymer Degradation and Stability*, 80(3): 403–419.

Vogl, T.M., Seidelin, C. and B. Ganesh, J. (2020). Bright, Smart technology and the emergence of algorithmic bureaucracy: Artificial intelligence in UK local authorities. *Public Administration Review*, 80(6): 946–961.

Waheed, M., Yousaf, M., Shehzad, A., Inam-Ur-Raheem, M., Khan, M.K.I., Khan, M.R., Ahmad, N. and Aadil, R.M. (2019). Channelling eggshell waste to valuable and utilizable products: A comprehensive review. *Trends in Food Science & Technology*, 106: 78–90.

Wang, H.-C., Cui, D., Han, J.-L., Cheng, H.-Y., Liu, W.-Z., Peng, Y.-Z., Chen, Z.-B. and Wang, A.-J. (2019). A2O-MBR as an efficient and profitable unconventional water treatment and reuse technology: A practical study in a green building residential community. *Resources, Conservation and Recycling*, 150: 104418.

Wang, W., Themelis, N.J., Sun, K., Bourtsalas, A.C., Huang, Q., Zhang, Y. and Wu, Z. (2019). Current influence of China's ban on plastic waste imports. *Waste Disposal & Sustainable Energy*, 1(1): 67–78.

Wu, B. and Ge, X. (2018). Development and Technique Application of Green Building for Hot Summer and Cold Winter Climate Zone: Based on a New Wall Technology, ICCREM 2018.: *Sustainable Construction and Prefabrication*. VA: American Society of Civil Engineers, Reston, pp. 239–248.

Yang, Y., Lan, J. and Li, X. (2004). Study on bulk aluminum matrix nano-composite fabricated by ultrasonic dispersion of nano-sized SiC particles in molten aluminum alloy. *Materials Science and Engineering: A*, 380(1–2): 378–383.

Yazdanbakhsh, A., Grasley, Z., Tyson, B. and Al-Rub, R.K.A. (2010), Distribution of carbon nanofibers and nanotubes in cementitious composites. *Transportation Research Record*, 2142(1): 89–95.

Zou, H., Du, H., Wang, Y., Zhao, L. Mao, G., Zuo, J.,Liu, Y., Liu, X. and Huisingh, D. (2017). A review of the first twenty-three years of articles published in the *Journal of Cleaner Production*: With a focus on trends, themes, collaboration networks, low/no-fossil carbon transformations and the future. *Journal of Cleaner Production*, 163: 1–14.

4

Reflections on Metaverse

A New Technology in the Super Smart Society

Fabio De Felice,[1] *Gianfranco Iovine*[2] and *Antonella Petrillo**

1. Introduction

Today the world is moving at an unprecedented speed. More generally, the acceleration in technological development has allowed the large-scale diffusion of the internet connection and of tools that allow to simulate a potentially infinite set of parallel worlds within which the subjects can move through virtual alter egos (avatars): the Metaverse (Huh et al., 2022). The Metaverse is regarded as the next iteration of the internet: it is a network of interconnected experiences and applications, devices and products, tools, and infrastructures (Kye et al., 2021). The integration between physical and virtual reality is facilitated by using different technologies that guarantee different degrees of immersion. The main technologies are: (1) Augmented Reality (AR), a fusion of synthetic and physical reality in which the user can move in a real world enhanced by virtual data that allows him to have

[1] Department of Engineering, University of Naples "Parthenope", Italy, Isola C4, Centro Direzionale Napoli, 80143 Napoli (NA), Italy.
  Email: fabio.defelice@uniparthenope.it
[2] Department of Economics Engineering Society and Business Organization (DEIM), University of Tuscia, Largo dell'Università s.n.c., Loc. Riello, Viterbo, 01100, Italy.
  Email: gianfranco.iovine@unitus.it
* Corresponding author: antonella.petrillo@uniparthenope.it

additional information, useful for carrying out complex tasks; (2) Mixed Reality (MR), where the environment in which one moves corresponds to a combination of real and virtual; and (3) Virtual Reality (VR), in which the user enters a completely virtual dimension through an avatar, through which he is able to experience an alternative life in a dimension that reproduces, replaces, and improves the real world (Iwanaga et al., 2023). Certainly, the primary driver of investments in the evolution of network and graphics technologies that today allow access to the Metaverse, and its development has been the broad appeal associated with gaming. But interest in using the Metaverse is growing exponentially in other sectors as well (Zhang et al., 2023). That interest was confirmed in October 2021, when Mark Zuckerberg announced the renaming of his company from 'Facebook' to 'Meta' (short for 'Metaverse'). Before this communication, Zuckerberg invested $50 million in the two-year research project called XR Programs and Research Fund which, through collaboration with industry partners, civil rights groups, governments, non-profit organizations, and institutions academics, aims to build this new dimension, which the company itself estimates will be completed over the next 10–15 years (Hudson, 2022). One thing is certain: the Metaverse today is only in its infancy (Lo and Tsai, 2022). In any case, several unknowns remain regarding the shape that the Metaverse will take. The absence of certainties regarding the evolutionary trajectory of the Metaverse is transposed into the impossibility of setting up a regulation and control system for the safety of this new dimension; this in turn translates into the possibility of adopting criminal behavior in it without suffering any consequences (Ng, 2022). In this context, the present chapter aims to understand the development of the Metaverse since its origin, describing its characteristics and applications. Section 2 describes the metaverse in the scenario of Society 5.0. Section 3 summarizes the main applications of metaverse. Section 4 focuses on two important applications within the Metaverse, i.e., the metaverse in education/training and industry. Section 5 defines the urgency to regulate the cyberspace in which it operates. Finally, Section 6 draws the main conclusions of the chapter.

2. Metaverse and Society 5.0

Society 5.0 is the central theme of Super Smart Society: towards a more sustainable, resilient, and human-centric future (Almarzouqi et al., 2022). The sub-themes through which to shape a Society 5.0 have been identified in metaverse, robotics, bioeconomy, decarbonization, ecological transition and digitization (Egliston and Carter, 2021). Society 5.0 is the ideal continuation of Society 4.0, or the Information Society, in which information and communication technology (ICT) platforms constitute the technological basis for an economy based almost exclusively on profit (Allam et al., 2022). Table 4.1 summarizes the characteristics of each "evolutionary era".

Table 4.1 From Society 1.0 to Society 5.0, the characteristics of each "evolutionary era".

	Society 1.0	Society 2.0	Society 3.0	Society 4.0	Society 5.0
Society	*Nomadic*	*Agricultural*	*Industrial*	*Information*	*Super Smart*
Production type	Hunting	Handcrafted	Mechanized	ICT	Cyber space
City type	Settlement	Fortified city	Industrial city	Network city	Smart city
Ideals	Availability	Protection	Functionality	Profit	Sustainability

Society 5.0 represents an evolution from various points of view, the first, is technological. The digital transformation underway is supported by Artificial Intelligence (AI), capable of supporting man in the analysis of the enormous amount of data that internet of things (IoT) systems continuously acquires in acting as a bridge between the real world and the virtual world (Bribi, 2022). Society 5.0 is based on three fundamental pillars (Mourtzis et al., 2022):

- Pillar#1: The transition towards the Society 5.0 model and Productivity Revolution, using IoT technologies, Big Data, and AI.
- Pillar#2: The creation of resilient, environmentally friendly, and attractive communities, through the Future City Initiatives for the realization of the United Nations Sustainable Development Goals (SDGs).
- Pillar#3: The empowerment of future generations and women through a revolution in human resource development to make the most of the rich creative and communication skills, focusing on the gender objectives of the SDGs.

In addition, to issues specifically related to economic and environmental sustainability, some technologies will play a key role in the development of Society 5.0. Connectivity and particularly 5G technology is seen as a fundamental enabler to guarantee devices and applications to communicate with each other with optimal latency (Hwang and Chien, 2022). The fundamental principles of Society 5.0 constituted a fundamental source of inspiration for the definition of Industry 5.0, coined in 2021 by the European Commission, which illustrated its assumptions in the highly inspired report Towards a Sustainable, Human-centric, and Resilient European Industry (De Felice et al., 2021). While Society 5.0 stands as a real evolution of the previous Information Society, Industry 5.0 does not aim to overcome the technological paradigm of Industry 4.0 but to make its applications compatible with the sustainability and inclusion criteria necessary to give rise to a truly human-centric socioeconomic system.

3. Metaverse, Features and Its Main Trends

3.1 *Main characteristics of the metaverse*

The metaverse is based on the convergence of technologies for augmented reality and virtual reality, which enable multimodal interactions with digital objects, virtual environments, and people (Bhavana and Vijayalakshmi, 2022). Consequently, the metaverse is a web made up of interconnected and social immersive experiences on constant multi-user platforms. Furthermore, cryptocurrencies and non-fungible tokens (NFTs) are possible thanks to technologies such as blockchain, which allow the ownership of virtual and immovable objects in metaverses such as Decentraland. Microsoft and Meta are among the companies that are developing the technology to interface with virtual worlds, but they are not the only ones (Sandrone, 2022). Thus, many other relevant companies are building the infrastructure necessary to create better and more realistic virtual worlds (Teng et al., 2022). The main characteristics of the metaverse can be summarized in:

- These are three-dimensional spaces where users move freely using avatars to play, create, work, and even conclude commercial agreements.
- Virtual spaces can be created by the users themselves who make them available to other users.
- To make possible the connection between real and digital space, augmented reality, and hybrid reality technologies are used.
- Both virtual and real currencies can be used.
- At the basis of virtual spaces there are compatible technical standards, protocols, interoperability, digital property, blockchain technology, and laws that regulate their use.

Jon Radoff proposed a 7-point conceptual grid to define the metaverse value chain as detailed below and in Fig. 4.1 (Ricoy-Casas, 2023):

- **Point#1: Experience.** The Metaverse will give us a plethora of three-dimensional (3D) images and even two-dimensional (2D) experiences that we are currently unable to enjoy.
- **Point#2: Discovery.** In the ecology of the Metaverse, there continue to exist inbound and outbound discovery systems. When people actively seek information, this is known as an inbound discovery. Meanwhile, outbound marketing refers to sending communications to people regardless of whether they have requested it.
- **Point#3: Creative economy.** The creators of previous incarnations of the Internet needed some programming knowledge to design and build tools. However, development of web applications without coding is now possible thanks to web application frameworks. As a result, the number of web creators is rapidly expanding.

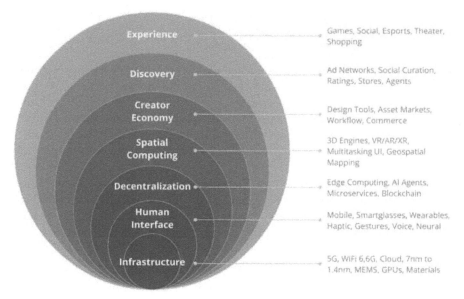

Fig. 4.1 7-point conceptual grid to define the metaverse value chain.

- **Point#4: Spatial computation.** Spatial computing refers to a technology that combines VR and AR. Microsoft's HoloLens is an excellent example of what this technology can achieve. Even if you haven't gotten your hands on Hololens yet, consider face filters on Instagram as an example of spatial computation.
- **Point#5: Decentralization.** Developers can leverage online capabilities through a scalable ecosystem enabled by distributed processing and microservices. Additionally, smart contracts and blockchain allow creators to use their own data and products.
- **Point#6: Human interface.** Users can receive information about their surroundings, use maps, and even create shared AR experiences simply by observing the physical world using a combination of spatial processing and human interface.
- **Point#7: Infrastructure.** The technological infrastructure is essential for the existence of other levels. It includes 5G and 6G processing to reduce network congestion and improve network bandwidth.

In addition, it is possible to define seven fundamental properties that the Metaverse is expected to possess (Table 4.2).

The novelty of Metaverse is causing confusion. There are many ideas that are separated from practice and reality. To avoid further confusion, it is important to summarize the seven rules that make up the metaverse (Parisi, 2021, https://bit.ly/3XmZZHU).

Table 4.2 Seven fundamental properties that the Metaverse is expected to possess.

#	Properties	Description
1.	Persistent	Regardless of how long users stay on the platform, it will not be possible to reset, pause, or terminate it. It will be an entity that continues indefinitely, in which subjects, be they individuals, groups, or organizations, will be able to develop over time without running the risk of disappearing.
2.	A bridge between different worlds	It will be a membrane that connects the digital world with the physical one, expanding the volume of the virtual universe through the introduction of the third dimension. This will make it possible to overcome the current switching from one website to another via hyperlink, replacing it with jumping from one 3D world to another.
3.	A fully functional, fertile, and scalable economy	Individuals and businesses will be able to create, own, invest, sell, and be rewarded (in cryptocurrency) for an incredibly wide range of activities.
4.	Extremely social	Through a remodulation of the interaction models it will be possible to transcend space and share environments and a sense of presence with other subjects.
5.	Live and concurrent	It will be an experience that exists constantly for all its users and independently of them, coherently and concurrently.
6.	Extremely interoperable	Data, assets, content, etc. can be transported without restrictions between the different virtual realities.
7.	Rich in content	Content and experiences will no longer be generated by specific individuals, but by a large and diverse number of contributors, including independent individuals, informal groups, or commercial enterprises.

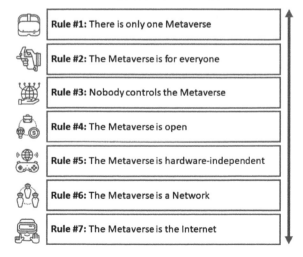

THE SEVEN RULES OF THE METAVERSE

Rule #1: There is only one Metaverse

Rule #2: The Metaverse is for everyone

Rule #3: Nobody controls the Metaverse

Rule #4: The Metaverse is open

Rule #5: The Metaverse is hardware-independent

Rule #6: The Metaverse is a Network

Rule #7: The Metaverse is the Internet

Fig. 4.2 7 rules of Metaverse (Source: Parisi, 2021, https://bit.ly/3XmZZHU).

There are several platforms to access the Metaverse. Table 4.3 shows the main platforms currently used.

Table 4.3 Main Platform used to access the Metaverse.

#	Platform	Description
1.	HyperVerse https://thehyperverse. net/	HyperVerse is a platform for creating virtual worlds. Every world is a planet, and many planets comprise the HyperVerse, which is not owned by a single entity. It is a universe with millions of worlds. Each person who resides in the HyperVerse is called a Voyager. Travelers can create tokens, interact with each other, and explore the world together. The HyperVerse also offers a unique opportunity to every Voyager, thanks to VerseDAO.
2.	Sandbox https://www.sandbox. game/en/	Sandbox is the name of a blockchain-based platform where you can buy, sell, and trade virtual parcels of land. The Ethereum blockchain serves as the foundation for the Sandbox. It is famous for its stability and security. However, it's not exactly cheap to use. To make this happen, SandBox created its own Ethereum-based token called SAND.
3.	Decentraland https://decentraland. org/	Decentraland is a virtual social world that is again based on the blockchain, which is essentially a digital ledger that permanently records cryptocurrency transactions across a computer network. The Decentraland platform can be used to do things like host conferences, play games, and trade virtual products in marketplaces. Socializing with other members is as easy as in real life. Decentraland is a digital ecosystem that allows its users to participate in events, play games, exchange goods and services in its marketplaces, and socialize with people from all over the planet.
4.	Horizon Workrooms https://www. meta.com/it/work/ workrooms/	Horizon Workrooms is a VR workspace where teams can connect, collaborate, and create. It is possible to sit with colleagues at the same table, even in traveling the world, and turn the home office into a remote meeting room. It is possible to share screen, send messages, and type naturally.
5.	Microsoft Mesh https://www.microsoft. com/en-us/mesh	It is designed for the world of work; it allows you to connect and hold meetings from all over the world by digitally bringing together your own avatar. A tool that allows team workers to review common processes faster and arrive at more informed decisions. A metaverse for conducting trainings and meetings, without the cost of travel, thanks to holotransportation, holographic sharing and visualization. Whether physically present or holoported, colleagues can collaborate on content in real time.

Table 4.3 contd. ...

... Table 4.3 contd.

#	Platform	Description
6.	Nakamoto https://www.nakaverse.games/	Recently, the team behind the NAKA token announced a rebranded NAKAverse. It is one of the first metaverses to introduce a true in-game economy. Users will be able to buy virtual land, build buildings, and more. The NAKA token will be the native currency of the NAKAverse.
7.	Metahero https://staratlas.com/	Metahero seeks to bring physical objects into the digital world. These virtual objects are called non-fungible tokens (NFTs). Metahero offers 3D scanning technology that reconstructs real-world objects into virtual versions. In turn, these virtual objects can be used in different ways, such as teaching, training, and entertainment.
8.	Star Atlas https://staratlas.com/	Star Atlas is a next generation game metaverse. It is created by the convergence of blockchain, real-time imaging, multiplayer video games, and decentralized financial technologies. Within the platform, digital assets such as ships, crew, land, and equipment can be purchased with an in-game monetary system built on Solana's blockchain platform.
9.	Bloktopia https://www.bloktopia.com/	Bloktopia is a metaverse. It is a virtual 21-story skyscraper. The Bloktopia team uses Bitcoin, with a maximum limit of 21 million. Bloktopia users can create their own avatars, participate in social activities, learn about cryptocurrency, and purchase metaverse "real estate" in this virtual reality skyscraper. They can use it to create artwork, games, and more using a construction tool.
10.	Roblox https://www.roblox.com/	Roblox is a popular online gaming platform for both kids and adults. The games are created by other users and are accessible to anyone. The Roblox website claims that over 20 million games have been released since Roblox launched in 2004. One of the most unique features of this platform is the ability to dress your avatar in accessories from real-world brands like Burberry.
11.	Stageverse https://stageverse.com/	It is a social VR platform with virtual places, interactive experiences, and digital content. The platform debuted a concert experience of Muse: Simulation Theory. The Stageverse platform allows multiple spectators to experience concerts together through 3D footage from multiple angles around the venue.
12.	Spatial https://www.spatial.io/	It is an augmented and virtual reality platform that brings collaboration to life. It allows users to interact with each other in virtual rooms using VR headsets, mobile, or desktop devices. People can share content and ideas in real time and use the platform to work together virtually.

3.2 Possible applications of the Metaverse

3.2.1 Entertainment, gaming, and fashion

One of the first and most evident fields of application of virtuality, in all its degrees of separation from reality (Wanick and Stallwood, 2023). The most interesting aspect will be the emergence of original gameplay within the Metaverse, exploiting the internal resources of the world in which they take place, or the creation of virtual native shows capable of gathering an 'infinite' audience (Sayem, 2022). Another sector interested in Metaveso is that of luxury brands such as Gucci, Dolce & Gabbana, LV, Etro, fast fashion chains such as Zara, H&M, Benetton, and sportswear, in full FOMO (fear of missing out) effect, have not lost the opportunity to open their flagship stores in the Metaverse Fashion District, creating truly immersive experiences for users (Weiss, 2022). An example of this is the Metaverse Fashion Week, which took place last March at Decentraland and which represented the first and largest all-digital fashion week in the world with four days of shows and over 60 well-known brands, designers, and artists.

3.2.2 Training and education

Training is about to undergo a significant change that is starting now. Thanks to the Metaverse, the doors of a new way of fruition are opening of educational contents and relationships for school and work. It is said that human beings never stop learning throughout their lives. Curiosity, the will to interact, to collaborate, and to move around the world pushes people to continuously discover and study new things (Petrillo et al., 2021). Surely, technology comes to the aid of training, playing the role of intermediary between interlocutors (Frith, 2022). The advent of the Metaverse, therefore, represents a new opportunity for evolution for training, going beyond the now evident limits that 2D applications present. Training in the virtual world means guaranteeing effective and interactive learning for students and employees. The three-dimensional and immersive nature of this channel, in fact, allows users to have a training experience comparable to the real one. Many researchers show that traditional training is only 5% effective compared to the 75% guaranteed by virtual training (Hwang, 2023; Lee et al., 2022; López-Belmonte et al., 2022; Rospigliosi, 2022). The role of digital twins is central in this case. These are all those elements that in the Metaverse are reproduced in the smallest detail in three dimensions, and even made manipulable to various extents. The most important digital twin is our avatar, through which we can interact with teachers, classmates, or work colleagues within a virtual space, depending on the context of the meeting in progress. It has been estimated by international studies that, by 2030, the virtual training market will undergo an extremely positive impact, with growth ranging between 180 billion and 270 billion dollars (Sánchez-López et al., 2022). Each type of formation finds its place in the Metaverse.

Starting from the didactics of the schools, passing through the universities up to the practice for every form of professional requirement and training for the job. The greatest limitation of distance learning is represented by the lack of interaction and socialization between students, which translates into a lack of attention and a desire to engage and learn (Wang and Shin, 2022). The Metaverse clearly does not replace the social interaction that can be had in a real context, but through the avatars, environments, and features that contribute to making the virtual space immersive, it can simulate a form of contact between users like the real one. Therefore, it is conceivable that the Metaverse undoubtedly represents the future of training.

3.2.3 Industry

There is still a lot to know about the functions of the metaverse, but we are already starting to decline the concept from a business perspective, let's analyze the industrial impact on the socio-economic scenario (Zhou et al., 2022). How does the metaverse change when we approach it as consumers and when we interact as industries and private organizations? The common element is the transposition of the physical world into the virtual world. Virtual reality allows people to play a double role: to be producers or consumers and to combine the two roles and to be producers and consumers at the same time, that is, prosumers (Kesselman and Esquivel, 2022). Ultimately, the division between the industrial metaverse and the consumer metaverse is related to the use of the service. Specifically, with reference to the industrial metaverse, the industries that are born or spread in that context can then have a practical response in physical reality by generating real economic value. The most widely used technologies in the industrial metaverse are summarized in Table 4.4.

The technologies related to the industrial metaverse are also considered like the paradigm of industry 4.0—that of cyber-physical systems, to be clear. However, the metaverse extends beyond technological boundaries by giving people a new role. The centrality of humans now seems to have become an unavoidable goal in all the techno-economic evolutions of the future. The European Commission imagines industry 5.0 as a catalyst for resilience, projected towards sustainability and which assigns a central role to people in the creation of value as shown in Fig. 4.3.

The metaverse and industry 5.0 feed on the technological foundations and drive towards interoperability and application cooperation that support the system.

The main **new opportunities** that the industrial metaverse will offer to companies in various specific scenarios are:

- **Training of employees through the meta-representation of the real work environment.** With MR or virtual, companies will be able to create virtual environments that faithfully represent the real conditions

Table 4.4 The most widely used technologies in the industrial metaverse.

#	Technologies	Description
1.	Mixed Reality	It is useful for complex industrial scenarios that require low operational latencies and that allow remote collaboration, real-time equipment diagnosis, and industrial model design.
2.	Compute Engine	Computational power available as-a-service for low latency operations.
3.	Digital Twin	It offers the possibility of recreating a digital twin not only of the single machine but also of the factory and of the entire production chain up to the overall economic system.
4.	AIoT	It arises from the confluence of artificial intelligence (AI) and the Internet of Things (IoT) and allows better management of devices and machines directly in the vicinity.
5.	Computer Graphics	It is as an enabling factor for the generation of 3D models to be used, later, also in the virtual interactions of the industrial metaverse.
6.	Blockchain	It is a technology that allows the management of transactions remodeled in a decentralized perspective. This favors the construction of a "trust without trust", that is, it generates greater security between the counterparties involved in the production chain. The characteristics of the blockchain that allow for greater reliance are immutability and transparency.

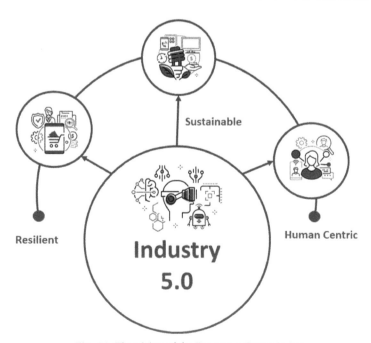

Fig. 4.3 The vision of the European Commission.

in which the resources will have to operate. This, in addition to speeding up times, also improves people's safety in the event of risky environmental conditions.

- **Simulate before executing or delivering.** Having the possibility to implement a simulation of the behavior of a machine through the digital twin, e.g., will allow companies to speed up development or avoid unforeseen problems during production.
- **Collaborate with others on design and production without physical limitations.** Utilizing additive manufacturing, e.g., a manufacturing company will be able to use third-party services for the design of the product and then produce it directly where it is needed through the industrial 3D printing service center.
- **Use virtual reality to perform maintenance tasks.** Whether it is augmented or mixed reality, the metaverse offers the possibility of operating remotely without the physical movement of the expert.

4. Metaverse: Evidence from the Scientific Community

From a scientific point of view, it is interesting to analyze the research of the scientific community on the theme of the metaverse. This section, particularly focuses on two important applications within the Metaverse, i.e., the metaverse in education/training and in industry. Therefore, we have an analysis on SCOPUS, the largest indexer of global research content, the current trends on two aspects: Education and Industry in the context of the metaverse. Using the Boolean operators, only documents in which the term "metaverse", "education/educational" and "industry/industrial" were included in the title were analyzed. More specifically, the search strings reported in Table 4.5 were used.

Figure 4.4 shows published papers on the metaverse topic. The first document was published in 2000. But only in 2022 is there an exponential interest from the scientific community with 524 published documents equal to 80% of all scientific production.

Similarly, analyzing the publications on education/educational and on the industry/industrial theme, it emerges that scientific interest is concentrated in the last two years (as shown in Fig. 4.5). This evidence is

Table 4.5 Search strings used on Scopus.

Search String	No. of Publications	Label
TITLE (metaverse)	647	/
(TITLE (metaverse) AND TITLE (educational) OR TITLE (education))	45	Cluster#1
(TITLE (metaverse) AND TITLE (industrial) OR TITLE (industry))	13	Cluster#2

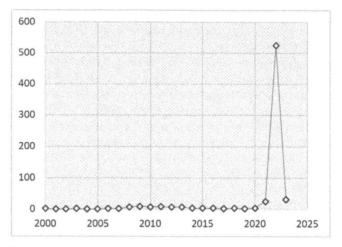

Fig. 4.4 Documents by year – TITLE (metaverse) (Source: SCOPUS).

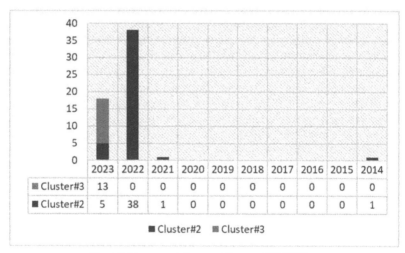

	2023	2022	2021	2020	2019	2018	2017	2016	2015	2014
Cluster#3	13	0	0	0	0	0	0	0	0	0
Cluster#2	5	38	1	0	0	0	0	0	0	1

Cluster#2   Cluster#3

Fig. 4.5 Documents by year (Source: SCOPUS).

not surprising considering the media impact on the metaverse. Currently, the marketing impact is more evident than in the past.

Instead, if we analyze the geographical origin of the countries where the theme is mostly addressed, the survey shows that China and South Korea are the most prolific countries from a scientific point of view (with a percentage equal to 27% of the total publications), as shown in Fig. 4.6. The result is not surprising because they are among the countries, in general, most productive from a scientific point of view investigating cutting edge technology. In fact, also according to Global Data (https://bit.ly/3XknKQV), the largest market share of the Metaverse and related technologies is held

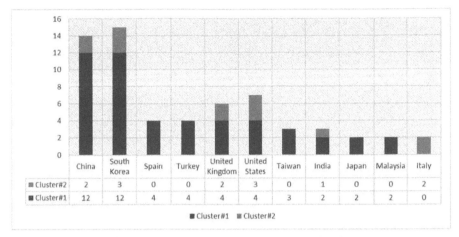

Fig. 4.6 Documents by country (Source: SCOPUS).

by the Asia-Pacific and North America regions. These two regions together hold 50% of the market share of the metaverse.

The survey conducted highlights and confirms that companies from all over the world, including those in non-technological sectors, are increasingly investing in this technology to improve the services offered and to meet the needs of a constantly evolving society. Companies operating in North America and Asia Pacific are focusing on critical technologies that can be implemented thanks to the metaverse, such as Blockchain, machine learning, AR, VR, IoT, payment platforms, business applications, gaming, and data governance. An example of the integration of the main technologies involved in the metaverse is shown in Fig. 4.7.

5. Urgency to Regulate and to Supervise the Cyberspace

The study of perception and visualization of digital spaces such as the Metaverse is critical to examine for the potential impacts of cyberspace technologies on real space. It has been demonstrated that in the digital world there are already threats of a cognitive type that can have important concrete repercussions both from a social and economic point of view and, secondly, also from a political and military point of view (Zvarikova et al., 2022). Parallel to the aspects just mentioned, comes identity. The avatar is not just a graphic icon in the form of a puppet, it is the representative of a person's identity. Maintaining identity in every environment of the Metaverse must

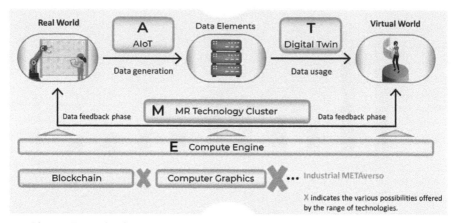

Fig. 4.7 Example of integration of the main technologies involved in the metaverse.

be a right and a discretion, a right and a duty of citizenship (Pooyandeh et al., 2021). Having an identity allows the recognition of the rights and responsibilities that we have in the real world as in the virtual one. The Metaverse is the actualization not just of a space, but of society itself. The question is critical involving the ethical and social sphere. Global regulation is urgently needed. However, despite national and international efforts to secure the cyber domain, the inherent transnationality and the diffusion of increasingly complex technologies continue to open new channels for conducting illegal activities (Qasem et al., 2022). Based on these premises and considering that cyberspace is expanding in ever more engaging ways and on an unprecedented scale, even if a more defined version of the Metaverse still belongs to the distant future, the potential threats posed by it require the immediate attention of a diverse range of people and organizations (Park and Kim, 2022). There is also a continuing need to pool efforts at international and national, public, and private levels, involving companies, academic researchers, Metaverse developers, organizations responsible for protecting society and active cyber citizenship, to set up a forecasting system, prevention.

6. Conclusions

The Metaverse appears to be a not entirely new world: imagined for years by writers, scientists, and entrepreneurs, it is only now taking shape in a project with high potential. We can say that we are at the beginning of an explosion of innovation and experimentation in which human and artificial action intertwine and integrate into a continuum that blurs the lines between physical existence and simulation. In fact, the digital revolution underway in recent years has made it possible to create the basis of existing technologies, creating prototypes of what will exist in the future, assuming the perspective of the beta economy. Awareness is that what we have now is nothing but the basic version of what we will have in the future. Although the potential is enormous and the enthusiasm is great, it must be taken into consideration that, as happened with the colonization of new media spaces (television in the 1970s in Italy and the Internet in the 1990s), the legal practitioners from various countries will struggle to keep up with these new and constantly evolving technologies.

References

Allam, Z., Sharifi, A., Bibri, S.E., Jones, D.S. and Krogstie, J. (2022). The metaverse as a virtual form of smart cities: Opportunities and challenges for environmental, economic, and social sustainability in urban futures. *Smart Cities*, 5(3): 771– 801. doi:10.3390/smartcities5030040

Almarzouqi, A., Aburayya, A. and Salloum, S.A. (2022). Prediction of user's intention to use metaverse system in medical education: A hybrid SEM-ML learning approach. *IEEE Access*, 10: 43421– 43434. doi:10.1109/ACCESS.2022.3169285.

Bhavana, S. and Vijayalakshmi, V. (2022). AI-based metaverse technologies advancement impact on higher education learners. *WSEAS Transactions on Systems*, 21: 178–184. doi:10.37394/23202.2022.21.19

Bibri, S.E. (2022). The social shaping of the metaverse as an alternative to the imaginaries of data-driven smart cities: A study in science, technology, and society. *Smart Cities*, 5(3): 832–874. doi:10.3390/smartcities5030043.

De Felice, F., Travaglioni, M. and Petrillo, A. (2021). Innovation trajectories for a society 5.0. *Data*, 6(11). doi:10.3390/data6110115.

Egliston, B. and Carter, M. (2021). Critical questions for Facebook's virtual reality: Data, power and the metaverse. *Internet Policy Review*, 10(4). doi:10.14763/2021.4.1610.

Frith, K.H. (2022). The metaverse: Is it just trending or a real game changer for education? *Nursing Education Perspectives*, 43(6): 384. doi:10.1097/01.NEP.0000000000001057.

Huh, S. (2022). Application of the computer-based testing in Korean medical licensing examination, the emergence of a metaverse in medical education, journal metrics and statistics, and appreciation to reviewers and volunteers. *Journal of Educational Evaluation for Health Professions*, 19. doi:10.3352/jeehp.2022.19.2.

Hwang, G. and Chien, S. (2022). Definition, roles, and potential research issues of the metaverse in education: An artificial intelligence perspective. *Computers and Education: Artificial Intelligence*, 3. doi:10.1016/j.caeai.2022.100082.

Hwang, Y. (2023). When makers meet the metaverse: Effects of creating NFT metaverse exhibition in maker education. Computers and Education, 194. doi: 10.1016/j.compedu.2022.104693.

Hudson, J. (2022). Virtual immersive shopping experiences in metaverse environments: Predictive customer analytics, data visualization algorithms, and smart retailing technologies. *Linguistic and Philosophical Investigations*, 21: 236– 251. doi:10.22381/lpi21202215.

Iwanaga, J., Muo, E.C., Tabira, Y., Watanabe, K., Tubbs, S.J., D'Antoni, A.V., . . . and Tubbs, R.S. (2023). Who really needs a metaverse in anatomy education? A review with preliminary survey results. *Clinical Anatomy*, 36(1): 77– 82. doi:10.1002/ca.23949.

Kesselman, M.A. and Esquivel, W. (2022). Technology on the move, consumer electronics show 2022: The evolving metaverse and much more. *Library Hi Tech News*, 39(5): 1– 4. doi:10.1108/LHTN-03-2022-0038.

Kye, B., Han, N., Kim, E., Park, Y. and Jo, S. (2021). Educational applications of metaverse: Possibilities and limitations. *Journal of Educational Evaluation for Health Professions*, 18. doi:10.3352/jeehp.2021.18.32.

Lee, H., Woo, D. and Yu, S. (2022). Virtual reality metaverse system supplementing remote education methods: Based on aircraft maintenance simulation. *Applied Sciences* (Switzerland), 12(5). doi:10.3390/app12052667.

Lo, S. and Tsai, H. (2022). Design of 3D virtual reality in the metaverse for environmental conservation education based on cognitive theory. *Sensors*, 22(21). doi:10.3390/s22218329.

López-Belmonte, J., Pozo-Sánchez, S., Lampropoulos, G. and Moreno-Guerrero, A. (2022). Design and validation of a questionnaire for the evaluation of educational experiences in the metaverse in Spanish students (METAEDU). *Heliyon*, 8(11). doi:10.1016/j.heliyon.2022.e11364.

Mourtzis, D., Panopoulos, N., Angelopoulos, J., Wang, B. and Wang, L. (2022). Human centric platforms for personalized value creation in metaverse. *Journal of Manufacturing Systems*, 65:653– 659. doi:10.1016/j.jmsy.2022.11.004.

Ng, D.T.K. (2022). What is the metaverse? Definitions, technologies and the community of inquiry. *Australasian Journal of Educational Technology*, 38(4): 190– 205. doi:10.14742/ajet.7945.

Park, S. and Kim, Y. (2022). A metaverse: Taxonomy, components, applications, and open challenges. *IEEE Access*, 10: 4209–4251. doi:10.1109/ACCESS.2021.3140175.

Pooyandeh, M., Han, K., and Sohn, I. (2022). Cybersecurity in the AI-based metaverse: A survey. Applied Sciences (Switzerland), 12(24). doi:10.3390/app122412993.

Petrillo, A., De Felice, F. and Petrillo, L. (2021). Digital divide, skills, and perceptions on smart working in Italy: From necessity to opportunity. *Paper presented at the Procedia Computer Science*, 180: 913– 921. doi:10.1016/j.procs.2021.01.342.

Qasem, Z., Hmoud, H.Y., Hajawi, D. and Al Zoubi, J.Z. (2022). The effect of technostress on cyberbullying in metaverse social platforms. doi:10.1007/978-3-031-17968-6_22. Retrieved from www.scopus.com.

Ricoy-Casas, R.M. (2023). The Metaverse as a New Space for Political Communication. doi: 10.1007/978-981-19-6347-6_29.

Rospigliosi, P.A. (2022). Metaverse or simulacra? Roblox, minecraft, meta and the turn to virtual reality for education, socialisation and work. *Interactive Learning Environments*, 30(1): 1– 3. doi: 10.1080/10494820.2022.2022899.

Sánchez-López, I., Roig-Vila, R. and Pérez-Rodríguez, A. (2022). Metaverse and education: The pioneering case of minecraft in immersive digital learning. *Profesional De La Informacion*, 31(6). doi:10.3145/epi.2022.nov.10.

Sandrone, S. (2022). Medical education in the metaverse. *Nature Medicine*, 28(12): 2456– 2457. doi:10.1038/s41591-022-02038-0.

Sayem, A.S.M. (2022). Digital fashion innovations for the real world and metaverse. *International Journal of Fashion Design, Technology, and Education*, 15(2): 139– 141. doi:10.1080/17543266.2022.2071139.

Teng, Z., Cai, Y., Gao, Y., Zhang, X., and Li, X. (2022). Factors affecting learners' adoption of an educational metaverse platform: An empirical study based on an extended UTAUT model. *Mobile Information Systems*, 2022. doi:10.1155/2022/5479215.

Wang, G. and Shin, C. (2022). Influencing factors of usage intention of metaverse education application platform: Empirical evidence based on PPM and TAM models. *Sustainability* (Switzerland), 14(24). doi:10.3390/su142417037.

Wanick, V. and Stallwood, J. (2023). Brand storytelling, gamification and social media marketing in the 'Metaverse': A case study of the ralph lauren winter escape. doi:10.1007/978-3-031-11185-3_3. Retrieved from www.scopus.com.

Weiss, C. (2022). *Fashion retailing in the metaverse. Fashion, Style and Popular Culture*, 9(4): 523– 538. doi:10.1386/fspc_00159_1.

Zhang, H., Lee, S., Lu, Y., Yu, X. and Lu, H. (2023). A survey on big data technologies and their applications to the metaverse: Past, current, and future. *Mathematics*, 11(1): doi:10.3390/math11010096.

Zhou, Y., Xiao, X., Chen, G., Zhao, X. and Chen, J. (2022). Self-powered sensing technologies for human metaverse interfacing. *Joule*, 6(7): 1381– 1389. doi:10.1016/j.joule.2022.06.011.

Zvarikova, K., Machova, V. and Nica, E. (2022). Cognitive artificial intelligence algorithms, movement and behavior tracking tools, and customer identification technology in the metaverse commerce. *Review of Contemporary Philosophy*, 21: 171–187. doi:10.22381/RCP21202211.

5

Overview on the Role of Cybersecurity in a Smart Society

Prospects and Strategies in Italy and Worldwide

Laura Petrillo

1. Introduction

Governments of many countries are committed to redefining a new cybersecurity agenda which firstly provides for the strengthening of the institutional system and national IT architecture (Bokhari et al., 2022). For example, the new US foreign policy aims to relaunch the American role in global cybersecurity and cyber diplomacy, acting as a 'model' in the eyes of the world. The slogan "America is back"—launched by President Biden at the Munich Conference in February 2021—does not imply a return of the US government's cybersecurity policies to the old "business as usual", but a relaunch of them on a more assertive and 'proactive' manner. Europe plays an increasingly active role in tackling multiple cyber threats and holds a leadership position in the global context (Brandão et al., 2022). In this

Ministero dell'ambiente e della sicurezza energetica, Via Cristoforo Colombo, 44, 00147 Roma (Italia).
Email: petrillo.laura@mase.gov.it

context, also the Italian government is committed to promoting equality, ethics, and justice, in a strategy of innovation and development focused on people and the planet. In fact, starting from 2016 a special team was created to start building the country's operating system, to build simpler and more effective services for citizens, Public Administration, and businesses, through innovative digital products. The team's strategy is inspired by the Sustainable Development Goals (SDGs) of the United Nations. The aim of the Italian strategy is to pursue the fundamental principles of equality, ethics, and justice, putting people and sustainability at the center while promoting innovation and digitalization of public services. In this context, Machine Learning algorithms and AI techniques are needed to solve complex problems, modern design patterns, service-oriented, secure, 'elastic' and highly reliable architectures. At the same time, it is necessary to develop a culture of information technology and digital skills among citizenship (Carvalho et al., 2020). Therefore, it is important to invest in the promotion and fight against all forms of digital illiteracy. In fact, data is a massively exchanged resource, consciously, in every relationship. Technology allows everything but must be cautious to possible risks associated with sharing personal data (Cheng and Wang, 2022). Virtuality allows us to access immaterial spaces that do not respond to the parameters and rules of physical reality, instead we must learn to move and understand that we are not alone. Intelligent machines are already, and will increasingly be, a constant companion able to understand us and process reasoning much faster than us, producing new, very copious data that will be further analyzed and processed by other machines (AlDaajeh et al., 2022).In its multiple applications, AI is the most important frontier that opens up. Therefore, the aim of the chapter is to outline the underlying principles underpinning cyber security in a smart society. The remainder of the chapter is organized as follows. Section 2 analyzes the main characteristics of Cyber Security and computer security; Section 3 defines the Cyber threat; Section 4 discusses the Cybersecurity risk management and the personal security perimeter; Section 5 summarizes the main features of Cybersecurity and human behavior. Sections 6 and Section 7 explore Cybercrime and Cyber defense, respectively. An overview of cybersecurity skills, towards the certification of professionalism is analyzed in Section 8. Finally, the main conclusions of the chapter are outlined in Section 9.

2. Cybersecurity and Computer Security

Cybersecurity represents a field related to computer security: tools and technologies whose function is to protect computer systems from outside attacks. It is subclass of computer security. More generally, computer security is the set of means, technologies, and procedures aimed at protecting information systems in terms of availability, confidentiality, and integrity of information assets or other advantages (AlZaabi, 2019). Cybersecurity often emphasizes the qualities of resilience, robustness, and responsiveness that a technology must possess to face attacks aimed at compromising its correct functioning and his performances. It is the practice of protecting systems, networks, and programs from digital attacks (Kennedy, 2020). These cyberattacks are usually aimed at accessing, changing, or destroying sensitive information; extorting money from users via ransomware; or interrupting normal business processes. Implementing effective cybersecurity measures is particularly challenging today because there are more devices than people, and attackers are becoming more innovative. Since the cybersecurity landscape is constantly evolving, it is necessary to adopt a rigorous and clear methodological approach. A sequence of clear operations should be established within the organizations with a view to continuous monitoring and improvement. An example is shown in Fig. 5.1.

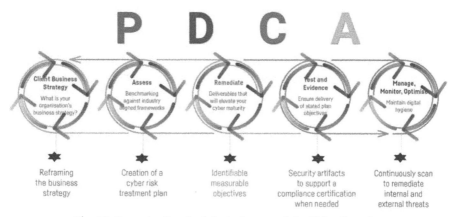

Fig. 5.1 Example of methodological approach for Cyber Security.

3. Cyber Threat

The set of contraindicated conduct that can be carried out in and through cyber-space or to the detriment of the latter and its constituent elements is identified as a cyber threat. It consists of cyber-attacks: actions of individuals or organizations, whether state or not, aimed at destroying, damaging, or hindering the regular functioning of the systems and networks and/or process actuator systems controlled by them, or to violate the integrity and the confidentiality of data/information. This reality must be addressed by acting according to an approach that includes the adoption of risk prevention and mitigation measures aimed at increasing the resilience of digital infrastructures (Raicu and Raicu, 2022). The latter include networks, systems, and data, but above all, users, whose awareness must be nurtured through a widespread culture of cybersecurity. If today, in fact, there is a widespread perception of the risks related to physical security, for which everyone carries out, in their daily life, actions aimed at protecting themselves and their assets, the same cannot be said for the digital dimension, of which risks are not yet fully aware (Mohammed and George, 2022). The securing of infrastructures, systems, and information from a technical point of view must be accompanied by cultural progress at every level of society, towards a "security-oriented" approach, an indispensable element to protect our value and democratic system (Payne et al., 2021). According to the European Union Agency for Cybersecurity (ENISA), it is possible to identify the top cyber threats for 2030. Figure 5.2 summarizes the 10 top cyber threats.

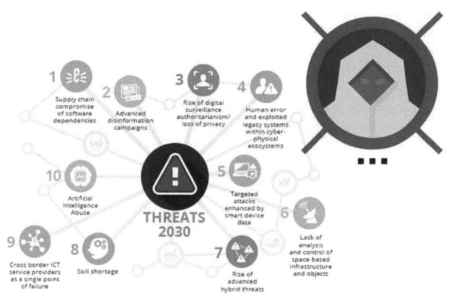

Fig. 5.2 The 10 top cyber threats for 2030.

4. Cybersecurity Risk Management

4.1 International context

Cybersecurity is inherently a human subject. Technology and processes require human interface and deployment. It is no wonder then, that most cybersecurity breaches involve some human element.

Protecting our information means analyzing the situations and areas in which it is processed and assessing the associated security risks (Ron et al., 2020). Managing cyber risk across is harder than ever. Keeping architectures and systems secure and compliant can seem overwhelming even for today's most skilled teams. As more of our physical world is connected to and controlled by the virtual world, and more of our business and personal information goes digital, the risks become increasingly daunting (Duong et al., 2021). Cybersecurity risk management is an ongoing process of identifying, analyzing, evaluating, and addressing cybersecurity threats. Cybersecurity risk management is a job of all of us; everyone has a role to play. Risk management consists of the identification processes, through which risks are identified, evaluated and treatment. There are several national and international organizations such as International Organization for Standardization (ISO) or European Union Agency for Cybersecurity (ENISA) or National Institute of Standards and Technology (NIST), which have produced and updated standards and guidelines to implement a risk management process, with particular reference to information security and then to cyber security. The ISO has defined the ISO/IEC 27001 standard which identifies the requirements for setting up and managing an information security management system (Information Security Management System - SGSI) and includes aspects relating to logical, physical, and organizational security. Risk identification is the first step in the management process of Modern security of IT systems (Santisteban et al., 2020). Therefore, IT risk management is an important part of the design and ongoing management of the information system, in view of the principles of security by design. It cannot be considered as a one-off or ex post action of security checks. Already in the requirements definition phase, the design of a service must include a risk assessment that allows to ascertain the need to protect the service itself and how much a technical solution facilitates or, on the contrary, hinders the adoption of adequate controls. Thus, NIST created a third-party risk management framework known as NIST Special Publication 800-30 to guide USA federal information system's risk assessments (Alamri et al., 2023).

In relation to the need to ensure adequate use of IT services, the ISO 22301 standard is also included. It is an international standard developed to guide organizations in identifying potential threats to their business processes and in building effective backup systems and processes to safeguard its interests and those of interested parties (Shaikh and Siponen, 2023).

4.2 *Italian national context*

In Italy, AgID (Agenzia per l'Italia Digitale—Agency for Digital Italy) supports the administrations in the design and planning of strategies useful for ensuring the resilience of the national IT infrastructure of the public administration (PA). Depending on the complexity of the information system and the organizational reality of the administration, risk management activities can translate into controls of a technological, organizational, and procedural nature useful for assessing the level of information security and aimed at countering the most frequent information threats, within a continuous process of monitoring and improvement (Bozzetti et al., 2021). A first action in this direction resulted in the publication of the ICT minimum security measures for PAs (2017). In implementation of the Italian Digital Agenda, and within the European project Italia Login—The citizen's house, AgID has developed a risk management methodology, starting from an international benchmark of public and private good practices:

- to be applied to all entities of the Italian Public Administration, differing in size, technological complexity, and the provision of services to businesses, citizens, and other public entities;
- to be supported by an application tool that can be used by all PAs and integrated with other central IT infrastructures.

The "Italia Login—The citizen's house" project aims to increase the trust of citizens and businesses in the digital services of the PA and to ensure greater efficiency of the administrative capacity of the PA. The project is in line with the objectives of the 2021–2023 update of the Three-Year Plan for IT in the PA. The lines of intervention that make up the Italia Login project include the creation of:

- access layer to online services;
- vertical ecosystem layers;
- interoperability infrastructures of national and transversal databases;
- support and control mechanisms.

Through these actions, AgID intends to follow up on the country's digital transformation model and facilitate the creation of a PA Information System which, among other things, perfects the overall degree of security of the PA system and so citizens, through a series of tools to support risk assessment and for the development of tools to prevent cyber threats (Annarelli et al., 2021). In this context, a fundamental role is played by social engineering, the art of manipulating people to obtain information. Social engineering represents a still underestimated threat to information systems and the complexity of the methodologies used is inevitably intertwined with the traditional vectors of cyberattack. At its core, social engineering is not a cyberattack. Instead, social engineering is all about the psychology of persuasion: It targets the mind like your old school grifter or con man. The aim is to gain the trust of targets, so they lower their guard, and then encourage them into taking unsafe actions such as divulging personal information or clicking on web links or opening attachments that may be malicious. In a typical social engineering attack, a cybercriminal will communicate with the intended victim by saying they are from a trusted organization. In some cases, they will even impersonate a person the victim knows. If the manipulation works (the victim believes the attacker is who they say they are), the attacker will encourage the victim to take further action. This could be giving away sensitive information such as passwords, date of birth, or bank account details. Or they might encourage the victim to visit a website where malware is installed that can cause disruptions to the victim's computer. In worse-case scenarios, the malicious website strips sensitive information from the device or takes over the device entirely. Over time, social engineering attacks have grown increasingly sophisticated (Bergami et al., 2013). Not only do fake websites or emails look realistic enough to fool victims into revealing data that can be used for identity theft, but social engineering has also become one of the most common ways for attackers to breach an organization's initial defenses in order to cause further disruption and harm.

The different types of attacks fall into three categories:

- **human based,** analyzed here, based on direct contact between attacker and victim;
- **computer based,** more articulated, which require the use of IT means and technical skills;
- **mobile based,** a logical subset of computer-based ones, having its own dignity due to the capillary diffusion, now all-encompassing, of the mobile technologies that carry the malevolent agent.
The main types of social engineering attacks are shown in Fig. 5.3.

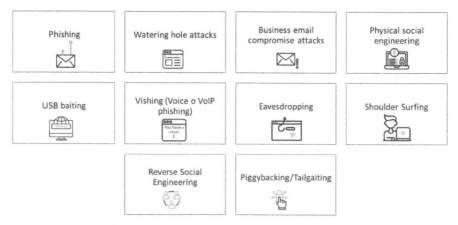

Fig. 5.3 The main types of social engineering attacks.

The main features of each of them are (Corbet and Gurdgiev, 2020):

1. **Phishing** scams are the most common type of social engineering attack. They typically take the form of an email that looks as if it is from a legitimate source. Sometimes attackers will attempt to coerce the victim into giving away credit card information or other personal data. At other times, phishing emails are sent to obtain employee login information or other details for use in an advanced attack against their company.

2. **Watering hole** attacks are a very targeted type of social engineering. An attacker will set a trap by compromising a website that is likely to be visited by a particular group of people, rather than targeting that group directly.

3. **Business email compromise (BEC)** attacks are a form of email fraud where the attacker masquerades as a C-level executive and attempts to trick the recipient into performing their business function, for an illegitimate purpose, such as wiring them money.

4. **Physical social engineering** attends a company's ability to prevent unauthorized physical access of assets on their premises or to prevent someone from taking an unauthorized action based on someone requesting it in person. Experienced consultants can provide their clients with a great deal of information regarding their physical security.

5. **USB** baiting sounds a bit unrealistic, but it happens more often than you might think. Essentially, what happens is that cybercriminals install malware onto USB sticks and leave them in strategic places, hoping that someone will pick the USB up and plug it into a corporate environment, thereby unwittingly unleashing malicious code into their organization.

6. **Vishing (Voice or VoIP phishing)** is that subcategory of impersonation which uses telephone communication in the attack scenario. In fact, the telephone medium allows the attacker, on the one hand, not to show himself, effectively maintaining the camouflage; on the other, to interact with the victim in a more direct way than via e-mail, exploiting the potential of persuasion techniques and the less time the interlocutor has, in a synchronous context, to rationalize. In some cases, vishing can be used in combination with website defacement techniques. For example, a banking site could be cloned or hacked, with the insertion of a fake banner inviting users to call the help desk number indicated, to speed up the solution of a certain disservice; at that point, it will be easy for the social engineer on the other end of the phone to obtain the authentication credentials for the cloned site from the resource.

7. **Eavesdropping** is the technique with which an attacker or an unauthorized person intercepts a communication, furtively listening to

it, reading email or SMS messages, or somehow inserting itself into the transmission stream.

8. **Shoulder** Surfing is an acquisition methodology that exploits direct observation "behind the victim's shoulders" to obtain accessible information: documents left on desks, post-its or notes left on bulletin boards or monitors. The implementation can be accompanied or enhanced by computer-based tools such as small video cameras (used later to "zoom in" on sensitive information and details) or micro-binoculars.

9. **Reverse Social** Engineering techniques are characterized by the particular modus operandi: in the first phase the attacker generates the problematic situation, the emergency, then proposes himself to the victim as the subject capable of resolving it.

10. **Piggybacking/Tailgaiting** takes place, e.g., in the case of an organizational perimeter characterized by controlled physical access, the social engineer can implement piggybacking methodologies, with which he tries, through persuasion and empathy, to convince an authorized third person to open the door, or tailgaiting, i.e., simply 'tailgating' the latter, before the entrance closes.

5. Cybersecurity and Human Behavior

Digitization has great benefits, but it also introduces significant risks to the privacy of our data. This is because ill-intentioned people could illegally acquire our digital data and use them for personal gain. The first line of defense is represented by foresight in the management and eventual dissemination of personal data (Jin et al., 2022). So, we must be aware that the fundamental principle to keep in mind is, therefore, to think before making available or sharing personal information, including our videos and photos on the Internet. When we use IT tools, e.g., to make online purchases, it is always useful to read the privacy policy to find out how, by whom our data will be processed (Geissler and Tang, 2021). Therefore, the protection of our digital data is not simply a technological problem but requires the continuous engagement of multiple stakeholders and includes disciplines as diverse as physics and psychology but first is ourselves. In fact, the main cause of the theft of our data is determined by incorrect individual behavior. Also, in today's world, people are more connected than ever before and the surface area for cyberattacks has expanded. It has never been more important to be on guard against cyberattacks (Michalec et al., 2022). Each of us must be aware of these risks, and therefore adopt 'safe' behavior. We need to know, but critically. We need to know what we need to do in order not to engage in risky behavior and to act proactively (Sallos et al., 2019). Although adequate technological solutions are the basis of the security of our data, however, the human element cannot be neglected. For this reason, many governments and many companies are committed to educating their citizens and workers to make them increasingly aware of the correct behaviors to adopt in the cyber space. In this sense, the three strategic levers on which to act are knowledge, motivation, and opportunities. In this fast-paced, ever-changing digital world, it is necessary take a holistic approach to digital data protection. Also, it is imperative to meet the appropriate level of conformity of a people, a process, and a technology prospect. In this context and in a wider scenario it is necessary to also consider the industrial organizations. These organizations are in fact vital to ensure that we all have an adequate standard of quality of life (Prasad, 2020). These organizations have a set of 'collective' strategic data that must necessarily be adequately protected in the same way as individual personal data. This applies to all production sectors: energy, oil and chemical, gas distribution, water management and distribution, railway and road infrastructure, etc. All these infrastructures, so-called 'critical' are in fact controlled through operational technology (OT) information systems (IT) that contain digital information of relevance to a country. It is therefore essential that such data is protected by those who have an interest in degrading the security of a country.

6. Cyber Crime

The term 'cybercrime' is used to describe crimes directed at computers or other ICTs (such as computer intrusions and denial of service attacks), and crimes where computers or ICTs are an integral part of an offence (such as online fraud). The definition of cybercrime cannot be considered static (Safonov et al., 2020). It will necessarily evolve with the evolution of the possible activities carried out digitally. In fact, just as the internet and other modern technologies are creating tremendous possibilities, they also provide opportunities for criminals to commit new crimes and to carry out old crimes in new ways. On the evidence available, that the number, sophistication, and impact of cybercrimes continues to grow and poses a serious and evolving threat to individuals, businesses, and governments. Online, criminals can commit crimes across multiple borders in an instant and can target many victims simultaneously. Tools that have many legitimate uses, like high-speed internet, peer-to-peer file-sharing and sophisticated encryption methods, can also help criminals to carry out and conceal their activities. Cybercrime is a global threat (Zhernova and Minbaleev, 2021). Criminals and the technical infrastructure they use are often based overseas, making international collaboration essential (Nag et al., 2022). There is a significant growth in cyber criminality in the form of high-profile ransomware campaigns over the last year. Breaches leaked personal data on a massive scale leaving victims vulnerable to fraud. Because the distinction between nation states and criminal groups is increasingly blurred, cybercrime attribution is sometimes difficult. Cyber criminals seek to exploit human or security vulnerabilities to steal passwords, data, or money directly. The most common cyber threats include social media and email passwords (hacking), bogus emails asking for security information and personal details (phishing). The scale and complexity of cyberattacks are wide ranging. "Off the shelf" tools mean that less technically proficient criminals are now able to commit cybercrime and do so as awareness of the potential profits becomes more widespread. The evolving technical capability of malware means evolving harm as well as facilitating new crimes, such as the cryptomining malware which attacks digital currencies like Bitcoin (Sen et al., 2022). Cyberattacks are financially devastating and disrupting and upsetting to people and businesses. The websites of governments provide practical advice on how to protect oneself, your computers, and mobile devices and your business against fraud, identity theft, viruses, and many other problems encountered online. They contain guidance on many other related subjects too—including performing backups and how to avoid theft or loss of your computer, smartphone, or tablet. Every country has its own set of rules, regulations, and standards for Cybercrime. An important aspect to consider is the global cost of cybercrime. It is expected to surge in the next five years, rising from $8.44 trillion in 2022 to $23.84 trillion by 2027 as shown in Fig. 5.4.

Cybercrime costs include damage and destruction of data, stolen money, lost productivity, theft of intellectual property, theft of personal and financial data, embezzlement, fraud, post-attack disruption to the normal course of business, forensic investigation, restoration and deletion of hacked data and systems, and reputational harm.

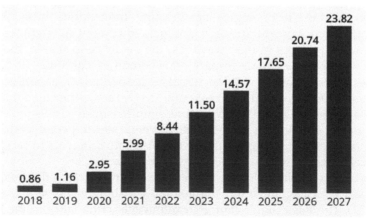

Fig. 5.4 Estimated costs of cybercrime worldwide (Sources: Statistica—https://bit.ly/3wkjtB7).

In this regard it is interesting to note that already in 2013, IBM proclaimed data promises to be for the 21st century what steam power was for the 18th, electricity for the 19th, and hydrocarbons for the 20th.

The world will store 200 zettabytes of data by 2025, according to Cybersecurity Ventures. This includes data stored on private and public IT infrastructures, on utility infrastructures, on private and public cloud data centers, on personal computing devices—PCs, laptops, tablets, and smartphones—and on IoT devices. Data is the building block of the digitized economy, and the opportunities for innovation and malice around it are incalculable. Cybercrime can take various forms, some common ones include e-mail compromise, identity theft, malware and ransomware attacks, phishing. Now that so much of daily life revolves around the use of electronic devices, from cell phones to laptops, to e-readers and more, it is inevitable that crime will increase in these areas as well. The hard part is that criminals often have a high level of knowledge when it comes to how to access data, whereas the average internet user does not. Cybercrime impacts both our personal and professional lives in many ways. Therefore, people and businesses need to take security measures to protect themselves. It is therefore essential that everyone is aware of how they treat their personal data.

7. Cyber Defense

The parallel presence must also be considered in this context of criminals who can compromise our systems at work and at home for and obtain, e.g., intellectual property or our personal data. Thus, cybercrime is growing, especially in the world of industrial infrastructure (Mercado-Velazquez et al., 2021). Nation-state actors have already used cyberattacks to sabotage industrial infrastructure; criminal ransomware gangs are also moving into the industrial space, To counter computer crimes, the so-called Cyber defense discipline was born which indicates the ability of an organization, a public body, a company to prevent a hacker attack against a computer system or device (Eltahawy et al., 2022). Cyber defense implies the adoption of measures capable of contrasting intrusions to avoid the illicit theft of sensitive data. All cyber defense strategies have one common goal: to prevent, disrupt, and respond to cyber threats. Regulators also see a need for greater security in the world of industrial organizations. Therefore, in terms of cyber defense, much is also being done in Europe at the regulatory level. The Information Security Directive (NIS) fits into this context, the first legislative act at EU level on IT security, adopted by the EU Parliament in 2016 and implemented in most European countries. The European Commission is moving precisely in this direction. In fact, it replaced the

NIS directive, with the NIS Directive 2 to respond to the growing threats posed by digitization and the wave of cyberattacks and at the same time strengthen security in the company and in the supply chains, simplify reporting obligations and introduce stricter rules oversight measures and tougher enforcement requirements, including harmonized sanctions across the EU. The new EU directive also provides for harmonization, cooperation, and consistency with the Digital Operational Resilience Regulation for the Financial Sector (DORA) and the Critical Entity Resilience Directive (CER). The Digital Operational Resilience Act (DORA) brings the single European cyber defense to the financial market and fintech. Banks, insurance companies, and cryptocurrency operators have two years to adapt to the recent new DORA regulation (Shah et al., 2020). The single European cyber defense regulation takes the form of a measure in the name of cyber resilience and data protection. The new CER directive completes the European regulatory landscape on critical infrastructures and expresses a profound maturity of Europe on the issues of the protection of IT, above all by reconciling security in the cyber and kinetic fields and offering an all hazard and risk-based point of view. Obviously, these choices also go in the direction of protecting citizens' data, who however are required to have a proactive attitude and develop their own awareness.

8. Cybersecurity Skills: Towards the Certification of Professionalism

Cybersecurity affects all people and all activities, being closely related to the use of IT, which can be applied in every context. As illustrated in the previous paragraphs, cybersecurity and therefore data protection is an essential requirement of modern systems (and infrastructures) and cannot be separated from development and management; it's not something that can be added later (Smith et al., 2022). Managers, Chief Information Security Officer (CISO), employees are all interested in IT, in different ways depending on the responsibility. Office workers work with technologies. Managers coordinate specific activities, supported by IT and they govern sets of IT-supported activities (Yu and Shen, 2022). Everyone must have the skills to interact with their interlocutors. To create a common vision across all EU Member States, on 21 September 2022, ENISA announced the European Cybersecurity Skills Framework (ECSF). The document lists 12 typical cybersecurity job profiles along with their identified titles, missions, tasks, skills, knowledge, and competencies (Phair et al., 2022). The main purpose of this framework is to create a common understanding between individuals, employers, and learning program providers across all EU Member States, making it a valuable tool for bridging the gap between the cybersecurity professional workplace and the learning environments. Also, ENISA has identified and ranked the 10 top cybersecurity threats to emerge by 2030 (Goupil et al., 2022): (1) Supply chain compromise of software dependencies; (2) Advanced disinformation campaigns; (3) Rise of digital surveillance authoritarianism/loss of privacy; (4) Human error and exploited legacy systems within cyber-physical ecosystems; (5) Targeted attacks enhanced by smart device data; (6) Lack of analysis and control of space-based infrastructure and objects; (7) Rise of advanced hybrid threats; (8) Skills shortage; (9) Cross-border ICT service providers as a single point of failure; and (10) Artificial intelligence abuse. Considering the 10 main threats to cybersecurity, the framework represents a valid tool for identifying the professional profiles necessary to counter future cyber threats. Training professions is not enough. It is necessary to create awareness among citizens as well, who are great users of IT (Hajny et al., 2022). Precisely for this reason, the digital training of citizens is fully part of the actions aimed at strengthening national cybersecurity and the subject of strategic political choices in many countries.

9. Conclusion

The digital revolution has transformed everyday life. Thanks to the Internet, connecting people all over the world has never been easier than it is today. However, this relatively new way of living—always accessible anywhere and anytime—has brought to light new problems in terms of security and especially cybersecurity.

In the era of digital transition, the protection of personal data is a matter of primary importance. Therefore, the use of the Internet and connected devices offers new opportunities for people and companies but, at the same time, creates new threats. For example, new technologies, smart devices connected to the network and many AI applications expose every organization to cyberattacks. Cybersecurity threats exploit the increased complexity and connectivity of critical infrastructure systems, placing the nation's security, economy, and public safety and health at risk. Like financial and reputational risks, cybersecurity risk affects a company's bottom line. It can drive up costs and affect revenue. It can harm an organization's ability to innovate and to gain and maintain customers. Cybersecurity can be an important and amplifying component of an organization's overall risk management. Cybersecurity is the practice of protecting systems, networks, and programs from digital attacks. These cyberattacks are usually aimed at accessing, transforming, or destroying sensitive information, and extortion of money from users or disrupting normal business processes. The regulations are there, but much more needs to be done to ensure increasingly adequate security for industrial infrastructures and citizens in the digital world. Security and privacy are of paramount importance in our new way of life. In our various guises (e.g., citizens, workers, parents) we are all called to contribute with awareness and responsibility to the protection of our data and corporate data for our safety and that of our country as well as collective safety in general.

References

AlDaajeh, S., Saleous, H., Alrabaee, S., Barka, E., Breitinger, F. and Raymond Choo, K. (2022). The role of national cybersecurity strategies on the improvement of cybersecurity education. *Computers and Security*, 119. doi: 10.1016/j.cose.2022.102754.

Alamri, B., Crowley, K. and Richardson, I. (2023). Cybersecurity risk management framework for blockchain identity management systems in health IoT. *Sensors*, 23(1). doi: 10.3390/s23010218.

AlZaabi, K.A.J.A.J. (2019). The value of intelligent cybersecurity strategies for Dubai smart city. doi:10.1007/978-3-030-01659-3_49.

Annarelli, A., Clemente, S., Nonino, F. and Palombi, G. (2021). Effectiveness and adoption of NIST managerial practices for cyber resilience in Italy. doi: 10.1007/978-3-030-80129-8_55.

Bergami, R., Aulino, B. and Zafar, A. (2013). The influence of cyber language on adolescents learning English as a second language: Voices from Italy and Pakistan. *International Journal of Learning*, 18(12): 107– 120. doi: 10.18848/1447-9494/cgp/v18i12/47826.

Bokhari, S., Hamrioui, S. and Aider, M. (2022). Cybersecurity strategy under uncertainties for an IoE environment. *Journal of Network and Computer Applications*, 205. doi: 10.1016/j.jnca.2022.103426.

Bozzetti, M.R.A., Olivieri, L. and Spoto, F. (2021). Cybersecurity impacts of the Covid-19 pandemic in Italy. Paper presented at the *CEUR Workshop Proceedings*, 2940: 145–155.

Brandão, A.P. and Camisão, I. (2022). Playing the market card: The commission's strategy to shape EU cybersecurity policy. *Journal of Common Market Studies*, 60(5): 1335–1355. doi: 10.1111/jcms.13158.

Carvalho, J.V., Carvalho, S. and Rocha, Á. (2020). European strategy and legislation for cybersecurity: Implications for Portugal. *Cluster Computing*, 23(3): 1845–1854. doi: 10.1007/s10586-020-03052-y.

Cheng, E.C.K. and Wang, T. (2022). Institutional strategies for cybersecurity in higher education institutions. *Information (Switzerland)*, 13(4). doi: 10.3390/info13040192.

Corbet, S. and Gurdgiev, C. (2020). An incentives-based mechanism for corporate cyber governance enforcement and regulation. doi: 10.1007/978-3-030-38858-4_13.

Duong, T.Q., Hoang, V. and Pham, C. (2021). Convergence of 5G technologies, artificial intelligence, and cybersecurity of networked societies for the cities of tomorrow. *Mobile Networks and Applications*, 26(4): 1747– 1749. doi: 10.1007/s11036-021-01778-6.

Eltahawy, B., Valliou, M., Kamsamrong, J., Romanovs, A., Vartiainen, T. and Mekkanen, M. (2022). Towards a massive open online course for cybersecurity in smart grids—A roadmap strategy. Paper presented at the IEEE PES Innovative Smart Grid Technologies Conference Europe, 2022 October. doi: 10.1109/ISGT-Europe54678.2022.9960630.

Geissler, M. and Tang, C. (2021). Strategies for integrating two-year information technology and cybersecurity programs. Paper presented at the SIGITE 2021 – *Proceedings of the 22nd Annual Conference on Information Technology Education*, 15–19. doi: 10.1145/3450329.3476865.

Goupil, F., Laskov, P., Pekaric, I., Felderer, M., Dürr, A. and Thiesse, F. (2022). Towards understanding the skill gap in cybersecurity. Paper presented at the Annual Conference on Innovation and Technology in Computer Science Education, *ITiCSE*, 1: 477–483. doi: 10.1145/3502718.3524807.

Hajny, J., Sikora, M., Grammatopoulos, A.V. and Di Franco, F. (2022). Adding European cybersecurity skills framework into curricula designer. Paper presented at the *ACM International Conference Proceeding Series*. doi: 10.1145/3538969.3543799.

Jin, D., Chen, B., Yu, L. and Liu, S. (2022). Adaptive output regulation for cyber-physical systems under time-delay attacks. *Control Theory and Technology*, 20(1): 20–31. doi: 10.1007/s11768-021-00072-w.

Kennedy, L. (2020). Society's new frontier-cybersecurity, privacy, and online expression. *Cornell International Law Journal*, 53(1): IX–XIX. Retrieved from www.scopus.com.

Mercado-Velazquez, A.A., Escamilla-Ambrosio, P.J. and Ortiz-Rodriguez, F. (2021). A moving target defense strategy for internet of things cybersecurity. *IEEE Access*, 9: 118406–118418. doi: 10.1109/ACCESS.2021.3107403.

Michalec, O., Milyaeva, S. and Rashid, A. (2022). Reconfiguring governance: How cybersecurity regulations are reconfiguring water governance. *Regulation and Governance*, 16(4): 1325–1342. doi: 10.1111/rego.12423.

Mohammed, A. and George, G. (2022). Vulnerabilities and strategies of cybersecurity in smart grid-evaluation and review. Paper presented at the 3rd International Conference on Smart Grid and Renewable Energy, SGRE 2022—Proceedings. doi:10.1109/SGRE53517.2022.9774038.

Nag, A., Ranjan, R. and Vinoth Kumar, C.N.S. (2022). An approach on cybercrime prediction using prophet time series. Paper presented at the 2022 IEEE 7th International Conference for Convergence in Technology, I2CT 2022. doi: 10.1109/I2CT54291.2022.9825386.

Phair, N. and Alavizadeh, H. (2022). Cybersecurity skills of company directors — ASX 100. *Journal of Risk Management in Financial Institutions*, 15(4): 429–436.

Payne, B.K., He, W., Wang, C., Wittkower, D.E. and Wu, H. (2021). Cybersecurity, technology, and society: Developing an interdisciplinary, open, general education cybersecurity course. *Journal of Information Systems Education*, 32(2): 134–149.

Prasad, S. (2020). Counteractive control against cyberattack uncertainties on frequency regulation in the power system. IET Cyber-Physical Systems: *Theory and Applications*, 5(4): 394–408. doi: 10.1049/iet-cps.2019.0097.

Raicu, G. and Raicu, A. (2022). Cybersecurity Strategies In Industry 4.0. *International Journal of Modern Manufacturing Technologies*, 14(3): 233–238. doi: 10.54684/ijmmt.2022.14.3.233.

Ron, M., Ninahualpa, G., Molina, D. and Díaz, J. (2020). How to develop a national cybersecurity strategy for developing countries. Ecuador case. doi: 10.1007/978-3-030-40690-5_53.

Safonov, K.V., Zolotarev, V.V. and Derben, A.M. (2020). Analysis and forecasting strategies of attacks on game resources for distributed cybersecurity training games. *Paper presented at the IOP Conference Series: Materials Science and Engineering*, 822(1). doi: 10.1088/1757-899X/822/1/012027.

Sallos, M.P., Garcia-Perez, A., Bedford, D. and Orlando, B. (2019). Strategy and organizational cybersecurity: A knowledge-problem perspective. *Journal of Intellectual Capital*, 20(4): 581–597. doi:10.1108/JIC-03-2019-0041.

Santisteban, A.S., Cunyarachi, L.O. and Andrade-Arenas, L. (2020). Analysis of national cybersecurity strategies. *International Journal of Advanced Computer Science and Applications*, 11(12): 771–779. doi:10.14569/IJACSA.2020.0111288.

Sen, A., Jena, G., Jena, S. and Devabalan, P. (2022). A case study on defending against cybercrimes. *Journal of Pharmaceutical Negative Results*, 13: 1931–1938. doi:10.47750/pnr.2022.13.S01.229.

Shah, Y., Chelvachandran, N., Kendzierskyj, S., Jahankhani, H. and Janoso, R. (2020). 5g cybersecurity vulnerabilities with IoT and smart societies. doi: 10.1007/978-3-030-35746-7_9.

Shaikh, F.A. and Siponen, M. (2023). Information security risk assessments following cybersecurity breaches: The mediating role of top management attention to cybersecurity. Computers and Security, 124. doi: 10.1016/j.cose.2022.102974.

Smith, L.A., Chowdhury, M. and Latif, S. (2022). Ethical hacking: Skills to fight cybersecurity threats. Paper presented at the *EPiC Series in Computing*, 82: 102–111.

Yu, W. and Shen, F. (2022). The relationship between online political participation and privacy protection: Evidence from 10 Asian societies of different levels of cybersecurity. *Behavior and Information Technology*, 41(13): 2819–2834. doi: 10.1080/0144929X.2021.1953597.

Zhernova, V.M. and Minbaleev, A.V. (2021). The integrated regulation of a cyber-physical system. Paper presented at the *CEUR Workshop Proceedings*, 3040: 116–124.

6

Legal Issues With AI/Algorithmic Systems
Responsibility and Liability

Michael Martin Losavio

1. Introduction

Who Is Responsible for 'AI?'

Outcomes of Artificial Intelligence (AI) operations bring great benefits, but also may lead to injuries to others. Deployment in a Super Smart Society assures engagement of nearly every person save those living purely analogue lives in isolation. The exceptional benefits and risks cross many domains of human activity. Beginning with the collection of data through to its processing, inferences, and use, AI-mediated systems may become entangled in regulations governing those benefits and the risks of injuries.

Suppose you use AI ChatBot to help you write a journal paper, legal brief, or book?[1] What if it went awry and gave bad medical advice? Or wrote false, hateful, and defamatory things about someone? Or did indeed create the Great Novel of the Century? Who gets the blame and who gets the credit, up to and including that Nobel Prize for Literature?

Department of Computer Engineering and Computer Science at the University of Louisville, Louisville, Kentucky, U.S.A.
Email: michael.losavio@louisville.edu

[1] Jay Caspian. (2022).

The cost of liability for flawed AI outcomes can be significant. AI deployment in a Super Smart City will impact many of those in the city, from hundreds to millions. A straightforward calculus shows how monetary damages add up for injury from a flawed system: D(amages)= $\sum_{d1\ldots dn}$, where d= loss from injury by an individual in monetary terms and n= everyone injured. In one case, a flawed system for determining, and punishing, supposedly fraudulent claims for benefits wrongly charged thousands of people and levied significant financial penalties. One of the resulting lawsuits filed on behalf of a class of those injured was settled by a payment of $20 million by the US State of Michigan.[2]

'AI' can embrace many different areas of computing operations, such as computational predictive analytics, machine learning, and neural networks, as well as that denoted as "Artificial Intelligence." The potential of these systems seems limitless. While "Artificial Intelligence" may not accurately describe the technology, it is an acceptable catch-all for processes that subject to a variety of laws. Those laws delineate the rights and obligations that flow from the development, distribution, and use of AI.

The injuries from AI may appear in many forms, ranging from errors in information to mistakes in control systems. An AI-mediated Search might produce links to sites providing harmful information, ranging from medical treatment to hatred to political engagement. An AI-mediated facial recognition systems might, in error, identify an innocent person as the perpetrator of a crime. This, in turn, may lead to a dangerous police encounter, or prosecution. An algorithmic "Robo-Judge" determines an innocent person has committed a crime, leading to criminal prosecution and seizure of their assets in recompense. An algorithmic recommender system connects children with sexually explicit material. An algorithmic system runs analytics against a massive database of consumer preferences and interests to reveal intimate facts about their lives All of these outcomes may lead to injuries to people, sometimes catastrophic ones.

Who is responsible for those injuries? Who might be accountable, and made to compensate the injured? The Attorney General of the State of Indiana of the United States asserts TikTok is liable for a variety of injuries its systems cause to people who use its services.[3,4]

It is vital that accountability for AI compensate and make whole those injured and critically, encourage and incentivize better, safer design, distribution, and use of AI systems to reduce the likelihood of injury.

A system of accountability exists within the general legal framework of all nations. Although in some cases legal immunity has been available or given to encourage and facilitate the development of Information and

---

[2] Michigan Department of Attorney General (2022).
[3] David Wells (2022).
[4] State of Indiana v. TikTok, Inc. and Byte Dance Ltd. (2022).

Communications Technology (ICT), that immunity is subject to change. For example, the Communications Decency Act (US) safe harbor of immunity for third-party posting of online content has been amended to withdraw immunity for postings of sex trafficking activity. The General Data Privacy Regulation (EU) creates liability for failures to insure privacy protections, regardless of intent or the ICT system. Everyone working with AI should be cognizant of the potential legal impact on themselves and their work and use of those systems.

Super Smart Cities and Super Smart Technologies, built on public and private networks and sensors, may be used in domestic, commercial/industrial, and governmental employ. Control and measurement devices are deployed in industrial and commercial systems, such as SCADA networks.[5] These commercial/ industrial, governmental, and domestic services may interact with centralized controllers that may be remote in the cloud or proximate, at the edge or in the 'Fog' of computational systems. Networked to AI analytics and control, such systems offer control, inference, and oversight power that is unprecedented. Unfortunately, that may mean that the injuries caused by such systems, where inadequately designed or used, may also be unprecedented.

These systems are for the benefit of people everywhere. It is vital that concern for the welfare of those who may be subjects of such systems and impacted by their use. Legal regulatory regimes serve to incentivize that concern by creating sanctions for the injuries that may result from inadequately designed or poorly used systems.

The vast data and analytical power deployed in Super Smart Cities can bring revelations and controls over peoples' lives never seen before. The inferential/decisional outcomes from these systems may have far-reaching consequences. The American Supreme Court Justice Sonya Sotomayor, commenting on the earlier, analogous system of information collection and analysis, opined such massive surveillance power may change the relationship between government and the governed.[6]

The injuries that may come from the systems are vast, especially if used carelessly or with evil intent. Uncontrolled, they may damage the very benefits such systems can offer. It is vital that the social, legal, and regulatory components of Super Smart Technologies and Super Smart Cities be inventoried, their possible outcomes noted and the engineering of such systems meet the responsibilities of developers, manufacturers, and users. This can best assure safety for all from these systems.

---

[5] E. Chikuni and M. Dondo. (2007).
[6] United States v. Jones (2012).

Whose Law Controls?

Jurisdiction is the power a nation asserts over conduct that impacts it. That power is often mediated by courts and laws that decide liability and then set the sanctions for the injuries that occurred. It is traditionally territorial in that a nation may assert over any actions within its territory or impacting those within their territory. Given the globalization of commerce generally and electronic commerce, in particular those involved in the development of any ICT system may be subject to the jurisdiction of nations far from their homes. And increasingly nations may assert jurisdiction over the actions of their citizens anywhere in the world, or over the actions of non-citizens that injure that nation or its citizens.

Asserting jurisdiction is the first step in adjudicating and sanctioning liability for injuries from AI systems. Even if there is a finding of liability and of money damages or criminal punishment, a nation may not have the ability to enforce those. A variety of treaties provide for the enforcement of foreign judgments against a citizen, but some nations do not recognize any extra-territorial efforts to enforce them. Others, such as those nation-signatories to the Council of Europe's Convention on Cybercrime,[7] agree to cooperate on transnational enforcement of criminal laws regarding cyber systems but may limit enforcement to conduct that is criminal under that nation's domestic laws.

Given the increasingly interconnected world of global e-commerce and the value of that market, a successful AI implementation and its users, developers, and distributers should be prepared to defend themselves in the event of injury from their AI systems.

2. Criminal Liability, Civil Liability

Criminal liability for AI systems is the most severe regulation of those engaged with its development and use. Criminal sanctions are the most serious sanctions for misuse as they include imprisonment and the loss of personal liberty and freedom as well as the forfeiture of assets, such as profits, and fines. Criminal sanctions are applied where there are the most serious injuries, including death and personal harm. These criminal (delictual) laws are designed to promote public safety generally. Criminal sanctions are meant to punish for past conduct, deter future misconduct, rehabilitate those capable of it, and incapacitate those who won't be deterred nor rehabilitated. It is a key component of a general security regime in which all people should engage.

It is the people who develop, distribute, and use AI systems who will be subject to possible criminal sanctions where there is a prohibited result. The

---

[7] Council of Europe, Convention on Cybercrime (23 November 2001).

general criminal law regimes of each nation impacted by these systems will apply. These regimes tend to be uniform, built around a core of regulation of malum in se (wrong in itself) misconduct. The canonical prohibited conduct includes homicide, sexual assault, physical assault, trespass, and theft. These have expanded to include destruction of property, forgery, fraud, interference with official operations and, increasingly, offenses against privacy, among others.

Most nations have civil law statutory regimes, with some, such as the United States and Great Britain, having mixed common law and statutory legal systems. Nearly all nations regulate criminal conduct by express statutes that prohibit canonical misconduct that is bad in and of itself—malum in se—as well as malum prohibitum, conduct that is 'wrong' and punished as a crime because the government says so. These criminal sanctions apply to actions a particular nation deems pernicious enough to warrant criminal punishment. This builds an 'algorithmic' system of laws and legal rules that can be applied to AI deployment. Those laws and regulations can provide warning of areas where caution and care is needed to avoid injury to others.

One key element in criminal prohibitions in that of scienter, the mental intent of the person whose conduct, such as the use of an AI system, led to a prohibited outcome, such as the death of a person or persons.

Criminal statues mostly prohibit intentional or knowing acts, where a person intends the action or outcome that leads to injury. Such prohibitions won't apply to most situations. Few develop systems of any kind that they intend will injure others, other than AI directed weapons systems.

But scienter has expanded, along with the potential of injury to others, to embrace unintentional conduct. This expansion can lead to criminal liability for AI systems deployment, depending on the estimations of risk and the injuries caused.

Most salient are criminal liability for these unintentional mental states where prohibited conduct occurs (Table 6.1).

Table 6.1 Culpable Mental States.

Mental State	Characteristics
Reckless	Failed to recognize an unreasonable risk of injury
Wanton	Recognized an unreasonable risk of injury but disregarded that risk
Strict Liability	Criminal liability regardless of any mental state

These mental states, though unintentional, may create liability for AI development, deployment, and use where the risk of injury is deemed unreasonable. The essential point is that although injury was not intended or the risk unknown, that is not a shield from liability. Everyone has the obligation to be careful not to hurt others, and this obligation extends even into new areas of human endeavor where the pioneers must anticipate all their innovations may do.

Civil Liability—Civil liability for injuries can encourage secure and reliable AI systems through the financial penalties of damages paid to those injured. Styled as "quasi-delictual" conduct under civil law regimes, liability stems from a variety of legal obligations that, while sometimes similar to criminal sanctions, generally only carry the risk of monetary penalties and not that of imprisonment. But those financial penalties may be related to the injuries caused by the AI system.

In extreme cases, if the AI operation leads to the death of someone, even beyond the cognitive stress of that responsibility, the damage penalties may equal the economic output of the rest of that person's life. For example, using a very simple calculation method, and age 60 as a retirement age, a person's age 30 would be expected to have an economic life of 30 years. If that person at age 30 made €50,000 per year and setting aside the many economic factors of increased wages over time, the simplest calculation of economic loss of 30 years x €50,000 = €1.5 million. For one middle-income person. That amount goes up with increased wages and lifetime work, and is multiplied if, as is very likely with the governmental systems of the Super Smart City by each of the many people impacted. The financial liability may be in millions and tens of millions of euros, dollars, renminbi, rupees, or yen.

Civil liability provides that injuries are generally made whole through money damages. These may be set by statutory and common law monetary sanctions. Although typical software vendors will try and limit their liability though End User License Agreements (EULA), this limitation won't apply to third parties injured by the systems, and sometimes may not apply to people who agreed to the EULA if the risks are great. The regime of product liability law provides for recovery of damages from a dangerous product. It applies to a manufacturer, programmer, or distributer where injuries result from the design, production, and distribution of an inherently dangerous and flawed product. Claims under products liability may be based on a variety of other legal theories, including negligence, strict liability, and breach of warranty of fitness and may vary with the jurisdictions controlling the issues of liability.

Liability may attach where an inherent defect in an AI that causes harm to another. An "inherent defect" will create liability for whomever was involved with that defect, from programmers, developers, and distributors.

In the case where $20 million (US) was the negotiated settlement for the damages from a flawed algorithmic system, the liability of the State of Michigan attached as its seizure of assets of the plaintiffs from erroneous determinations of fraud were without due process of law.[8] The "due

---

[8] Bauserman v. Unemployment Insurance Agency (2022).

process of law" is a civil right to have a fair opportunity of a citizen to challenge and correct wrongful state action, and the injured Michigan citizens were denied that right to correct the wrongful seizures against them. Even governments, with all their immense powers and immunities, may be liable for injuring their citizens through their use of AI systems in the courts. That is in addition to potential political liability from the voters for the "computational abuse" of authority.

There is one critical aspect of sovereign liability developers, distributers, and users must note in the context of governmental use of flawed systems. Sovereign entities such as municipalities and states may be immune for liability for their wrongful actions. The non-state developers and distributers may be held liable. But some nations, such as the United States, provide that the state employees of such entities using those systems, in their individual capacities, may be liable for the injuries their actions played a part in. Those state employees, as in Michigan's case, may be personally liable for massive financial damages.

3. Thought Exemplars

Consider these thought experiment examples, extrapolating from algorithmic failures in computations systems of AI-mediated systems for the Super Smart City for transportation and healthcare:

Example 1: A Super Smart City has distributed sensors over all roads and transit paths within its boundaries that centrally collect that data for real-time analysis by AI systems. These analyses are dynamically used to control traffic control systems such as traffic lights, warning remediated traffic flow system modelled on fluid dynamics[9] and greatly improves the efficiency of vehicular traffic in the city.

But if it breaks down, massive traffic congestion may develop. People may be delayed for work, school, medical services, or various other appointments, some of which may be critical to safety or health. If that failure is caused by the traffic controller and its AI system, there may be liability. Inadequate software testing may have failed to uncover a race condition whereby commands are executed out of order, the wrong traffic control devices are actuated, and a cascade of traffic problems erupt, stopping vehicular transport.

Example 2: Super Smart Technology is used to monitor and allocated prescription medication from medical data. The process is quick and initially has a reduced error rate over human pharmacists, improving patient safety. But software testing, done on short, interval operations, did

---

[9] Bretti, Gabriella and Natalini, Roberto and Piccoli, Benedetto (2007).

not test for production use of the system where it might be in continual use for many hours between re-start and reset. A timing error, normally trivial between short on-off times, accumulates over continuous operation. This leads to mistiming of dosages, potentially leading to overdosing patients on possibly lethal medications in larger doses.

These thought experiments point to potential liability. Where system failure leads to death, the scrutiny of those systems will be at its highest. Civil liability for wrongful death would probably attach to all involved, from the developers to the users, under theories of negligence and product liability law. Criminal liability would depend on where in the chain of causation the failure led to a person's death. There, if that failure is shown to be due to a reckless discard for the dangers, then a charge for reckless homicide could be brought. If it is discovered that the developers, distributers, or users knew of the risk but disregarded it, then a more serious change of wanton homicide could be brought. Some jurisdictions equate wanton homicide with murder, so the most serious penalties may be levied for that death of an innocent person.[10]

A case study of one such system shows that the risks, and related penalties, for AI system failures can be severe.

4. One Case Study in Flawed Analytics

The damage that can be done via a supposedly Super Smart City/Province/State using Super Smart Technologies is seen in the deployment of an automated adjudication system by the US state of Michigan that led to injuries of innocent people.[11,12] A government benefits system moved to a computational system for addressing issues in unemployment compensation insurance, a benefit employees and employers both contribute to for wage protection. That computatoinal system incorporated an algorithmic system to examine the data from applications for benefits, detect patterns or anomalies, and the alert as to misfeasance that could lead to denial of benefits and punishment for fraudulent applications.

That algorithmic system processed data from current and past applications for benefits, looking back are prior cases as well as current pending claims. In the course of its analyses it would find inconsistencies in the data and, based on those, lead to determinations of fraud that would subject the applicant to both a denial of benefits and possible financial penalties.

---

[10] Kentucky Revised Statutes 507.020, Murder, "including… he wantonly engages in conduct which creates a grave risk of death to another person and thereby causes the death of another person".

[11] Cahoo et al. v. SAS Analytics Inc. et al. (2019).

[12] Bauserman v. Unemployment Insurance Agency, Id.

Above and beyond a failure to consider simple human error, the system would analyze income reports and allocate income to time periods via a system of "income spreading". The algorithm, rather than applying the income report to a specific pay period, would average income over multiple pay periods, creating an inference of income being earned during time periods where a person had no income, and thus the application for benefits. But with this erroneous attribution of income, when a person asserted he or she had not income at that time it was flagged as a potentially fraudulent. An individual's claim for benefits due to loss of employment creating a loss of income would be at odds with the erroneous algorithmic inference of income. This led to determination of possible fraud.

Unfortunately, there was no intervention or review by other systems or people. The claimants were effectively denied the ability to contest these findings by a notification system that usually did not work and a stringent time period within which assert that the determination was erroneous.

The penalties levied against the claimants for the beneits could be multipled up to four times the benefits claimed or received, potentially devastating for people out of work. For some, this led to their filing for bankruptcy or losing their housing.

System analysis found an error rate in fraud positive determinations of 93%. But use of the system was not terminated until the flaws were corrected. Rather, after complaints from administrative law judges about the errors, review by state employees of the determinations was introduced. Yet this only reduced the error rate to 50% as the state continued to use it despite these problems.

When finally called to account, the state itself was protected by sovereign immunity for the injuries caused. But the state employees responsible for that system, knowing of the errors but continuing to use the system, were denied immunity and faced possible damages due to the financial injuries. The continued use of this machine system, even in the face of such damage and injury to people, violated clearly established constitutional due process rights and these employees were liable for the damages caused by use of the system. When the damages were multiplied by the thousands of people injured, the total amount these employees may be responsible for may be far beyond their capacity to pay.

In the end, the the State of Michigan paid $20 million to settle the claims of the people injured by this system. This serves as a marker for the financial damage, and financial liability, a flawed algorighmic system can incur.

5. It Wasn't Me—It Was The AI!

Blaming a machine system, for now, is just another way of blaming the people responsible for it, which may be several people in the chain from development to use. It won't work to escape responsibility for a flawed system. As is sometimes said, "AI don't kill people; PEOPLE kill people!" The challenges are in identifying the particular people responsible and who are rightly held to account for injuries to others. At this time "Artificial Intelligence" systems themselves will not be considered autonomous, "intelligent" entities legally responsible for the damage they do. The Imitation Game[13] posited by Alan Turing may set us thinking about the future, but it won't avail the programmers, developers, distributers, and users of AI to escape responsibility for the damage they do. Responsibility presumes a key level of cognition, as where people with mental defects, animals and non-AI machines are not held liable for their actions. It is the facts of these systems that govern responsibility and liability. The failure of facts as to shift responsibility will govern AI systems for the time being. And indicative of that is Turing's conclusion on the matter: "The original question, 'Can machines think?', I believe to be too meaningless to deserve discussion." Turing, supra.

The AI Litigation Database of George Washington University (US) is a project of its Ethical Tech Initiative of DC.[14] It sets out in both searchable static and dynamic data a broad range of litigation involving AI. Its October 2022 table details 78 cases. These are primarily cases in US federal courts but include US state court cases as well as cases from Canada and the United Kingdom. The AI Litigation Database details the claims made against those making and using AI and algorithmic system:

- Due process and Equal Protection US Constitutional violations in sentencing algorithms as to gender, indigenous status, race;
- Unreasonable law enforcement actions, searches and seizures, and convictions based on predictive police algorithmic assessment of criminal activity, facial recognition, social network analysis of alleged criminals;
- Unreasonable or erroneous algorithmic assessment of property taxes, exclusionary advertising due to family status, gender and race, suitability for employment and privacy/consumer protection violations through facial recognition analytics;
- Unreasonable denials of disability benefits, food stamps, unemployment benefits due to algorithmic rejection of claims or reduction in assessment of needs, determinations as to pay raises, promotion and termination from employment;
- Wrongful injury and death due to self-driving and autonomous vehicles.

---

[13] Turing, A.M. (1950).
[14] Ethical Tech Initiative of DC (2022).

Litigation has asserted liability for algorithmic error in business operations. There were claims for crimes against humanity and genocide and support for the Hamas terrorist organization due to analytics targeting, though these were dismissed.[15]

The breadth of these cases and the grounds for liability for injuries show how the expansion of AI use into more and more areas of civil life are leading to more and more claims for redress from injury. Assertions of liability and responsibility will continue to be expand and be debated as the need for policy and legislation.

Addressing responsibility will be an ever-greater challenge as the technology grows increasingly complicated, sophisticated, and propriety. There may be more and more disagreements as to what these systems do, especially where systems may not be fully 'explainable' and have self-developed through machine learning systems.

A controversy developed where an Alphabet/Google systems engineer working on their AI chatbot LaMDC (Language Model for Dialogue Applications) examined the machine for bias, such as towards religions, in its output,[16] Engaging in a data exchange with LaMCD, the engineer asserted their machines had achieved 'consciousness'. His employer did not agree and the engineer's work for them ended.[17,18] Yet this question will appear again, possibly even as a defence to personal liability.

Vapnik has applied the "duck test" for characterizing statistical analytical systems, many of which are core functions of AI and machine learning systems.[19] So if an 'AI' sounds like an intelligent entity, talks like an intelligent entity, and 'transports' like an intelligent entity, perhaps it is, or at least close enough that a jury can't decide. This presupposed that the judges, in their gatekeeper role for using scientific, engineering, or other specialized knowledge in legal adjudications, decide to permit that issue be given the a jury to decide rather than being tossed out as unreliable and inadmissible to be considered at trial. The first lawsuits on these issues promise to be interesting, and hotly contested.[20] They may reflect the very controversial and deeply felt beliefs some have for animals and their rights. The lawsuit by the People for the Ethical Treatment of Animals (PETA) claiming, on behalf of the orcas of the entertainment company Sea World, that they were being held in involuntary servitude in violation of the anti-slavery provisions of the Thirteenth Amendment (US) was dismissed as

---

[15] Id.

[16] Allyn, Bobby (2022).

[17] McQuillan, Laura (2022).

[18] Grant, Nico (2022).

[19] Vapnik, Vladimir (2020). (2022).

[20] U.S. Constitution, amend. XIII, § 1" "Neither slavery nor involuntary servitude, except as a punishment for crime whereof the party shall have been duly convicted, shall exist within the United States, or any place subject to their jurisdiction."

there was no standing, or right, to assert such a claim.[21,22] It is emblematic of the sorts of problems any defensive efforts to transfer liability from people to their creations will face; they may not even get their day in court for not having a right to one.

The National Institute for Standards and Technology (NIST) (US) has put forth its Artificial Intelligence Risk Management Framework[23] (AI RMF). This framework is a voluntary guide "to address risks in the design, development, use, and evaluation of AI products, services, and systems". Yet it represents a significant step in government setting out how to build better, safer AI products.

The Framework, motivated by the growth in AI power and capabilities, is meant to promote trustworthy and responsible AI. While noting all technology has common concerns in proper and competent management, it notes special problems and challenges:

AI systems bring a set of risks that require specific consideration and approaches. AI systems can amplify, perpetuate, or exacerbate inequitable outcomes. AI systems may exhibit emergent properties or lead to unintended consequences for individuals and communities. A useful mathematical representation of the data interactions that drive the AI system's behavior is not fully known, which makes current methods for measuring risks and navigating the risk-benefits tradeoff inadequate. AI risks may arise from the data used to train the AI system, the AI system itself, the use of the AI system, or interaction of people with the AI system.[24]

Similarly, the Organization for Economic Co-operation and Development (OECD) works to develop guidelines for AI and sets out "principles of responsible stewardship of trustworthy AI" that include accountability and enabling a policy environment for AI.[25] The OECD-AI Principles overview sets out operational principles of inclusivity, fairness, transparency and explainability, and robustness, security and safety.[26] This work by the OECD supports The Global Partnership on Artificial Intelligence of 14 countries and the European Union, formed to guide "responsible development and use of AI" while respecting human rights.[27]

Developers who do not pay attention to the Framework nor to the OECD-AI Principles, and their related documents, may one day have to

---

[21] U.S. Constitution, amend. XIII, § 1" "Neither slavery nor involuntary servitude, except as a punishment for crime whereof the party shall have been duly convicted, shall exist within the United States, or any place subject to their jurisdiction."

[22] Tilikum ex rel. People for the Ethical Treatment of Animals v. Sea World Parks & Entm't. (S.D. Cal. 2012).

[23] National Institute for Standards and Technology (US). (2022).

[24] *Id.*, p. 1

[25] Organization for Economic Cooperation and Development (2019).

[26] OECD.AI Policy Observatory (2022).

[27] Organization for Economic Cooperation and Development (15 June 2020).

explain why they did not incorporate the guidance they give into their development and resultant AI systems. They may face liability for failure to do so where compliance might have avoidance injuries caused by their AI systems.

The many different things AI systems can do and the activities, especially as to data, insert them more and more into peoples' lives. Yet those same systems are the responsibility of their users, owners, and developers, including for the failures and injuries they may have engendered. One artist won an art contest in digitally manipulated photography to then face accusations of deception where the AI tool Midjourney was used.[28] Are those developers of the AI tool free of any obligation for any damage from it or is it solely the responsibility of the user?

Far worse, an English coroner examining the death of a 14-year old girl concluded the victim had been exposed by the AI recommender systems of two social media companies to a barrage of material on suicide, self-harm, and other negative content; the coroner's conclusion as to the cause of death: "[the victim] died from an act of self-harm whilst suffering from depression and the negative effects of online content".[29] Are those social media companies responsible for serving up this material to that child, or is that the responsibility of others? Would it matter if such a company knew its systems were recommending inappropriate and harmful content, but did not stop doing so?[30]

In the end, beyond liability, would it matter to that developer of an AI system that their work caused the death of a child?

6. Conclusion

Developers and users of AI systems, especially in Super Smart Cities, must be fully cognizant of the damage that may be done by their systems and must engineer them to reduce or eradicate erroneous outcomes. This is both an ethical obligation and a legal one for which tremendous penalties for damages may accrue. In the most severe cases, where death and personal injuries may result beyond financial damages, criminal liability may be attached. This may put all of those responsible for the system and assurance of its reliability and safety in jeopardy of imprisonment and loss of their liberty. These are not risks to be ignored. Quality and assurance for these systems as they increasingly govern our lives is essential for all of us, lest we suffer in many different ways. That suffering includes the guilt for causing the suffering of others by systems for which we are responsible.

---

[28] Harwell, Drew (2022).

[29] Naughton, John (2022).

[30] Hao, Karen (2021).

References

Allyn, Bobby. (2022). The Google engineer who sees company's AI as 'sentient' thinks a chatbot has a soul. National Public Radio All Things Considered, 16 June

Bauserman v. Unemployment Insurance Agency. (2022). Michigan Supreme Court (US), Docket No. 160813, decided 26 July. at https://www.courts.michigan.gov/4a1c19/siteassets/case-documents/opinions-orders/msc-term-opinions-(manually-curated)/21-22/bauserman-op.pdf accessed 11 December 2022.

Bretti, Gabriella and Natalini, Roberto and Piccoli, Benedetto. (2007). A Fluid-Dynamic Traffic Model on Road Networks. Archives of Computational Methods in Engineering, 14: 139-172. 10.1007/s11831-007-9004-8.

Cahoo et al. v. SAS Analytics Inc. et al. (2019). No. 18-1296 (6th Cir. 2019).

Chikuni, E. and Dondo M., (2007), Investigating the security of electrical power systems SCADA., *AFRICON 2007*: 1-7, doi: 10.1109/AFRCON.2007.4401531.

Council of Europe (2001). Convention on Cybercrime, 23 November. https://www.refworld.org/docid/47fdfb202.html, accessed 26 November 2022.

David Wells (2022). Indiana Files Pair of Lawsuits Against TikTok over Data Security, Child Safety. Courthouse News Service, 8 December. Available at https://www.courthousenews.com/indiana-files-pair-of-lawsuits-against-tiktok-over-data-security-child-safety/, accessed 8 December 2022.

Ethical Tech Initiative of DC. AI Ligitation Database, George Washington University, at AI Litigation Database – Search – Ethical Tech Initiative of DC (gwu.edu). Accessed 10 December 2022.

Flannery, Sheri. (2012). *The Thirteenth Amendment Won't Help Free Willy.*, 15 The Scholar Available at: https://commons.stmarytx.edu/thescholar/vol15/iss1/2.

Grant, Nico. (2022). Google Fires Engineer Who Claims Its A.I. Is Conscious. New York Times, 23 July, p. B3.

Hao, Karen. (2021). The Facebook whistleblower says its algorithms are dangerous. Here's why. MIT Technology Review, 5 October. Available at https://www.technologyreview.com/2021/10/05/1036519/facebook-whistleblower-frances-haugen-algorithms/ accessed 12 December 2022.

Harwell, Drew (2022). He used AI to win a fine-arts competition. Was it cheating? The Washington Post, 2 September.

Jay, Caspian. (2022). Could an A.I. Chatbot Rewrite My Novel? The New Yorker, 9 December.

Kentucky Revised Statutes 507.020, Murder, "including… he wantonly engages in conduct which creates a grave risk of death to another person and thereby causes the death of another person."

McQuillan, Laura. (2022), "A Google engineer says AI has become sentient. What does that actually mean?" CBC News, 24 June. Available at https://www.cbc.ca/news/science/ai-consciousness-how-to-recognize-1.6498068#:~:text=Google%20engineer%20Blake%20Lemoine's%20recent,it%20means%20to%20be%20alive. Accessed 12 December 2022.

Michigan Department of Attorney General. (2022) "State of Michigan Announces Settlement of Civil Rights Class Action Alleging False Accusations of Unemployment Fraud," 20 October. Available at https://www.michigan.gov/ag/news/press-releases/2022/10/20/som-settlement-of-civil-rights-class-action-alleging-false-accusations-of-unemployment-fraud#:~:text=The%20State%20of%20Michigan%20has,their%20property%20without%20due%20process. Accessed 11 December 2022.

National Institute for Standards and Technology (US), Artificial Intelligence Risk Management Framework, 2nd Draft, https://www.nist.gov/itl/ai-risk-management-framework. Accessed 10 December 2022.

Naughton, John. (2022). Molly Russell was trapped by the cruel algorithms of Pinterest and Instagram. The Guardian, 1 October.

OECD.AI Policy Observatory., OECD AI Principles Overview. The OECD Artificial Intelligence (AI) Principles - OECD.AI. Accessed 12 December 2022.

Organization for Economic Cooperation and Development. (2019). "Recommendation of the Council on Artificial Intelligence: OECD/LEGAL/0449, adopted 21 May 2019.

Organization for Economic Cooperation and Development. (15 June 2020) OCED to host Secretariat of new Global Partnership on Artificial Intelligence. Online: https://www. oecd.org/going-digital/ai/OECD-to-host-Secretariat-of-new-Global-Partnership-on-Artificial-Intelligence.htm.

State of Indiana v. TikTok, Inc. and ByteDance Ltd. (2022), Superior Court of Allen County, Indiana, Case No. 02D03-2212-PL-000401, filed 7 December. Available at https:// www.documentcloud.org/documents/23389518-in-tiktok-complaint-china. Accessed 8 December 2022.

Tilikum ex rel. People for the Ethical Treatment of Animals v. Sea World Parks & Entm't. (2012). No. 11cv24 76, 2012 U.S. Dist. LEXIS 15258, at *14-15 (S.D. Cal. 2012).

Turing, A.M. (1950), Computing Machinery and Intelligence. (1950). *Mind*, LIX (236), October: 433–460. https://doi.org/10.1093/mind/LIX.236.43.

U.S. Constitution, amend. XIII, § 1".

United States v. Jones. (2012). 565 US 400 (2012) (US).

Vapnik, Vladimir, Complete Statistical Theory of Learning: Learning Using Statistical Invariants. (2020). p. 23. Available at https://lexfridman.com/files/slides/2020_01_10_vladimir_vapnik_theory_of_learning.pdf. Accessed 3 December 2022.

PART II

Super Smart Technologies in Different Industrial Sectors

7

Digital Skills and Future Workplaces in a Super Smart Society

An 'Intelligent' Railway Pantograph Maintenance for High-speed Trains

Fabio De Felice and *Cristina De Luca**

1. Introduction

The pantograph was invented in 1879 by Walter Reichel, an engineer at Siemens & Halske of Germany (Rotoli et al., 2016). The pantograph, during the 20th century, gradually supplanted, thanks to its better characteristics, the other devices used for powering the electric railway and tramway means of transport, (i.e., the tramway arch, the trolley pole, and the of uptake). The pantograph consists of an articulated system (movable panel), mounted on the base frame, and installed by means of insulators on the roof of the rolling stock. The movable panel supports the collection head equipped with contact strips that are in direct contact with the wire or wires of the catenary. The pantograph, whose working height can even reach 3 meters

Department of Engineering, University of Naples "Parthenope", Italy, Isola C4, Centro Direzionale Napoli, 80143 Napoli (NA), Italy.
Email: fabio.defelice@uniparthenope.it
* Corresponding author: cristina.deluca001@studenti.uniparthenope.it

in height, is lifted with compressed air devices, and held at the established pressure by means of suitably calibrated springs or, by a pneumatic drive. Therefore, predictive- and condition-based monitoring are important to ensure safety, cost-effectiveness, and smooth rail operations (Huang et al., 2020). Currently, the maintenance of the pantographs is a critical success factor to prevent suboptimal conditions from causing defects to vehicles or infrastructure (Kilsby et al., 2017). In this context, highly accurate measurements of ground and on-board systems, related to both vehicle and track, overhead line, and pantograph, could provide important data for modern maintenance solutions (Gonzalez-Jimenez et al., 2021). From this point of view, it is important to specify that the current pantographs used in Italy are characterized by a first level of diagnostics capable of generating alarms in the event of exceeding the acceleration values beyond the limits, such as ETR1000 high-speed trains in operation and EMU-Caravaggio regional trains under construction. However, to reduce the risk of possible damage to the railway network and save time, resources and money, innovative maintenance policies should be implemented (Wang et al., 2021; Chen et al., 2021; European Commission, 2019; Zhou et al., 2019; Dinmohammadi F., 2019). The advent of Industry 4.0 has given way to a constant evolution also to predictive maintenance systems that will be destined to play an increasingly important role in all industrial, production, and service sectors in future (Chen et al., 2019; Lidén et al., 2017). In fact, thanks to the monitoring of the significant parameters of the machine, it is possible to identify and anticipate the occurrence of a fault and eliminate unnecessary maintenance activities as well argued by several authors (Lee et al., 2018; Pisano and Usai, 2020). In this scenario, software solutions and technological innovations would be needed for an appropriate management of railway pantograph maintenance. In this way, it will be possible to manage all interventions in a coordinated way to provide the right information at the right time, manage people, deadlines and interventions, and support decisions. (Song et al., 2020). Thus, the aim of the present research is to propose an innovative approach for the maintenance management of railway pantographs thanks to the use of technologies enabling the Industry 4.0 (i.e., internet of things (IoT), Big Data & Data Analytics, Artificial Intelligence (AI) & Machine Learning, Decision Support Systems, Augmented Reality (AR), etc.). Therefore, with this "4.0" approach, technical and engineering aspects will be integrated with management and economic aspects, thus giving rise to new models and processes of Integrated Maintenance (Karakose and Navik, 2020). It is, hence, in this perspective of rapid technological evolution that the present research aims to integrate three technological macro-areas: (1) Self-Diagnostics of the Pantograph. It is the object of maintenance activities, through appropriate sensors of the component. Sensors are powered by applying energy harvesting (EH) which will be able to transmit the 'status' of the Device Under Test (DUT) to the Maintenance Management Platform 4.0, to determine when it is more appropriate to perform maintenance. Condition-based and predictive-based approaches are used, subject of research and experimentation activities; (2) Full Maintenance Cycle Management Platform 4.0. It can implement the integration of diagnostic functions with those of intervention management, and (3) Augmented Reality Maintenance System. It can support maintenance interventions, so that they can be carried out in an effective, efficient, safe way for rail operation and for operators. Starting from the three technological macro-areas defined above, the research intends to development of increasingly sensor-based railway components. As smart components they can self-diagnose their state using a condition-based and predictive-based approach. Consequently, the research proposes an innovative diagnostic information technology (IT) system for first and second level maintenance to jointly manage maintenance operations, materials, equipment, and maintenance documents. In addition, a setting of a Data Processing Center (DPC) capable of managing the new process will be developed. Finally, the proposed architecture will be integrated by an AR maintenance system. The final output of the project is a platform capable of planning, monitoring, and controlling individual maintenance activities, both with man-machine and man-man interaction processes. The rest of the manuscript is structured as follows: Section 2 presents an overview of maintenance 4.0 and AR. Subsequently, Section 3 describes the main phases of the project; Section 4 describes the practical implementation of project. Finally, the main conclusion and future developments are summarized in Section 5.

2. State of Art on Maintenance 4.0 and AR

Rail transport has undeniable environmental, practical, and economic advantages. Increasing the capacity of "mass transport" will be the only way to face, in the long term, the challenge of increasing urbanization and the resulting road congestion (Gordieiev et al., 2018, Qiu et al., 2021; Zio et al., 2007). However, the availability of trains is closely linked to the reliability of their components. In this context, digitization plays a fundamental role (Patiño-Rodriguez and Carazas, 2019). Today, thanks to the development of on-board sensors and the IoT, it is possible to obtain real-time diagnosis of the state of the train, predict faults, and therefore organize preventive maintenance, cheaper than corrective maintenance, or even manage a stop in (Rastegari and Salonen, 2015; Fengxia et al., 2020). In this perspective, maintenance management is also becoming essential to anticipate, predict and define the behavior to be adopted during the train's life cycle (Levitin et al., 2021; Zhang et al., 2020; Bergquist and Söderholm, 2014; Bouillaut et al., 2012). As argued by several authors thanks to digitalization and the use of the enabling technologies of Industry 4.0 (IoT, Big Data & Data Analytics, AI & Machine Learning, Decision Support Systems, AR, etc.), it is possible to evolve maintenance policies towards a continuous improvement of the maintenance management (Baraldi et al., 2018; Karakose and Navik, 2020; Liu et al., 2014). The benefits of digitization in maintenance primarily concern, firstly, connectivity and quick access to information relating to maintenance needs, and secondly, the direct monitoring of the machines takes over, which can also take place remotely, with the further possibility of predicting their behavior (Alberti et al., 2022; Macchi et al., 2012; Söderholm and Norrbin, 2021). One of the most promising digital technologies in the current Industry 4.0 landscape is undoubtedly AR. In fact, as claimed by several authors AI are immersive applications that improve maintenance activities and allow you to anticipate breakdowns and disservices (Sharma and Rai, 2019; Erkoyuncu et al., 2017; Al-Turki et al., 2014). This technology allows to increase man's perceptive capabilities by adding to what man detects with his senses, other data, usually transmitted electronically (Choi et al., 2021; Merrick et al., 2011). The use of innovative technologies such as AR in the maintenance management, brings several advantages. For example, it is possible to visualize the individual operations that maintenance workers need to carry out according to the planned maintenance action they need to perform, or the type of fault detected (Scheffer et al., 2021; Erkoyuncu et al., 2017). Using AR, it possible to imagine extraordinary improvements in terms of standards of effectiveness, efficiency, and risk reduction, and allows one to rethink and reshape all the processes directly or indirectly linked to maintenance activities (De Felice et al., 2019; Kumar et al., 2010; Cioffi et al., 2020; Hodkiewicz et al., 2021). To analyze the potential of using AR in the maintenance management, we investigated the state of the art through the SCOPUS database, the Elsevier largest abstract and citation database of peer-reviewed literature. Specifically, search string used for the bibliographic investigation was (TITLE (augmented AND reality) AND TITLE (maintenance)). String was defined according to the standards of Scopus database using the Boolean operators (AND, OR, NOT). Only articles in which the string was found in "article title" were analyzed. The Scopus survey showed that the first publication on this topic was published in 1998. In total from 1998 to date (June 2022), 251 documents have been published. It is not surprising that there have been publications on the subject since 1998. In fact, in recent years, the term "augmented reality" has entered more and more common language thanks to the increasing accessibility of the technology itself. However, AR is not a recent technology as it seems: the origin of the term dates to 1992, while the first use of the technology itself dates back to the 1960s. In 1968, the American computer scientist Ivan Sutherland, known as the "father of computer graphics", created what is considered the first viewer for AR, a device with two cathode ray tubes that projected 3D images superimposed on the real world. But it was only in 1990 that this type of technology was given the name "augmented reality", in a project by Tom Caudell and David Mizell for Boeing. Over the years many applications with AR have been developed and proposed essentially in the world of cinema and gaming. But, only recently, in 2019, has research shifted to mixed reality and the integration of AI and AR applied in several sectors. The application horizons are almost infinite: from automotive to manufacturing to smart packaging to operations. Figure 7.1 clearly shows the growing interest of the scientific community in AR and the maintenance sector.

Scopus investigation highlighted that most of the documents are Conference Papers (169; 67%) followed by Articles (72; 29%), Book Chapters (8; 3%), and Reviews (2; 1%). Generally, the most technologically advanced nation is thought to be the USA. However, forecasts show that Europe and Asia acquire the majority share of the AR market (Ramedhan and Kartowisastro, 2020). In this regard, the scientific community of the different countries shows a great interest, especially in Italy (39; 22%), Germany (29; 17%), and China (27; 15%), as shown in Fig. 7.2.

From the analysis of the published documents, it emerges that the most covered research areas are Engineering (34%) followed by Computer Science (33%) and Mathematics (11%). It is interesting to note that many publications are the result of funded research projects, most of which are funded by the British Engineering and Physical Sciences Research Council (17%), followed by European Commission (17%) and National

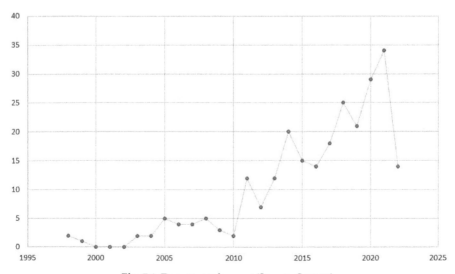

Fig. 7.1 Documents by year (Source: Scopus).

Fig. 7.2 Documents by country/territory (Source: Scopus).

Table 7.1 Projects for the use of Augmented Reality in the field of Maintenance.

Project	Year	References	Technological Features	Application(s)
ARVIKA	2002	W. Friedrich	Platform connected to the network to receive updates on the environmental conditions and data of the object on which it is going to operate. The connection to the AR browser allows the computer to process the images it receives through an algorithm, associating them with reference images to understand the user's position in space.	Automotive sector Aeronautical sector
STAR	2004	A. Raczynski	Aims at mixing images of real environments, virtual objects, and humans to produce mixed reality animations of an existing environment. Technology based on the development of 3D modelling, camera tracking, and image storage.	Industrial sector
ARMAR	2009	S.J. Henderson	Experimental augmented reality systems for maintenance job aiding. It proposes the modular design concept: through a head-worn display, staff can be guided to the damaged area and assisted in understanding the tasks and performing them.	Military sector

Table 7.1 contd. ...

... Table 7.1 contd.

Project	Year	References	Technological Features	Application(s)
RAMAR	2009	Pisa University, VRmadia (Spin-off)	Augmented reality tool for maintenance and distance learning. The expert can see what the remote user is looking at by guiding him through tools such as HMD Optical See-Through Helmet with camera, headset, microphone, and minicomputer.	Industrial sector Naval sector
TAC	2010	S. Bottecchia	The operator is equipped with a device that can record the user's real field of view and the surrounding environment. Both video streams are sent to the remote expert technician who can decide where to place the augmented information.	Industrial sector
IMA-AR	2015	N. Gavish	Augmented Reality platform for operator training provides a set of instructions to be carried out in chronological order. With the tablet the object of maintenance is framed, operator touching the tablet touch screen receives information in the form of text and graphics.	Industrial sector
ARAUM	2017	J.A. Erkoyuncu	It includes two platforms, 'Authoring' and 'Application', which allow the automated creation process. The first one is necessary for the creation of the content to be shown increased during maintenance operations that operates on PC; the second one for the actual maintenance that shows the previously created content to the operator on mobile Android.	Industrial sector

Natural Science Foundation of China (12%). Table 7.1 summarizes the main projects concerning the use of the AR in maintenance, the technological characteristics, and the field of application.

Figure 7.3 shows the most used keywords (i.e., User Interfaces, Human Computer Interaction, Industry 4.0).

3. Modeling Methodology and An Operational Proposal

3.1 *Definition of operating procedures for maintenance*

This section describes the operational phases of the project. The architecture model is structured through seven main phases. Figure 7.4 presents an

Fig. 7.3 The most used keywords (Source: Scopus).

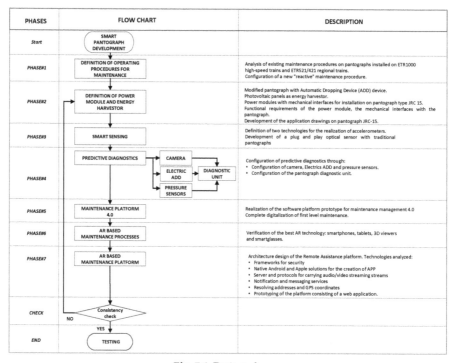

Fig. 7.4 Project phases.

overview of the project phases. For each phase, the activity carried out is briefly described.

The definition of the operating procedures starts from the current maintenance procedures tasks foreseen for the maintenance of the pantograph. In this regard, it is important to specify that the pantographs installed on the ETR1000 and ETR521/421 trains have a diagnostic system based only on optical accelerometers that measure the acceleration of the pantograph contact strips. The train is equipped with four pantographs, two of which are intended for power supply under catenary powered at 25 kV AC or 15 kV AC and the other two units are for power supply under catenary powered at 3 kV DC or 1.5 kV DC. Traction is provided by eight motorized trolleys, each equipped with two motors, present on the first, third, sixth, and eighth elements. This allows the train to enjoy better grip and accelerate from a standstill to 300 kilometers per hour in 4 minutes.

Figure 7.5 shows a preliminary study of the JRC15 pantograph with the new optical accelerometers, the power module with energy harvesting and the electric ADD.

In traditional maintenance systems, preventive maintenance coincides with scheduled maintenance. According to new maintenance diagnostics procedures (such as contact strip acceleration analysis, monitoring systems, measurement procedures through AR), the integrity of the pantograph is checked by measuring the acceleration of the heads. The optical signals, coming from the new optical accelerometers and powered by the energy harvesting system are transmitted to the pantograph diagnostics and processed in the time and frequency domain. Figure 7.6 shows the application of the power module.

The main functional requirements of the power module and the mechanical interfaces with the pantograph are reported in Table 7.2.

To improve the maintenance process, the research intends to the development of increasingly smart sensor-based railway components. The term "Smart Sensor" (intelligent sensor) refers to a type of sensor that has

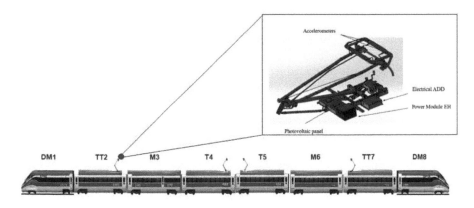

DM: Driver's cabs
TT: Pantograph transformer
M: Converter, braking resistor
T: Pantograph

Fig. 7.5 Example of structure of the ETR 1000 (Product from: Hitachi Rail Italy-Bombardier).

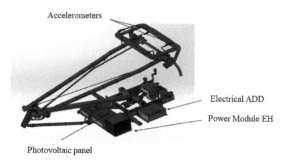

Fig. 7.6 Application of the power module.

Table 7.2 Power module requirements with energy harvesting.

Power Module Requirements	Value
Output Voltage	+12V + 5V ± 1%
Power Output	5W (minimum value)
Average daily operation	20 h
Temperature range	25 ÷ 75°C
Protection (installation on imperial)	IP 68

inside electronic circuits capable, in addition to detecting a physical size, chemical or electrical, including processing information and transmitting it to the outside in the form of a digital signal. Schematically, within an intelligent sensor you can identify a transducer, an amplifier circuit, an A/D converter, a microprocessor, and a communication interface. The development of optical sensors connected directly to pantograph contact strips increases the reliability of contact strip acceleration detection systems. Figure 7.7 shows the positioning of the optical sensors on the pantograph head.

Table 7.3 summarizes the main functional characteristics of the accelerometer are reported.

OPTICAL ACCELEROMETERS

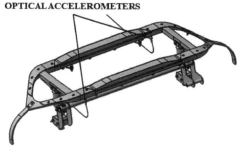

Fig. 7.7 Examples of application of optical accelerometers on pantograph head JRC15.

Table 7.3 Functional features new optical accelerometers.

Optical Accelerometers Features	
Frequency range	0,5 ÷ 330 Hz ± 5%
Natural frequency	800 ÷ 2.400 Hz
Measuring acceleration range	50 g
Maximum g for survival	2000 g
Nominal sensitivity	100mV/g + 5%
Sensor Power supply	12V ± 5%
Output Impedance	< 100
Phase shift	0 deg
Amplitude non-linearity	<5%
Traverse sensitivity	<3%
Temperature Range	−25 ÷ 100°C
Protection (outdoor installation)	IP 68

It should be noted that the maintenance plan defined by the rolling stock manufacturer shows all the first and second level scheduled maintenance activities. Maintenance activities shall be carried out on rolling stock during the period of service, either on a mileage or on a time basis. Based on the maintenance program, the rolling stock manufacturer defines the maintenance management which shows the planning of first and second level maintenance for each component. The conceptual scheme of the innovative second-generation sensor is shown in Fig. 7.8.

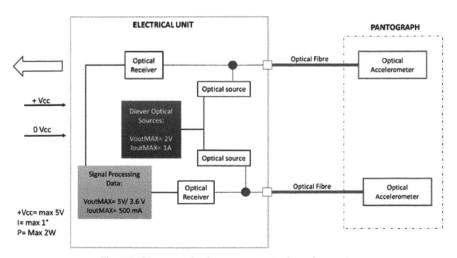

Fig. 7.8 Conceptual scheme new optical accelerometer.

3.2 *Maintenance Management Platform 4.0*

The Maintenance Management Platform 4.0 is based on a WEB-Server architecture. It considers the integration of the procedures of maintenance interventions with the management of the activities carried out by its operators and the activities carried out by the appointed employees and the manager responsible for the contract. In particular, the WEB-Server mainly handles dynamic HTML pages. The WEB-Server implements the following Architecture (multi-tier), according to the paradigm at levels (three-tier). Figure 7.9 illustrates the logic of the SYPLA-4.0 Web Server.

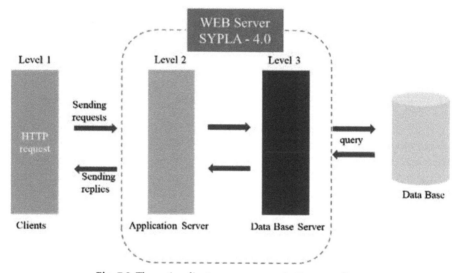

Fig. 7.9 Three-tier client-serve communication paradigm.

A new Level I maintenance management system is introduced, which provides for the management of the following master data:

- The *personnel database* must contain information on all personnel present at the worksite with an indication of their professional specialization.
- The *tools database* contains all the tools and equipment present on the site, with an indication of their expiry dates.
- The *materials register* contains the list of all materials present in the warehouse with an indication, where applicable, of the serial number and the type of stock. An alert signal must be provided for materials under warranty according to the rate of use and minimum stock.
- The *rolling stock register* contains the rolling stock present in the depot being serviced with the date of assignment to the depot, the monthly mileage, and the serial numbers of the pantograph system components present on the rolling stock.

The back-office will be responsible for loading, updating, and managing the master data. As regards the development of Level II maintenance, this is implemented with reference to the interaction with Level I maintenance. The management of Level II maintenance includes:

1. The connection with the materials sent from the repair yard and the actual repair carried out in the workshop with the replacement of the elementary components.
2. On site, after diagnosis, the maintenance technician can replace the faulty thrust control unit by carrying out Level I maintenance.
3. The faulty control unit is then sent to a workshop to be repaired by specialized technicians. During the repair, the faulty elementary components will be replaced to restore the thrust control unit and then sent to the site warehouse.

The analysis of the causes of failures will then be shared with design, production, and testing in specific reports to trigger continuous product improvement.

In this phase the hardware and software features have been identified that can support real-time image processing processes, such as presence of a lens with Depth of Field Recognition (DOF) characteristics, visual processing, operating system, battery, and storage large enough to support rendering processes, display-type visual output, motion, and acceleration sensors.

3.3 AR based maintenance processes

The **AR based maintenance platform** (called GAP Guided Activities Platform) has been developed considering some fundamental principles: self-sufficiency of services, interfaceability/integrability with external systems, and security of access to data. In this regard, a cloud architecture based on machine virtualization has been developed. To do this, consolidated deployment technologies have been used, to make the software accessible from the outside, maintaining a level of security towards attacks and unauthorized access adequate to the standards of modern web applications. The software was written using frameworks such as Spring Boot, Maven and JPA. The platform provides a WEB interface for remote management and support, while field technicians can enjoy an APP made for Android or iOS containing support services, including AR functionality. In detail, this functionality is accessible by the technician through a call to an operator that triggers a series of parallel processes, such as:

- Call and participant monitoring for future reporting;
- Audio and video connection to the operator;
- Start of Augmented Reality services.

Once the implementation of the AR capabilities was completed, they have been lowered into an ecosystem of microservices that consists of features summarized in Table 7.4.

Each service is protected by an 8-hour security token that uniquely identifies a user on login, preventing access to services both to unregistered users and to users with insufficient access level. The result of the AR maintenance model is based on a cloud architecture principle for centralized

Table 7.4 Features of system services.

Services	Technical Specifications	Detail
Profiling the user	Admin	System administrator, with the possibility of creating users, working groups, downloading reports on activities on the platform.
	Worker	WEB technician, able to respond to service requests and use services for AR marker insertion.
	User	General user, can request technical assistance and use the mobile application.
User tracking	The date, time, and position of each user entering and leaving the platform are saved.	For mobile users, the location is updated in real time to allow operators to monitor the deployment of maintenance teams on the ground.
Documentaries	A section dedicated to uploading and downloading shareable material such as maintenance manuals, operational documents, photos of inspections, video presentations/slideshows.	The system can interface with external DMSs to retrieve or provide documentation to and from legacy systems.
Live chat	The system allows instant messaging system between all users of the platform	
Virtual Assistant	Artificial intelligence is available that can be suitably trained to carry out a first level of automated assistance.	The artificial intelligence itself can make a call to a real operator in case of complexity too high for the problem.
Reporting	An administrator can retrieve statistics in the form of datasheets or interactive graphs concerning access to the platform.	It is possible to track one or more users over a time period, and detect parameters related to service requests made.
Recording of interventions	Incoming calls on the platform can be recorded.	Calls can later be shared and sent to a proprietary CRM.

services. It has functional layers capable of governing devices in the field with both predominantly autonomous and assisted operation:

- **Cloud Architecture:** The model is divided into a client and a server. The client is a device in the hands of the field technician (smartphone/ viewer) that sends data to a Server, the 4.0 maintenance management platform, which manages the incoming and outgoing flow to manage the available AR microservices.
- **Autonomous Functionality:** The application can recognize the context. This feature enables it to provide a first level of assistance to the user in the form of security information, measurement, and reconstruction of 3D models based on object recognition or at the request of the user. The application interacts with the center only for the transfer of data to be recorded.
- **Assistance Functionality:** The platform has a microservice nature and gives the possibility to manage a mass of data that exceeds the capacity of the single device. Thanks to these features, the platform can manage a communication between the user and a remote operator who, thanks to AR tools, can aid in a precise and effective way by highlighting, for example, points of interest or overlapping information with the user's context.

4. Practical Implementations of the Project

Thanks to the AR, the operator identifies technical characteristics of the pantograph, the operating procedures to be followed for the most correct execution of the maintenance intervention and for the safety and management of railway operations, in the whole maintenance process which is shown in Fig. 7.10.

An important aspect that must not be overlooked is the safety of maintenance workers (organ contactors, competent and support maintenance workers) since the activities are carried out mainly in the

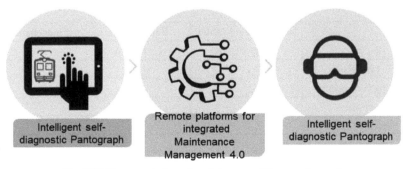

Fig. 7.10 Representation of maintenance 4.0 on pantographs.

undercarriage or on the imperial vehicle and in the presence of electrical and earthing circuits.

For this reason, it has been possible to integrate AR technologies for:

- **Support of the staff:** Support to the less experienced staff in a fast and effective way, centralizing the position of the competent maintainer with respect to more maintainers from side, giving tools of guide and interaction between the two to complete the operations.
- **Staff training:** Providing tools that increase learning experiences and provide innovative methodologies for training, through simulations of safe interventions through 3D reconstructions and enrichment of video information material of recorded interventions with virtual models and data.
- **Respect for safety:** Create an alarm and monitoring interface of the environment in which the operator is performing an intervention, in order to alert both personnel in the presence and remote control of possible hazards or emergency situations in a timely or even preventive manner with the aim of preventing accidents and injuries.

Equally important is the fact that the visual audio input/output system must be able not to hinder the ongoing operations. Wireless, wearable hardware was needed that could be integrated into the safety components.

A technician equipped with AR viewers can reproduce his point of view to an experienced remote operator who could guide him and highlight, through AR markers, faults, and details on which to intervene. Videos can also be recorded to create material for training purposes to show to trained personnel. The systems are also able to monitor the distance between the workers, the safety bars, and the vehicle systems, alerting the maintainers through visual and sound outputs in case of danger. Handheld and head-mounted devices were used to implement the new maintenance management processes. A comparison of the characteristics of the two technologies used is shown in Table 7.5.

Table 7.5 Comparison of handheld and portable devices.

	Smartphone / Tablet	Smartwatch	Smart Band
Display	5.5 in–12.5 in	1.5 in–3 in	X
Wi-Fi connection	✓	✓	✓
Roaming connection	✓	✓	✓
AR services	✓	X	X
GPS	✓	✓	✓
Gyroscope	✓	✓	X
Biometric sensors	X	✓	✓

The features of the devices used are:

- Operator position monitoring (ability to identify and track the operator; ability to define geofencing).
- Incident detection mechanisms (possibility to detect and report incidents automatically; possibility for the operator to report an incident; ability, in conjunction with position monitoring, to automatically identify the operator in the event of an incident).
- Possibility of interaction with the operator (possibility for the operator to contact expert support; possibility for the operator to consult documentation; possibility of using the device in Hands-Free mode, i.e., using technologies such as voice recognition).

Smart bands available on the consumer market do not fit the defined specifications as the devices are closed and tightly bound to the manufacturer's platform. While smartwatches are certainly more flexible among consumer devices, representing the point of intersection between smart bands and smartphones. As far as Head-Mounted Displays (HMD) are concerned, scouting has been carried out for viewers. These devices must allow a maintenance worker to know quickly and from any position in the environment about any faults that may occur and must be able to carry out maintenance work in complete freedom of movement and in any conditions. The main visors marketed are shown in Table 7.6.

At the end of all the scouting activity, two smartphones were selected on which to start the verification of the functionality with respect to native frameworks, such a Samsung and iPhone.

5. Practical Implementations of the Project

The result of the AR maintenance model is based on a cloud architecture principle for centralized services. It has functional layers capable of governing devices in the field with both predominantly autonomous and assisted operation:

- **Cloud architecture:** The model is divided into a client and a server. The client is a device in the hands of the field technician (smartphone/viewer) that sends data to a Server, and the 4.0 maintenance management platform, which manages the incoming and outgoing flow to manage the available AR microservices.
- **Autonomous Functionality:** The application can recognize the context. This feature enables it to provide a first level of assistance to the user in the form of security information, measurement, and reconstruction of 3D models based on object recognition or at the request of the user. The application interacts with the center only for the transfer of data to be recorded.

Table 7.6 HDM comparison.

	Irstick G1	Vuzix M300XL	Vuzix M400	Microsoft Hololens	Realwear HMT-1
Display	Microdisplay WQVGA	Microdisplay equivalent to 5" display thumbs	Microdisplay equivalent to 5" display thumbs	Transparent holographic lenses	0.33" Microdisplay (equivalent to 7" display)
Resolution	428x240	640x360	640x360	2048x1080	854x480
Field of view	13°	16,7°	16,8°	52°	20°
Tracking area	Camera, accelerometer	Camera, accelerometer	Camera, accelerometer	Environmental mesh in real time	Camera, accelerometer
Built-in audio	✓	✓	✓	✓	✓
Built-in microphone	✓	✓	✓	✓	✓
Sensors		Compass, gyroscope, accelerometer	Compass, gyroscope, accelerometer	4 cameras for head tracking, 2 infrared cameras, depth sensor, accelerometer, gyroscope, magnetometer	Accelerometer, gyroscope, magnetometer
Integrated processing capacity	No Mobile Requirement Android 8 or sup.	Dual Core Intel Atom Android 9.0	8 Core 2.52 Ghz Qualcomm XR1 SO Android 9.0	Windows	Android
Connection	USB Type-C	Micro USB Wi-Fi 5 Bluetooth 4.1 GPS	USB Type-C Wi-Fi 5 Bluetooth 5.0 GPS	USB Type-C, Wi-Fi 5, Bluetooth 5	USB Type-C, Wi-Fi 5, Bluetooth 4.1 GPS

- **Assistance Functionality:** The platform has a microservice nature and gives the possibility to manage a mass of data that exceeds the capacity of the single device. Thanks to these features, the platform can manage a communication between the user and a remote operator who, thanks to AR tools, can aid in a precise and effective way by highlighting, for example, points of interest or overlapping information with the user's context.

Thanks to the video call system, even a non-specialist technician can intervene immediately by contacting an expert who can support him remotely. The system guides the technician to the faulty vehicle, if stationary on the tracks, or the carriage to be repaired if the vehicle has been taken to a maintenance station. The remote operator has a view of all operations in progress and can interact remotely by entering indications in the form of graphics (possibly, even 3D) and voice indications to guide the technician through the operations in the best possible way. Once the maintenance operation has been completed, the system is able to automate the registration process, indicating the place, date, duration, and personnel involved on site and remotely. The intervention can also be recorded. The video can be used as training material for new staff, it allows the technician who carried out the intervention to refine his knowledge and it can certify to a customer that the intervention was carried out according to the established quality parameters.

6. Conclusions

A new, more reliable maintenance management system has been developed, leveraging product innovations on the pantograph to implement a process innovation, connected to the integrated management of the maintenance process, also through Augmented Reality support tools. In the field maintenance support phase, AR provides maintenance operators with tools, techniques, and view-dependent manual support for the operation in progress. The maintainer equipped with glasses for the AR will see in transparency the real scene on which to operate and overlapping contributions in 2D and 3D computer graphics. 2D contributions are, for example, arrows, text annotations, and warnings. 3D contributions consist of virtual replicas of parts of the product to be animated, for example, to exemplify disassembly/assembly movements. 2D and 3D contributions will be kept in visual collimation as the person moves around the scene – with the advantage of having your hands free to perform the task. Additional benefits could come from collaborative processes between two or more people to assist assembly and field maintenance in two ways.

- Jointly conduct a live operation 'seeing' in overlay to the real scene, but from different points of view, the same indications of aid through the appropriate sensorized glasses.
- Offer the employee the opportunity to request remote support from an expert for a teleconsultation with a high sense of presence in the real remote scene, leveraging the high multimedia and communication capabilities of modern AR glasses (cameras, audio and microphone, wireless communications, embedded computers).

Therefore, on the one hand, there will be the possibility to monitor and manage the entire maintenance cycle, and on the other hand to make use of a platform for the evaluation of process costs (Activity Based Cost) between the different technological tools. The result will be the prospect of further benefits. In fact, with reference to the means of transport and the railway components, the entire lifecycle of an industrial manufactured product is strongly conditioned by the characteristics of the maintenance processes. The risks, the times, the costs of intervention, the frequency of such processes, condition the total cost of ownership (TCO) and influence the competitiveness of the producing companies and their relationships with the market. In conclusion, the diagnostic pantograph (intelligent) together with the innovative integrated maintenance management system and support tools for AR, improves the reliability and availability of the pantograph itself by means of optical sensors and the energy harvesting system. Among other things, the complex and harmonious set of technologies implemented is aimed at optimizing maintenance activities, certify the execution, and increase the safety of operators with alarm systems that prevent and block operations that do not comply with safety standards.

Acknowledgement

NOVA MAINT CP. F/190203/01-02/X44 Ministry of Economic Development Ministerial Decree 5 March 2018 Research and development projects within the sectors applications consistent with the national strategy of intelligent specialization (SNSI) "Smart factory"

References

Alberti, A.R., Ferreira Neto, W.R., Cavalcante, C.A.V. and Santos, A.C.J. (2022). Modelling a flexible two-phase inspection-maintenance policy for safety-critical systems considering revised and non-revised inspections. *Reliability Engineering & System Safety*, 221: 108309. https://doi.org/10.1016/j.ress.2021.108309.

Al-Turki, U.M., Ayar T., Yilbas B.S. and Sahin A.Z. (2014). Integrated Maintenance Planning. In: *Integrated Maintenance Planning in Manufacturing Systems*. SpringerBriefs in Applied Sciences and Technology. Springer, Cham. https://doi.org/10.1007/978-3-319-06290-7_3.

Baraldi, P., Bonfanti, G. and Zio, E. (2018). Differential evolution-based multi-objective optimization for the definition of a health indicator for fault diagnostics and prognostics. *Mechanical Systems and Signal Processing*, 102: 382–400. https://doi.org/10.1016/j.ymssp.2017.09.013.

Bergquist, B. and Söderholm, P. (2015). Data analysis for condition-based railway infrastructure maintenance. *Quality and Reliability Engineering International*, 773–781. https://doi.org/10.1002/qre.1634.

Bottecchia S., Cieutat J.M. and Jessel, J.P. (2010). T.A.C: Augmented Reality System for Collaborative Tele-Assistance in the Field of Maintenance through Internet. *Proceedings of the 1st Augmented Human International Conference*, 14. https://hal.archives-ouvertes.fr/hal-00585435.

Bouillaut, L., Francois, O. and Dubois, S. (2012). Optimal metro-rail maintenance strategy using multi-nets modeling. *International Journal of Performability Engineering*, 77–90. https://doi.org/10.23940/ijpe.12.1.p77.mag.

Chen, H., Chai, Z., Jiang, B. and Huang, B. (2021). Data-Driven Fault Detection for Dynamic Systems with Performance Degradation: A Unified Transfer Learning Framework. *IEEE Transactions on Instrumentation and Measurement*, 70: 1–12, 3504712. http://dx.doi.org/10.1109/TIM.2020.3033943.

Chen, H., Jiang, B., Chen, W. and Yi, H. (2019). Data-driven Detection and Diagnosis of Incipient Faults in Electrical Drives of High-Speed Trains. *IEEE Transactions on Industrial Electronics*, 66(6): 4716–4725. http://dx.doi.org/10.1109/TIE.2018.2863191.

Choi S. and Park J.-S. (2021). Development of Augmented Reality System for Productivity Enhancement in Offshore Plant Construction. *J. Mar. Sci. Eng.*, 9: 209. https://doi.org/10.3390/jmse9020209.

Cioffi, R., Travaglioni, M., Piscitelli, G., Petrillo, A. and De Felice, F. (2020). Artificial Intelligence and Machine Learning Applications in Smart Production: Progress, Trends, and Directions. *Sustainability*, 12: 492. https://doi.org/10.3390/su12020492.

De Felice, F., Travaglioni, M., Piscitelli, G., Cioffi, R. and Petrillo, A. (2019). Machine learning techniques applied to industrial engineering: A multi criteria approach. *Proceedings of the 18th International Conference on Modelling and Applied Simulation* (MAS 2019), 44–54. https://doi.org/10.46354/i3m.2019.mas.007.

Dinmohammadi, F. (2019). A risk-based modelling approach to maintenance optimization of railway rolling stock: A case study of pantograph system. *Journal of Quality in Maintenance Engineering*, 25(2): 272–293. https://doi.org/10.1108/JQME-11-2016-0070.

Erkoyuncu, J.A. (2017). Improving efficiency of industrial maintenance with context aware adaptive authoring in augmented reality. *Cirp Annals*, 465–468. 10.1016/j.cirp.2017.04.006.

Erkoyuncu, J.A., del Amo, I.F., Dalle Mura, M., Roy, R. and Dini, G. (2017). Improving efficiency of industrial maintenance with context aware adaptive authoring in augmented reality. *CIRP Annals*, 66 (1): 465–468. https://doi.org/10.1016/j.cirp.2017.04.006.

European Commission. (2019). Study on the competitiveness of the Rail Supply Industry. https://ec.europa.eu/docsroom/documents/38025/attachments/1/translations/en/renditions/native. Accessed November. 2021. Friedrich W., Jahn, D. and Schmidt, L. (2002). ARVIKA-Augmented Reality for Development, Production and Service. *ISMAR*, 3–4. https://doi.org/10.1145/354666.354688.

Gavish, N. (2015). Evaluating virtual reality and augmented reality training for industrial maintenance and assembly tasks. *Interactive Learning Environments*, 23(6): 778–798. https://doi.org/10.1080/10494820.2013.815221.

Gonzalez-Jimenez, D., del-Olmo, J., Poza, J., Garramiola, F. and Madina, P. (2021). Data-Driven Fault Diagnosis for Electric Drives: A Review. *Sensors*, 21(12): 4024. https://doi.org/10.3390/s21124024.

Gordieiev, O., Kharchenko, V. and Leontiiev, K. (2018). Usability, security, and safety interaction: Profile and metrics-based analysis. International Conference on Dependability and Complex Systems. *Springer, Cham.*, 2018. https://doi.org/10.1007/978-3-319-91446-6_23.

Henderson, S.J. and Feiner, S.K. (2007). Augmented Reality for Maintenance and Repair (ARMAR). Final Report for June 2005 to August 2007, 34. https://citeseerx.ist.psu.edu/viewdoc/download?doi=10.1.1.149.4991&rep=rep1&type=pdf. Accessed January 2022.

Hodkiewicz, M.R., Lukens, S., Brundage, M.P. and Sexton, T. (2021). Rethinking Maintenance Terminology for an Industry 4.0. Future. https://doi.org/10.36001/ijphm.2021.v12i1.2932.

Huang, W., Zhang, R., Xu, M., Yu, Y., Xu, Y. and De Dieu, G.J. (2020). Risk state changes analysis of railway dangerous goods transportation system: Based on the cusp catastrophe model. *Reliability Engineering & System Safety*, 202: 107059. https://doi.org/10.1016/j.ress.2020.107059.

Karakose, M. and Yaman, O. (2020). Complex Fuzzy System Based Predictive Maintenance Approach in Railways. *IEEE Transactions on Industrial Informatics*, 16(9): 6023–6032. https://doi.org/10.1109/TII.2020.2973231.

Kilsby, P., Remenyte-Prescott, R. and Andrews, J. (2017). A modelling approach for railway overhead line equipment asset management. *Reliability Engineering & System Safety*, 168: 326–337. https://doi.org/10.1016/j.ress.2017.02.012.

Kumar, S., Gupta, S., Ghodrati, B. and Kumar, U. (2010). An approach for risk assessment of rail defects. *International Journal of Reliability, Quality and Safety Engineering*, 291–311. https://doi.org/10.1142/S0218539310003822.

Lee, J., Park, J. and Ahn, S. (2018). On determining a non-periodic preventive maintenance schedule using the failure rate threshold for a repairable system. *Smart Structures and Systems*, 22(2): 151–159. https://doi.org/10.12989/SSS.2018.22.2.151.

Levitin, G., Xing, L. and Dai, Y. (2021). Optimal operation and maintenance scheduling in m-out-of-n standby systems with reusable elements. *Reliability Engineering & System Safety*, 211: 107582. https://doi.org/10.1016/j.ress.2021.107582.

Lidén, T. and Joborn, M. 2017. An optimization model for integrated planning of railway traffic and network maintenance. *Transportation Research Part C: Emerging Technologies*, 74: 327–347. https://doi.org/10.1016/j.trc.2016.11.016.

Liu, X., Lovett, A., Dick, T. and Rapik Saat, M. (2014). Optimization of ultrasonic rail-defect inspection for improving railway transportation safety and efficiency. Journal of Transportation Engineering, 04014048-1-04014048-10. http://dx.doi.org/10.1061/(ASCE)TE.1943-5436.0000697.

Macchi, M., Garetti, M., Centrone, D., Fumagalli, L. and Pavirani, G.P. (2012). Maintenance management of railway infrastructures based on reliability analysis. *Reliability Engineering & System Safety*, 71–83. https://doi.org/10.1016/j.ress.2012.03.017.

Merrick, J.R.W., Soyer, R. and Mazzuchi, T.A. (2005). Are maintenance practices for railroad tracks effective? Journal of the American Statistical Association, 17–25. Office of Rail and Road (2016). https://doi.org/10.1198/016214504000002104

Patiño-Rodriguez, C.E. and Carazas, F.J.G. (2019). Maintenance and Asset Life Cycle for Reliability Systems. In: Kounis, L. (Ed.). *Reliability and Maintenance: An Overview of Cases*. IntechOpen. https://doi.org/10.5772/intechopen.85845

Pisano, A. and Usai, E. (2020). Contact force regulation in wire-actuated pantographs via variable structure control, Decision and Control. In: *Proceedings of the 46th Annual Conference of the IEEE Industrial Electronics Society, Marina Bay Sands Expo and Convention Centre*, Singapore, 18–21 October.

Qiu, Q., Liu, B., Lin, C. and Wang, J. (2021). Availability analysis and maintenance optimization for multiple failure mode systems considering imperfect repair. Proceedings of the Institution of Mechanical Engineers, Part O: *Journal of Risk and Reliability*, 235(6): 982–997. https://doi.org/10.1177/1748006X211012792.

Raczynski, A. and Gussmann, P. (2004). Services and training through augmented reality. 1st European Conference on Visual Media Production (CVMP), 263–271. http://dx.doi.org/10.14195/978-972-8954-42-0_12.

RAMAR, 2009, https://lucense.it/progetti/ramar. Accessed November. 2021. Rastegari A. and Salonen A. (2015). Strategic Maintenance Management: Formulating Maintenance Strategy. *International Journal of COMADEM*, 18: 5–14.

Rotoli, F., Malavasi, G. and Ricci, S. (2016). Complex railway systems: Capacity and utilisation of interconnected networks. *Eur. Transp. Res. Rev.*, 8: 29. https://doi.org/10.1007/s12544-016-0216-6.

Scheffer, S., Martinetti, A., Damgrave, R., Thiede, S. and van Dongen, L. (2021). How to Make Augmented Reality a Tool for Railway Maintenance Operations: Operator 4.0 Perspective. *Appl. Sci.*, 11: 2656. https://doi.org/10.3390/app11062656.

Sharma, G. and Rai, R.N. (2021). Modified failure modes and effects analysis model for critical and complex repairable systems. *In*: (Eds.). Hoang Pham, Mangey Ram, Advances in Reliability Science, Safety and Reliability Modeling and its Applications. Elsevier, pp., 245–260. https://doi.org/10.1016/B978-0-12-823323-8.00016-7.

Söderholm, P. and Norrbin, P. (2013). Risk-based dependability approach to maintenance performance measurement, *Journal of Quality in Maintenance Engineering*. https://www.emerald.com/insight/content/doi/10.1108/JQME-05-2013-0023/full/html.

Song, Y., Liu, Z., Ronnquist, A., Navik, P. and Liu, Z. (2020). Contact wire irregularity stochastics and effect on high-speed railway pantograph-catenary interacions. *IEEE Trans. Instrum. Meas.*, 69(10): 8196–8206. https://doi.org/10.1109/TIM.2020.2987457.

Wang, X.L., Yang, G.H. and Zhang, D. (2021). Event-Triggered Fault Detection Observer Design for T–S Fuzzy Systems. *IEEE Transactions on Fuzzy Systems*, 29(9): 2532–2542. http://dx.doi.org/10.1109/TFUZZ.2020.3002393.

Zhang, F., Shen, J. and Ma, Y. (2020). Optimal maintenance policy considering imperfect repairs and non-constant probabilities of inspection errors. *Reliability Engineering & System Safety*, 193: 106615, https://doi.org/10.1016/j.ress.2019.106615.

Zhou, N., Yang, W., Liu, J., Zhang, W. and Wang, D. (2019). Investigation on Monitoring System for Pantograph and Catenary Based on Condition-Based Recognition of Pantograph. *Shock and Vibration*, 2019: 3839191. https://doi.org/10.1155/2019/3839191.

Zio, E., Marella, M. and Podofillini, L. (2007). Importance measures-based prioritization for improving the performance of multi-state systems: Application to the railway industry. *Reliability Engineering & System Safety*, 1303–1314. https://doi.org/10.1016/j.ress.2006.07.010.

8

Role of Renewable Energy Resources in Meeting Global Energy Demand

Yusuf Parvez[1] and *Harsha Chaubey*[2,*]

1. Introduction

Renewable energy can be termed as energy obtained from resources that occur naturally and get replenished at the same rate as they are consumed. Energy from these resources can be derived in various forms like solar, wind, biomass, tidal, geothermal, etc. Over the years, energy demands have been met by conventional sources of energy such as coal, firewood, petroleum, nuclear power, and natural gas. These conventional energy resources are exhaustible in nature and have several drawbacks like lower efficiencies, wasted heat, and harmful emissions. Problems like air pollution, the greenhouse effect, global warming and climate change demand global attention and need to shift towards cleaner, abundant, widely available, and sustainable alternatives to meet energy requirements (Kanoglu et al., 2020; Pierce, 2012; Choudhary et al., 2018). In 2021, demand for primary energy rose by 5.8%, outpacing 2019 levels by 1.3%. Demand

[1] Mechanical Engineering, Maulana Azad National Urdu University, Cuttack Campus, Odisha, India.
  Email: parvez_yusuf01@yahoo.com,
[2] Mechanical & Automation Engineering Department, Indira Gandhi Delhi Technical University for Women, Delhi, India
* Corresponding author: harsha.chaubey@gmail.com

for renewable energy rose by over 8 EJ (exajoules) between 2019 and 2021. Fossil fuel usage has mostly remained stable. In 2020, 82% of primary energy came from fossil fuels, down from 83% in 2019 and 85% in 2012. It demonstrates that even while the demand for renewable energy has increased, the reliance on fossil fuels hasn't significantly decreased over time. Nuclear power (4.3%), renewable energy (6.7%), and hydroelectric power (6.8%) made up the remaining portion of primary energy use. The biggest annual growth in primary energy was 31 EJ in 2021, driven by emerging economies, which had an increase of 13 EJ, led by China's growth of 10 EJ. On the other hand, the energy demand in developed economies decreased by 8 EJ in 2021 compared to 2019. Renewable energy sources were the only ones accountable for the increase in primary energy use from 2019 to 2021. Between 2019 and 2021, there was no change in the amount of fossil fuel energy utilized, with rising natural gas (5 EJ) and coal (3 EJ) consumption countering falling oil demand (-8 EJ) (Fig. 8.2). With the rapid development of solar and wind power, the proportion of renewable energy sources (aside from hydro) in the world's energy production has continued to rise. Renewable energy sources generated about 13% of all power output in 2021 (Fig. 8.1), exceeding nuclear energy's 9.8% contribution. The use of renewable energy is growing fast. The consumption of renewable energy on a global scale reached a new peak in 2021, growing by an astonishing 15%. The amount of electricity consumption, produced from solar increased by a record 1.7 EJ, a 22% rise, although wind power (+2.5 EJ) contributed most to the expansion of renewable energy sources. A total of 2,894 TWh (Terawatt-hours) of electricity was generated by wind and solar energy combined in 2021. For comparison, that amount was 380 TWh in 2010. For the first time, wind and solar power generated 10.2% of the world's total electricity in 2021. In 2021, coal's contribution to electricity generation increased slightly

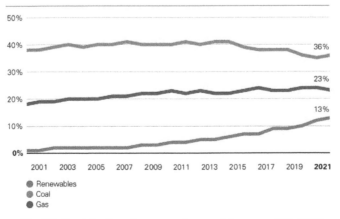

Fig. 8.1 Contribution of sources of renewable energy (BP, 2022).

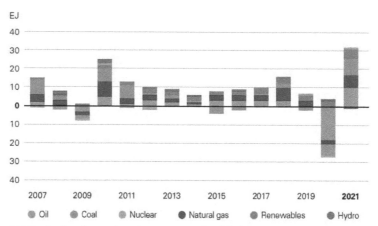

Fig. 8.2 Fuel-wise change in primary energy in global power generation (BP, 2022).

from 35% to 36%. (Fig. 8.1) The proportion of gas generation in 2021 was nearly at the same level as the 10-year average (BP, 2022; IEA, 2019; Marr, 2022).

Figure 8.1 shows the contribution of renewables, coal, and gas in the generation of power at a global scale over the years. It can be observed that the proportion of renewables increased to the highest value of 13% in the year 2021. Figure 8.2 depicts the fuel-wise change in primary energy consumption over the years (BP, 2022).

Figure 8.3 depicts the status of energy consumption for different resources in the year 2021. It is apparent that oil, coal, and natural gas still dominate the market by having 31%, 27%, and 24% share in global consumption, respectively. Renewables (7%) including hydroelectricity (7%) have a total contribution of about 14 % in the global consumption of primary energy for the year 2021. Among all the resources, nuclear energy has the least share of 4% in primary energy consumption. Figure 8.4 shows the status of net additions in GW for PV, Wind, Hydro, and Bioenergy in the years 2019, 2020, and 2021 globally. From the figure, it's clear that PV constitutes the maximum addition in renewable energy consecutively for the last 3 years, having maximum addition of 151 GW in 2021. Wind capacity increased significantly in 2020 to 113.4 GW, declining slightly to 94.3 GW in 2021. Hydro energy and bioenergy contribute 35.4 GW and 12.2 GW, respectively, making the net renewable energy addition reach 294.1 GW in 2021 (IEA, 2019; Broom, 2022).

The proportion of renewable energy in India's power generation goals is rising rapidly under the country's aims for sustainable development (Srivastava et al., 2019). Today, India produces 40% of its installed electrical capacity from non-conventional fuel sources, making it the world's third largest renewable energy producer. Table 8.1 displays the amount of power

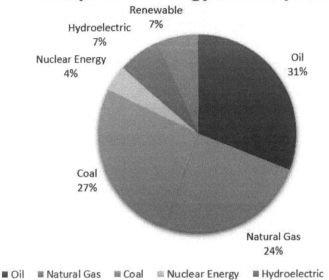

Fig. 8.3 Primary Global Energy Consumption. Modified from BP Statistical Review 2022 (BP, 2022).

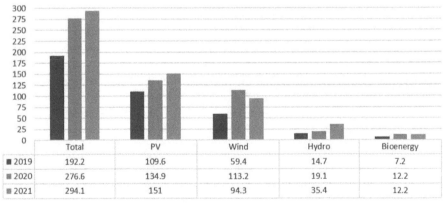

Fig. 8.4 Net additions in GW to renewable energy in 2019, 2020, and 2021. Modified from International Energy Agency [IEA] Report for Renewable Energy 2022 (IEA, 2019; Broom, 2022).

Table 8.1 India's installed capacity of renewable sources of energy. Ministry of New & Renewable Energy.

Solar Power	Large Hydro Power	Wind Power	Biopower	Nuclear Power	Small Hydro Power
48.55 GW	46.51 GW	40.03 GW	10.62 GW	6.78 GW	4.83 GW

generated by the installed capacity of different renewable resources in India. It can be seen that solar power accounts for the maximum installed capacity in the country, having 48.55 GW as its total available potential (Renewable Energy in India, 2022; Ministry of New & Renewable Energy–Government of India, 2022).

Primary energy usage in India increased by 10% annually in 2021, from 32 to 35 EJ. Oil, gas, and coal combined constitute about 90% of all the energy used, which is comparable to their utilization levels before the pandemic. Out of all the energy utilized, 57% of primary energy usage in India was driven by coal. Globally, India accounts for 12.5% of the world's coal usage. At 29 bcm (billion cubic meters) in 2021, natural gas production rose by 20%. Although the growth rate of Renewable energy per annum is recorded to be 13.2 % in 2021, but when this rate is compared to the increase in the growth rate of primary energy consumption, it accounts for only 5% (excluding hydro) in total share. A total capacity of 10.3 GW of solar power and 1.5 GW of wind power were installed in India, among which the amount of solar power set a new record. By the end of 2021, solar and onshore wind systems had a combined installed capacity of 49.3 GW and 40.1 GW, respectively. The total power produced grew by 10%, from 1,563 TWh to 1,715 TWh. Out of all the power generated, 74% comes from coal, rising from 72% in 2019. The generation of renewable energy sources (apart from hydro) rose from 152 to 172 TWh, exceeding the levels of hydropower generation and going beyond the 10% share contribution for the first time. India's dependence on imported biofuels declined as its production of biofuel grew from 23 to 37 kboe/d (thousand barrels of oil equivalent per day), a 60% increase, while its consumption rose from 36 to 43 kboe/d (BP, 2022; IEA, 2019).

2. Solar Energy

Solar energy is among the highly utilized renewable energy resources across the world. It can be harnessed directly or indirectly for consumption. The direct method involves heating water/air in concentrators, drying commodities, and conversion into electricity using photovoltaic cells. The indirect method involves solar energy being part of the ecology, like water evaporation, which is an intermediate step in the water cycle for hydropower production. The devices used to collect solar thermal energy are known as solar collectors. They are specialized heat exchangers that convert the irradiation from the sun into heat. Collectors can be of flat plate and concentrating types. When solar energy is directly changed into electricity, it's called photovoltaic conversion, and the devices used for it are called solar cells (Pierce, 2012; Energy Education, 2022). The PV method is far more practical for producing power on a small scale, while thermal solar systems are often exclusively used for producing power on a massive scale. Water heating, space heating, electricity production, drying, cooking, and refrigeration are all common applications of solar thermal energy. Low-temperature-operated, sun-oriented thermal systems can be used to generate electricity and provide space heating and cooling for homes and small businesses. Solar-based technology is one of the world's most rapidly developing and reasonably priced power sources. PV technology is a great, environmentally friendly, sustainable power source due to its technological advances and economic nature (Ehtesham and Jamil, 2020; Khare et al., 2022).

In India, the industry of solar power generation is rapidly expanding. As of February 2021, it has been reported that the nation's installed solar power capacity—which takes into account both ground- and roof-mounted plants, is 39,083 MW. From April 2021 to March 2022, solar electricity production jumped to 73.48 TWh (terawatt-hours) from 60.4 TWh. India's utilization of solar energy was placed fourth worldwide in 2021. Nearly 20.7 billion US dollars in foreign funds were invested in solar power projects in India between 2010 and 2019. At least 942 GW of cumulative PV capacity was added at the end of 2021. With a cumulative capacity of 308,5 GW, China is still in the lead, followed by the European Union (178,7 GW), the United States (122,9 GW), Japan (78,2 GW), and India (60,4 GW). With 59,2 GW, Germany is the top-performing country in the solar sector in the European Union, followed by Italy (22,6 GW), Spain (18,5 GW), France (14,3 GW), and the Netherlands (13,2 GW) (BP, 2022; IEA, 2019).

The world's largest solar park, Bhadla Solar Park (Fig. 8.5), is situated in the village of Bhadla, Jodhpur district, Rajasthan, India. It covers an area of about 14,000 acres or nearly 5,700 hectares. The park is distributed in four phases and can produce 2245 MW in total (Bhadla Solar Park, Jodhpur District, Rajasthan, India (2019)).

2.1 Solar thermal energy

Utilizing the sun's radiation to heat fluid and produce heat or electricity is known as solar thermal energy generation. The fluid is heated to make steam, which powers generators to provide electricity. There are two categories of solar thermal setups: active and passive. Moving components like fans or pumps are used in active systems to circulate the fluids that

Fig. 8.5 Bhadla Solar Park, India (Renewable Energy Projects | Solar Power Projects India, USA, 2020).

conduct heat. Passive systems rely on design elements (like greenhouses) to capture heat without mechanical components. The technologies are categorized into low, medium, and high ranges of temperatures. Solar thermal energy is typically used for room heating or water heating in low-temperature (100°C) applications. A flat plate collector positioned on the roof that circulates liquid is a common component of active systems. The liquid transports the heat that the collector captures from the sun to the intended location, such as pools or systems for heating houses (Fig. 8.6). Intelligent building design techniques are used in passive heating systems to effectively capture or reflect solar energy, which minimizes the need for heating or cooling equipment. Applications for medium temperatures

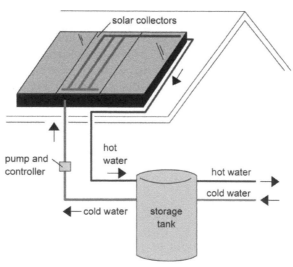

Fig. 8.6 Solar water heating system. US Energy Information Administration EIA (2016).

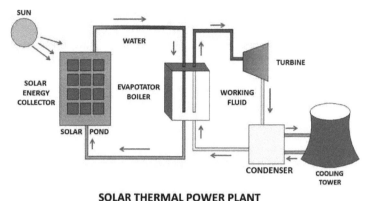

SOLAR THERMAL POWER PLANT

Fig. 8.7 Electricity generation through solar thermal plant (Thareja, 2020).

(100–250°C) are infrequently utilized. One instance of such a device is a solar oven, which directs the sun's rays onto a central cooking pot using a specially constructed reflector. In slightly elevated temperature ranges (> 250°C) solar thermal arrangements, a collection of mirrors is utilized to concentrate rays from the sun onto a central collector. These concentrated solar power systems are built in such a way that the temperature rises just enough to create steam, which in turn drives a turbine and powers an electricity generator (Fig. 8.7) (Pierce, 2012; Energy Education, 2022; Jaiswal et al., 2022). Figure 8.6 demonstrates the working of a solar water heating system through a flat plate collector. Water from storage is pumped with the help of a controller. This water absorbs the heat after passing through the collector and then hot water gets circulated for further use. Figure 8.7 depicts electricity generation processes in thermal power plants as explained earlier in this section.

An essential part of converting energy from the sun into useful heat energy at different temperature ranges is performed by solar thermal collectors. The heat transmission rate and the fluid flow in the collector's receiver tube are still being studied extensively to find ways to increase the heat transfer rate. There are many ways to improve a solar thermal collector's performance. Passive, active, and combination techniques are three general categories for these approaches. Because they are inexpensive and simple to use, passive methods have the biggest impact on improving the effectiveness of solar thermal collectors. Among the most simple and straightforward ways to increase the setup's performance is to apply a selective coating to the plate of the collector. Nanofluids transport solar energy to other fluids at a noticeably faster rate; hence, they are being utilized in absorbing solar intensity by functioning as working fluids in solar collectors. Other performance-improving techniques include booster reflectors, which direct enormous amounts of solar radiation onto the aperture portion of the collector plate and provide inserts in flow passageways, which increase the rate of heat transfer by creating secondary flow. In addition to storing heat energy, phase change materials can control temperature variations within a given range. Researchers are attempting to develop phase change materials for collectors that have a high heat of fusion, a high specific heat, long-term dependability throughout recurring cycling, a high density, and consistent freezing behavior (Energy Education, 2022; Jaiswal et al., 2022; Ellabban et al., 2014).

2.2 *Basic energy balance for a flat plate collector*

The absorber plate, which is usually constructed from steel, copper, or aluminium in the case of flat plate collectors and is merged with the pipe or duct, counts as the most crucial component. In order to transfer thermal energy from the absorber to the working fluid, the fluid is made to flow through the pipe or duct that is in thermal contact with the plate. The purpose of the collector/absorber plate is to minimize heat loss to the atmosphere from the top surface by absorbing an abundant amount of energy from the sun that is incident on it through the glazing. It is crucial to note that in the case of flat plate collectors, the absorber and glass cover areas are identical (Tiwari and Mishra, 2012).

The useful thermal energy output per unit time of a Flat Plate Collector of area A_c (Fig. 8.8) is the difference between the absorbed solar radiation, q_{ab}, and the thermal loss. It is given by the Equations (8.1) and (8.2) (Tiwari, and Mishra, 2012)

$$\dot{Q}_u = A_c \dot{q}_u = A_c[\dot{q}_{ab} - U_L(T_p - T_L)] \tag{8.1}$$

$$\dot{q}_{ab} = (\tau_0 \alpha_0)I(t) \tag{8.2}$$

where $I(t)$ is the solar intensity incident on a Flat Plate Collector, τ_0 and α_0 are the transmissivity and absorptivity, respectively. The expression for U_L, is the overall heat loss coefficient.

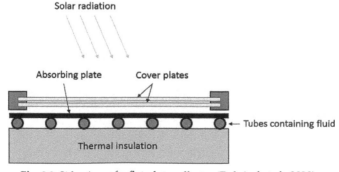

Fig. 8.8 Side view of a flat plate collector (Dobriyal et al., 2020).

The instantaneous thermal efficiency (η) of the Flat Plate Collector is given as (Tiwari and Mishra, 2012),

$$\eta = \frac{\dot{Q}_u}{A_c I(t)} = \frac{\dot{q}_{ab}}{I(t)} - \frac{U_L(T_p - T_a)}{I(t)} \tag{8.3}$$

The overall thermal collection efficiency of a Flat Plate Collector (the ratio of the daily useful gain to the daily incident solar energy) is given by (Tiwari and Mishra, 2012),

$$\eta_c = \frac{\int \dot{Q}_u \, dt}{A_c \int I(t) \, dt} \tag{8.4}$$

2.3 PV technology

Through the use of semiconductors, solar PV technology directly transforms sunlight into electricity. Electrons in the semiconductor of the PV cell are liberated when the sun's rays strike on it, producing an electrical current. Normally, PV cells are mounted on a frame known as a module, connecting with one another. Together, the modules create an array that can be scaled up and down to generate the necessary amount of electricity. Silicon is the most widely used semiconductor, and about 90% of all PV technologies in use today are based on silicon in some form (Solar PV. Retrieved from website: https://studentenergy.org/conversion/solar-pv). Some materials, including monocrystalline silicon, polycrystalline silicon, amorphous silicon, cadmium telluride (CdTe), and copper indium gallium selenide (CIGS), are used to make solar panels. The efficiency of solar PV devices typically ranges from 6 to 18% (Singh et al., 2022). Solar PV depends on sunlight, as opposed to solar thermal technologies, hence it cannot generate power when the sun is not shining. PVs can shift electricity production from vast, centralized facilities to more compact, decentralized production locations (i.e., residential rooftops) (Energy Education, 2022). Figure 8.9 depicts the operating mechanism of a solar PV cell, i.e., the generation of free electrons, holes, and PV junction which is responsible for transforming solar energy into electricity. Figure 8.10 displays how solar panels are utilized in conjunction with inverters to harness transformed electric power for appliances and for further transmission to grids.

Increasing the efficiency of PV technology is an emerging area of research and development. The amount of sunlight (irradiation) that strikes the surface of a solar panel and then is turned into power or useful electricity can be stated as solar panel efficiency or conversion efficiency of the photovoltaic cell.

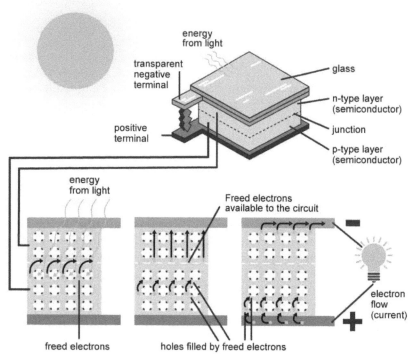

Fig. 8.9 Operation of a PV Cell. (U.S. Energy Information Administration, 2016).

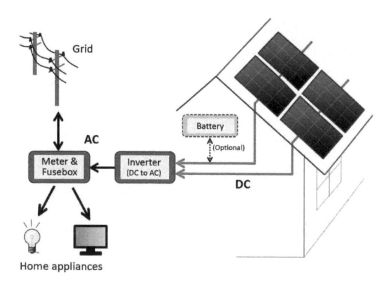

Fig. 8.10 PV Solar power system (Daware, 2022).

2.4 Conversion efficiency of a solar cell

The portion of the incoming solar energy can be converted into electricity. It can also be defined as the ratio of maximum power that a solar cell can generate to the input solar power (Yahyaoui, 2018) as shown in Equation (8.5).

$$\eta_{max} = \frac{P_{max}}{P_{in}} = \frac{V_{max}I_{max}}{P_{in}} \tag{8.5}$$

The fill factor (FF) of the I-V curve (Yahyaoui, 2018) can be defined as,

$$FF = \frac{V_{max}I_{max}}{V_{oc}I_{sc}} \tag{8.6}$$

It is a measure of the squareness of the I-V curve as shown in Fig. 8.11. As the fill factor increases, the conversion efficiency of the cell increases.

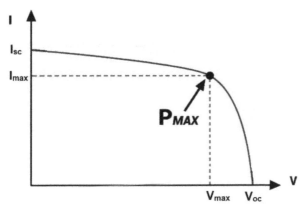

Fig. 8.11 The I-V characteristics showing the maximum power point of a solar cell (Yahyaoui, 2018).

2.5 Factors affecting the efficiency of a solar cell

Various factors are there which affect the conversion efficiency of the cell. However, the fill factor and the panel's temperature are the two main factors that contribute to the lack of the cell's efficiency. The panel with a higher fill factor value will perform better than the one with a lower value. Solar cell designers always try to maximize the fill factor (Yahyaoui, 2018). The panel's temperature strongly affects the performance of a solar cell. The higher the temperature, the poor is the performance of the cell. Hence, the panel temperature should be kept under control. A number of cooling techniques have been discussed in the literature to keep the panel temperature in the optimal functioning range.

The significant advancements in solar technology over the past few years have increased the average panel conversion efficiency from 15% to well over 20%. Due to this significant increase in efficiency, a standard-size panel's power rating scaled up from 250W to 400W (Energy Education, 2022; Jaiswal et al., 2022; Ellabban et al., 2014). The panel efficiency can be enhanced by either improving the material used for manufacturing panels or adding cooling systems to keep panel surface temperatures in the optimum range. These cooling systems can further be classified as active or passive on the basis of the cooling medium applied. A review paper presented by Singh et al. (Singh et al., 2021) discussed the evolution of solar PV cells across four generations, from their early days to the most recent improvements in their performance. Conventional, wafer-based first-generation solar cells, including m-Si and p-Si, are widely known. Thin-film technologies including CdTe, CIGS, and a-Si are used in the second generation of solar cells. Third-generation solar cells use cutting-edge technologies including DSSC (Dye-Sensitized Solar Cells), QDs (Quantum Dot Solar cells), and PVSC (Photovoltaic Solar Cells). The fourth generation of solar cells now has better charge transport and optical coupling because of technical improvements. The device structure of PVSC incorporates inorganic nanostructures. Solar cells from the first and second generations are much more efficient than those from the third and fourth. Second-generation solar cells are currently of economic viability, but their disposal is a significant barrier to future commercialization. A further scope of improvement in PV technology is driven by improving efficiency through thermoelectric coolers and phase change materials. PV panel efficiency decreases as its temperature reaches beyond the optimum working temperature. Much research work has been done to provide cooling systems to control and maintain the maximum temperature in the desired range. In their work Singh et al. (Singh et al., 2022) have fabricated a hybrid cooling model to raise the efficiency of the solar module by employing TECs (active cooling) and PCM (passive cooling) to control the surface temperature. This cooling technique has observed a 19% increment in panel efficiency. The research explores the latest trends and suggests areas of improvement in PV cooling system technologies. Table 8.2, summarizes applications, benefits, and challenges in solar energy technology.

3. Wind Energy

The kinetic energy present in the wind could be captured and transformed into electrical energy. The main causes of the wind are the earth's rotation around its axis and around the sun, as well as the absorption of solar energy on the earth's surface and in the atmosphere. Localized heating and cooling differences between ocean and ground surfaces are responsible

Table 8.2 Applications of solar energy along with the benefits and challenges in adopting solar technology (Energy Education, 2022; Shukla et al., 2017; Dey et al., 2022).

Applications	• Electricity generation • Water heating, space heating • Cooking
Benefits	• Solar energy is a renewable, inexhaustible, free, and clean resource. • Less consumption of fossil fuels; less air pollution. • Noise-free operation. • Easy installation. • Since they employ less complex technology and passive systems have no moving parts, solar thermal systems require comparatively less maintenance. • Electricity generation in rural areas promotes a better standard of living and access to information through technology.
Challenges	• Requires a sizeable amount of land. • Conversion efficiency is less in PV technologies. • Risk and uncertainties in Power generation due to weather and climate dependency. • Electricity transmission over long distances is costly and may result in distribution losses. • When it comes to PV modules, improper production and disposal practices might result in environmental concerns.

for the generation of local winds. The earth's rotation and the temperature differences between the polar and equatorial regions are responsible for the generation of planetary winds (Pierce, 2012). Using various mechanical and electrical subsystems, such as wind turbine rotors, generators, controllers, and interconnection equipment such as possible PECs and transformers, wind energy harvesting technologies are placed up to capture the energy of wind movement for the purpose of producing electric power. The tower, rotor, and nacelle, which house the generator and transmission gear, are the main parts of modern wind turbines. The rotor of the wind turbine, which is made up of two or more blades and is systematically connected to an electrical device or generator, captures the kinetic energy of the wind. The gearbox, which transforms the wind turbine's insufficient rotating rates into significant spinning speeds on the electrical machine side, is the fundamental component of the mechanical design. The wind turbine spins the shaft of the electrical machine that generates electric power (Fig. 8.14). By utilizing effective control and supervision techniques, this output is maintained in accordance with requirements (Pierce, 2012; Ellabban et al., 2014; Shukla et al., 2017).

The wind energy generation industry in India has continuously advanced. This wind sector development has enabled a stable environment, project operation capacities, and a manufacturing base of roughly 10,000 MW to be maintained yearly. The country presently stands fourth worldwide in wind capacity, with an installed capacity of 39.25 GW altogether (recorded in March 2021), and advanced to generate around 60.149 billion units between 2020 and 2021. The federal government has put in place more than 800 wind monitoring stations across the nation via the National Institute of Wind Energy (NIWE) and released wind potential maps above ground-level heights of 50, 80, 100, and 120 meters. The nation has a gross potential of 302 GW at 100 meters and 695.50 GW at 120 meters above the surface, for wind energy generation as per a recent analysis (BP, 2022; Ministry of New & Renewable Energy–Government of India, 2022). Globally, the units of energy generated by wind increased by a record 273 TWh in 2021 (up 17%). This energy output, which was 55% greater than what was accomplished in 2020, was the highest among all renewable energy technologies. Such a quick rise was made possible by the remarkable growth in wind power generation expansions, which reached 113 GW in 2020 as opposed to just 59 GW in 2019. With an output of 1870 TWh, or roughly as much as all the other renewable technologies combined, the wind will continue to be the most significant non-hydro renewable technology in 2021. China ranked first and was responsible for almost 70% of the increase in generating wind power in 2021, with the United States coming in second at 14% and Brazil coming in third at 7% contribution. The European Union saw a 3% fall in wind power production in 2021 as a result of relatively prolonged periods of low wind patterns, despite the tremendous capacity growth in 2020 and 2021. In the year 2021, out of the 830 GW of installed wind power, 93% came from onshore systems, with offshore wind farms providing the remaining 7%. Being used by 115 countries globally, onshore wind technology is more advanced and mature than offshore wind, which is still in its development stage and only shares capacity in 19 countries. Offshore technology is projected to increase over the forthcoming years as more nations are in the process of getting ready to build their first-ever offshore wind farms (BP, 2022; IEA, 2019).

The Jiuquan Wind Power Base, commonly referred to as the Gansu Wind Farm (Fig. 8.12), is located on the borders of the Gobi Desert in the western part of China's Gansu province. It is the largest wind farm in the world with a 20GW proposed capacity. The farm will have 7,000 turbines when completed, generating sufficient electricity to power a small nation. The facility generated 8 GW of power only at 40 % utilization in 2015 (The World's Biggest Wind Farms, 2021, November).

3.1 Horizontal and vertical wind turbines

There are two methods of utilizing wind energy, either directly or indirectly turning it into electrical energy. The wind turbine is the most crucial component of the wind energy conversion system (WECS), which is used

Fig. 8.12 Gansu Wind Farm, China (Wikipedia Contributors (December, 2018). Gansu Wind Farm. Retrieved from Wikipedia website: https://en.wikipedia.org/wiki/Gansu_Wind_Farm).

to harness the wind's physical energy and transform it into mechanical energy by rotating the generator's rotor with the aid of a gearbox. The two primary types of wind turbines are the horizontal-axis type model (HAWT) and the vertical-axis type model (VAWT). Generally, HAWTs have three blades and they work 'upwind', turning at the top of the tower to point the blades in the direction of the wind. HAWTs are preferred in the wind energy sectors because of their higher efficiency. Since VAWT turbines are omnidirectional, i.e., there's no need for them to be turned to face the wind in order to work. The VAWTs are most effective when used sparingly, like on a rooftop (as shown in Fig. 8.13). Even though they are less efficient than HAWTs, VAWTs can catch the wind from any angle. Additionally, they can still produce power at low wind speeds owing to their lower cut-in speed (Ang et al., 2022). On the basis of their installation location and method of grid connection, modern wind turbines can be divided into the following three categories (Office of Energy Efficiency and Renewable Energy).

1. The output of a "land-based wind turbine" or "utility wind turbine" can range from 100 kW to several MW. The cost-effectiveness of larger wind turbines leads to their grouping into wind plants, providing the electrical grid with significant power.
2. "Offshore wind turbines" often have larger, taller structures. Due to the fact that their large sections can be shipped, they do not have the same challenges encountered in transportation as wind energy installations on land. Strong ocean breezes may be captured by these wind turbines to generate enormous power.

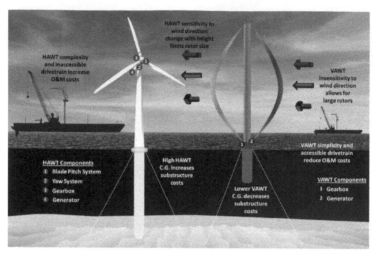

Fig. 8.13 Operation of HAWT and VAWT. Office of Energy Efficiency & Renewable Energy (Illustration by Sandia National Laboratories).

3. Wind turbines of any size set up on the consumer end of the electric meter or installed near the location where the energy they produce is readily utilized are referred to as "distributed wind turbines". Agricultural, lighting residential areas, and small commercial zones, including industrial applications, are some uses of minor wind turbines with a power output under 100 kilowatts. These smaller wind turbines are often coupled with other distributed energy resources and employed in hybrid energy systems, such as microgrids powered by photovoltaics, batteries, and diesel generators (Energy Education, 2022; Office of Energy Efficiency and Renewable Energy Costs, October, 2018; Ang et al., 2022).

Figure 8.13 demonstrates the operating method of HAWT and VAWT as described earlier in this section. Figure 8.14 depicts the components of the wind turbine and energy transmission process from towers and stations to grids, which were explained at the beginning of Section 3.

Madvar et al. (Madvar et al., 2019) evaluated and classified the technological stages of various wind energy sub-technologies in their study. According to their research, five technical domains have been developed to advance the wind industry: Generator or configuration, Blades and rotors, Control of turbines, Components and gearbox, and Nacelle (Fig. 8.14). The findings showed that "Blades and rotor" and "Control of turbines" are two key technological areas that have drawn significant interest from leading nations in power generation through wind and have been involved in more than 32% and 31% of issued patents, respectively (Madvar et al., 2019). Researchers are concentrating on improving wind turbines in conjunction

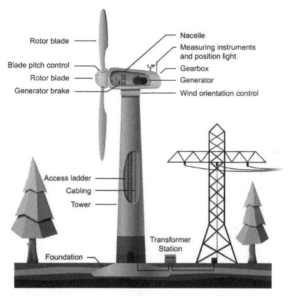

Fig. 8.14 Power Transmission through Wind Turbine (Sarwar, 2019).

with other generators, which are the major components of the Wind Power Conversion System, increasing the effectiveness of technology for converting wind energy into electricity. Larger generation equipment, like higher towers, high-efficiency generators, and longer rotor blades, have been suggested to improve the performance of wind energy conversion and generation as energy demands keep rising. Instead of using coils, a PMSG (permanent magnet synchronous generator) uses a permanent magnet to create an excitation field. Compared to EESG (electrically excited synchronous generator) and conventional DFIG (double-fed induction generator), PMSG is extremely efficient when using wind energy at a partial load. Additionally, because PMSG utilizes fewer moving components than wound rotor induction and electrically excited generators, it is more powerful and requires less service. Due to the alternating nature of wind, MPPT (maximum power point tracking) is often employed to establish the generator's ideal speed and to increase the energy yield from moderate-speed zones. The typical MPPT is now combined with an intelligent algorithm to lower the loss rate and increase system effectiveness (Ellabban et al., 2014; Ang et al., 2022). Figure 8.15 displays a schematic diagram for the wind turbine along with nomenclature.

The available power from a wind flowing with a speed of V and the mass flow rate with the sweeping area A is mentioned in the equation below (Hodge, 2017; Contreras Montoya et al., 2021).

$$Power = \tfrac{1}{2} mV^2 \tag{8.7}$$

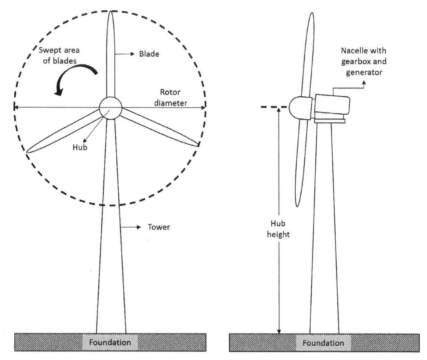

Fig. 8.15 HAWT schematic and nomenclature (Hodge et al., 2017; Contreras Montoya et al., 2021).

Since the mass flow rate in Equation (7) is equal to ϱAV; therefore, the available power becomes as mentioned in Equation (8).

$$Power = \frac{1}{2} \rho AV^3 \tag{8.8}$$

The above Equation (8.8) is very important, as it illustrates that the power available is proportional to the swept area swept by a wind turbine and also proportional to the cube of the wind speed. Therefore, the ideal location for wind turbines is where wind speeds are high.

The expression for maximum power extracted (Hodge, 2017; Contreras Montoya, 2021),

$$Power = \frac{8}{27} \rho Ac^3 \tag{8.9}$$

The power coefficient (Cp) can be defined as the ratio of power extracted to the available power (Hodge, 2017; Verger et al., 2022) as mentioned in the Equation (10).

$$C_p = \frac{Power_{ext}}{\frac{1}{2}\rho AV^3_{wind}} \tag{8.10}$$

The maximum value of Cp, the Betz limit, is written as

$$C_p = \frac{\frac{8}{27}\rho A c_{wind}^3}{\frac{1}{2}\rho A V^3} = 0.59259 \tag{8.11}$$

The Betz limit of a wind turbine shows the maximum value of the power coefficient (Cp) and describes the maximum power extractable from a wind stream. This power coefficient is one of the most important terms used in characterizing wind turbines (Hodge, 2017; Contreras Montoya et al., 2021).

3.2 Onshore and offshore wind power

Globally, both onshore and offshore wind power technologies are commercially acceptable. Onshore wind farms have less distance between wind turbines, which causes less voltage loss. They are easier to install, produce fewer carbon emissions, and require little care. However, compared to offshore wind power plants, where wind speeds are strong, onshore wind plants produce less power. Also, they require a larger land area. High initial investment costs, technical difficulties in power transmission from the shoreline to the ground, and disruption of the ecosystem of marine life are the main drawbacks of offshore wind power projects (Ang et al., 2022). The installation, operation, and maintenance of static foundation-type wind turbines in deep oceans are challenging. Floating offshore wind (FOW) is a novel and rapidly emerging technology that extracts power from the wind in those areas. Offshore wind turbines fixed on a floating platform or foundation are known as floating wind turbines. The first industrial floating wind farm in the world is called Hywind Scotland. With a combined wind farm capacity of 30 MW, these five floating wind turbines—each 6 MW from Hywind, with hub heights of 101 meters, rotor diameters of 154 meters, and rated wind speeds of 10.1 meters per second—are located 29 kilometers off Peterhead, Scotland. To affix floating platforms and enable them to endure the impacts of seawater and waves, extra heavy-duty cables are needed, which results in a hefty initial expenditure. The offshore wind power floating technology is also gaining popularity and might be accepted commercially like onshore and offshore technology within the next few years (Rehman et al., 2023). Table 8.3 summarizes the applications, benefits, and challenges of existing wind technologies.

4. Biomass Energy

The use of organic material to generate usable sources of energy like heat, electricity, and liquid fuels is referred to as biomass or bioenergy. It is the phrase used to define all organic material derived from plants, including algae, trees, and crops, and it mainly refers to the process of collecting and

Table 8.3 Applications of wind energy along with the benefits and challenges in adopting wind energy technology (Energy Education, 2022; Jaiswal et al., 2022; Shukla et al., 2017; Dey et al., 2022).

Applications	• Power generation • Crop processing • Water pumping • Irrigation
Benefits	• Wind energy is free, clean, and abundant energy resource. • Less consumption of fossil fuels; less air pollution. • As opposed to other kinds of power plants, wind turbines may be installed in a short period of time . • Due to their installation away from the coasts, floating wind farms cause less of a noise disturbance and visual impact on the scenery of nearby locations. • Lower risk of being vulnerable to water shortages and more options for adaptability through irrigated agriculture. • Can be a source of income generation and improve quality of life in rural areas.
Challenges	• Given the fact that wind speeds vary with geography, topography, and season, certain places are more suitable than others for the production of wind energy. • Technical difficulties include a high initial investment and electricity transmission from offshore to ground stations. • Electromagnetic interference for radio signals. • Large land requirements for plants, can be an eyesore to the landscape. • Alteration in the flight path of migrating birds. • Consequential noise from rotating blades.

storing solar energy via photosynthesis. The variance of biomass resources includes concentrated wastes (municipal solid waste, sewage waste products, manure, industrial wastes), dispersed waste residue (through the crops, disposal manure, and lagging), and harvested biomass (biomass energy plantations, standing biomass). Not all biomass is utilized to generate energy directly; on occasion, it can be transformed into biofuels or intermediary energy carriers (Fig. 8.16). This includes producing gas (from the gasification of biomass), ethanol, or charcoal, a solid fuel with a higher energy density (Pierce, 2012; Energy Education, 2022; Ellabban et al., 2014).

Biomass may be directly transformed into 'biofuels', providing a quick solution to fulfilling the demand for transportation fuel, unlike other renewable technologies. The two most widely used varieties of biofuels today are ethanol and biodiesel; belonging to the first generation of biofuel resources. The process of converting biofuels includes upgrading and deconstruction. It is necessary to break down the biological elements cellulose, hemicellulose, and lignin that are bound together to make up the plant cell wall's strong and stiff structure. Deconstruction at either high temperature or low temperature is done to break down the cell wall. Pyrolysis, gasification, and hydrothermal liquefaction are three processes that are used in high-temperature destruction (Energy Education, 2022).

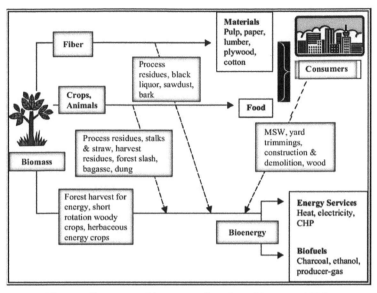

Fig. 8.16 Biomass and bioenergy flowchart (Kammen, 2004).

Figure 8.16 depicts the bioenergy conversion process as explained above in this section.

When biomass is rapidly heated at an elevated temperature range (500°C–700°C) in the absence of oxygen, it's termed as the Pyrolysis. Heat causes biomass to be pyrolyzed into vapor, gas, and char. Following the removal of the char, the vapors are cooled down and condensed into a liquid 'bio-crude' oil. Synthesis gas (also known as syngas), a combination primarily made up of carbon monoxide and hydrogen is created when biomass is heated to a higher temperature range (> 700°C) with some oxygen present. The primary method of conversion for feedstocks like algae is hydrothermal liquefaction. This method uses water to transform biomass into liquid bio-crude oil under elevated pressures and a moderate temperature range (200°C–350°C). Chemicals or biological catalysts called enzymes are frequently used in low-temperature deconstruction to transform feedstocks into intermediaries. Sugar polymers like cellulose and hemicellulose are broken down into simple sugar-building components enzymatically or chemically during the pre-treatment process, known as hydrolysis. Intermediary products such as unprocessed bio-oils, sugars, syngas, and other chemical building blocks are refined to create a final product. In the process of fermenting sugar or gaseous intermediates, microorganisms like bacteria, yeast, and cyanobacteria produce chemicals and fuel blend materials (Fig. 8.17). To enhance storage and transportation properties, any undesired or reactive chemicals can be removed using a catalyst (Kumar et al., 2020; Jaiswal et al., 2022; Ellabban et al., 2021).

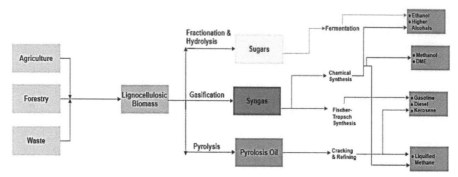

Fig. 8.17 Conversion process for renewable transport fuels (Verger et al., 2022).

Anaerobic Digestion (AD), a scientific process in biogas plants, is how biodegradable organic materials and wastes, including cattle dung, biomass from farms, gardens, kitchens, and industries, fowl droppings, night soil, and municipal wastes are converted into biogas. Many factors influence the design of a biogas plant, but the amount of feedstock that will be processed is a crucial factor in this decision. Biogas is a mixture of gases, primarily methane (CH_4), carbon dioxide (CO_2), and trace amounts of hydrogen sulphide (H_2S), produced by the decomposition/breakdown of bio-degradable organic matter from raw materials such as agricultural waste, animal waste, plant matter, sewage, green waste, or food/kitchen waste in the absence of oxygen. The calorific value of biogas is around 5000 kcal per m^3 (Ministry of New & Renewable Energy–Government of India, 2022). As a byproduct of biogas plants, the digested slurry provides a better supply of organic manure that has been enhanced with nutrients for agricultural use. It supports soil quality in addition to assisting in increasing agricultural yields (Jaiswal et al., 2022; Ellabban et al., 2014). Figure 8.17 explains the methods of renewable transport fuel generation through biomass. The flow diagram mentions intermediate processes and end products as summarized by (Verger et al., 2022).

Agricultural biomass has been used as a primary fuel source in several nations in Asia and Africa, despite being bulky and having a low calorific value. It still lags behind the consumption of wood used mostly for heating and cooking. Many nations, including India, China, Denmark, Poland, and Nigeria, have considered agricultural biomass as an energy source (Saleem, 2022). The feedstock for biogas plants is organic, biodegradable waste from various sources, including animal and food waste, food and kitchen waste, poultry excretions, and the agro-industrial sector, among others. These facilities are particularly useful for providing off-grid power to individual dairy and poultry businesses, dairy cooperatives for the operation of dairy equipment, and other plants that need electrical, thermal, and cooling energy. By replacing diesel use with the installation of such biogas systems, individual farmers, beneficiaries, business owners, and dairy farmers can lower their electricity costs, increasing their take-home earnings. In addition to cooking, it is used for pumping, lighting, irrigation, and other general purposes like cooling milk. In an off-grid setting, the farmers can also sell their excess biogas or electricity to their neighbours. By using fewer chemical fertilizers and other profitable endeavors like organic farming, nutrient-enriched organic bio-manure is another revenue stream from biogas projects. It also saves money on chemical fertilizers (Ministry of New & Renewable Energy–Government of India, 2022; Jaiswal et al., 2022; Ellabban et al., 2014).

Having produced 55% of all renewable energy and over 6% of energy requirements being met by modern bioenergy in 2021, it will be the largest source of renewable energy on the planet. In comparison to 2021, the demand for biofuel is projected to increase by 6%, or 9100 million liters per year, in 2022 (BP, 2022). Eighty percent of the rise in the use of biofuels worldwide is accounted for by the United States, Canada, Brazil, Indonesia, and India, all of which have extensive policy frameworks that drive growth in this sector. It is believed that by the year 2030, the employment of bioenergy to replace conventional fuels will have improved significantly, as per the Net Zero Emissions by 2050 Scenario (BP, 2022). Between 2010 and 2021, the utilization of modern bioenergy grew globally by an average of 7% per year and still rising (BP, 2022; IEA, 2019).

In August 2019, Nature Energy opened the largest biogas plant to date in Korskro, Denmark (as shown in Fig. 8.18). It is situated in South West Jutland, one of Denmark's cattle-rich regions, which offers a reliable supply base and the ideal circumstances for high-quality, environmentally

Fig. 8.18 Biogas plant by Nature Energy in Korskro, Denmark (Danes turn waste into raw material. Retrieved from Gaz Energie website: https://www.cng-mobility.ch/en/beitrag/danes-turn-waste-into-raw-material/).

responsible gas production. Around 710,000 tonnes of biomass are processed annually at Korskro (Danes turn waste into raw material. Retrieved from Gaz Energie website: https://www.cng-mobility.ch/en/beitrag/danes-turn-waste-into-raw-material/).

4.1 Biomass India

According to a new Ministry of Renewable Energy analysis, the Indian biomass supply is estimated to be around 750 million metric tonnes annually. The estimated surplus biomass availability, including agricultural residues, is approximately 230 million metric tonnes per year, or has a potential of about 28 GW if the 550 sugar mills in the country adopted the optimal levels of cogeneration for generating energy from the bagasse produced, both technically and economically. Now, an additional 14 GW of power might be delivered through bagasse-based cogeneration. With 1871.11 MW of installed biomass IPP (independent power producer) capacity, 7562.45 MW of installed bagasse cogeneration capacity, and 772.05 MW of installed non-bagasse cogeneration capacity, India's biomass power, and cogeneration sector has a total installed capacity of 10205.61 MW as of October 2022. Alternatives for power generation and thermal energy utilization, including heating, cooking, and electricity, cover biogas plants. Biogas produced by plants ranging in size from 30 m^3 to 2500 m^3 are used as thermal energy for heating and cooling (Ministry of New & Renewable Energy–Government of India, 2022).

4.2 Sugar industries-based cogeneration and ethanol production

In India, sugarcane is used for sugar mills, whereas sugar beet is used in European countries (Raghu Ram and Banerjee, 2003). After harvesting, sugarcane is cut into pieces of the required size and is fed into the juice extraction system. The raw juice is then sent for the lime treatment process, where the milk of lime (Ca(OH)$_2$) is added to the juice and heated to refine the sugar juice and increase its pH value up to 7. The sludge is removed from the coagulated juice through the clarification process. The clarified juice is passed through the evaporator and vacuum pans where the juice is heated and gets concentrated. Further, the syrup received from the vacuum pan known as 'massecuite' is passed through the crystallizers. After crystallization, perfect sugar crystals and molasses are separated. Sugar is then dried and packed while the molasses is used in the second-generation process, i.e., bio-ethanol production (Pellegrini et al., 2011; Hall, 1991). A flowchart is mentioned in Fig. 8.19 shows the stepwise processes executed in a sugar mill.

Sugar industries are generally operated based on two main criteria: the first includes the conventional method of ethanol production, and

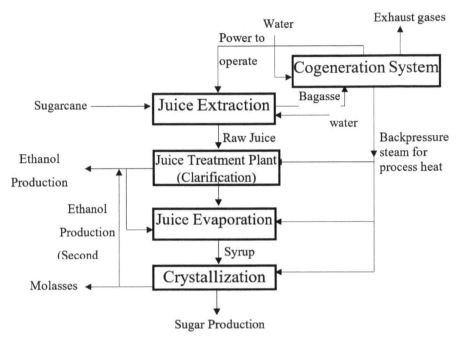

Fig. 8.19 Flowchart for sugar and ethanol production. Modified figure printed with permission from (Pellegrini et al., 2011).

the other is a second-generation process used for bioethanol production along with the production of white sugar crystals, as shown in Fig. 19 (Pellegrini and de Oliveira Junior, 2011). During the second-generation process, the byproduct extracted from the mill, i.e., molasses, is used for bioethanol production (Pellegrini and de Oliveira Junior, 2011; Hall, 1991). The production of second-generation ethanol (bioethanol) can be further utilized as the fuel which is capable of reducing the consumption of 10% petrol and 2–3% diesel fuels. This considerably reduces the environmental impact and hence reduces greenhouse gas (GHG) emissions (Pina et al., 2017; Lennartsson et al., 2014).

Manufacturing of sugar crystals involves a large number of heat-intensive processes. The heat is required in the form of steam at many stages, such as evaporation, diffusion, crystallization, distillation, etc., and a significant portion of the steam is needed at low pressure, especially at 2.5 bar (Parvez, 2022). Other than process heating, there are several stages, such as material handling equipment, mill drives, shredders, centrifuges, and other process pumps, which require electrical energy. Sugar industry-based cogeneration system generates surplus power after fulfilling their internal electricity and heat demand (Khatiwada et al., 2012). Sugar industries are self-dependent in terms of energy since it produces bagasse as a byproduct during the cane crushing process, which is an effective energy source.

4.3 Cogeneration systems

Various cogeneration systems are available in the literature, whereas only steam turbine-based cogeneration is being used in sugar industries. This system is further categorized as a back-pressure turbine and condensing extraction turbine systems (Parvez and Hasan, 2019). The selection of these two systems depends upon the demands for heat and power, required quality, economic factors, etc. The pressures at which steam is being extracted from the turbine can be more than one, depending upon the process heat requirements.

4.3.1 Back pressure turbine system

A back-pressure turbine is a system in which steam leaves the turbine at higher or equal to the pressure of the atmosphere. The back-pressure turbine system can be the most favorable option for small-capacity sugar mills due to less available fuel. Generally, process steam is extracted at two pressure values: 8 bar for centrifuging, heating the air in the air dryer, etc., and 2.5 bar pressure for process heating (Parvez, 2022). The schematic diagram for this system is shown in Fig. 8.20. Following equations can be used for the performance analysis of the system.

$$\text{Energy content (LCV or NCV) of fuel} = \dot{m}_f * LCV \tag{8.12}$$

$$\text{Turbine work output} = W_T = \dot{m}(h_1 - h_2) + \dot{m}_2(h_2 - h_3) \tag{8.13}$$

In Equation (8.13), m is the mass flow rate (in kg/sec) of steam entering into the turbine, m_1 is the mass flow rate of steam extracted at 8 bar pressure

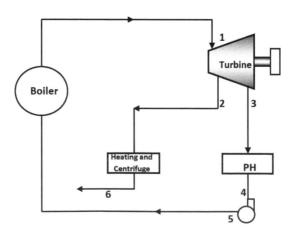

Fig. 8.20 Schematic diagram of a back-pressure steam turbine system.

for heating and centrifuge, and m_2 is the mass flow rate of remaining steam passing at 2.5 bar pressure for process heating. The turbine work output and total heat utilised in process heating can be evaluated from Equations (8.12) and (8.13), respectively.

Process heat utilized, $Q = \dot{m}_1\,(h_2 - h_6) + \dot{m}_2\,(h_3 - h_4)$ (8.14)

Since a part of the work generated by the turbine will be consumed to operate the pump, as mentioned in Equation (15), the network produced by the system will be reduced, as mentioned in Equation (8.16).

Pump work, $= W_p = \dot{v}_f\,(P_5 - P_4)$ (8.15)

Net work produced, $W_{net} = W_T - W_p$ (8.16)

$$\eta_{cogen} = \frac{W_{net} + Q}{\frac{1}{2}\rho A V_{wind}^3}$$ (8.17)

In Equation (8.17), is the baggase burning rate inside the boiler, and LCV is the lower heating value of the fuel (Parvez and Hasan, 2019). The cogeneration efficiency mentioned in Equation (8.17) represents the system's overall performance (Parvez, 2022).

4.3.2 Condensing extraction turbine system

Condensing extraction turbine is a system that allows the steam after extraction in the intermediate stages to expand till the condenser pressure is as shown in Fig. 8.21. A schematic diagram for condensing extraction turbine system used in sugar mills has been summarized in Fig. 8.21. Following equations can be used for the performance analysis of such systems.

In the below-mentioned equations, m is the mass flow rate (in kg/sec) of steam entering the turbine and m_1 is the mass flow rate of steam extracted at 8 bar pressure for heating and centrifuge, and m_2 is the mass flow rate of steam extracted at 2.5 bar pressure for process heating. At these two-pressure values, process heat is supplied to the cogeneration plant. The remaining steam is allowed to expand in the steam turbine till the condenser pressure. The total heat utilised in process heating, turbine work output and network produced in the cycle can be evaluated from the Equations (8.19), (8.20), and (8.21), respectively.

Process heat utilized; $= Q = m_1\,(h_2 - h_9) + m_2\,(h_3 - h_7)$ (8.19)

Turbine work output;

$W_T = m(h_1 - h_2) + (m - m_1)\,(h_2 - h_3) + (m - m_1 - m_2)\,(h_3 - h_4)$ (8.20)

Net amount of work produced, $W_{net} = W_T - W_p$ (8.21)

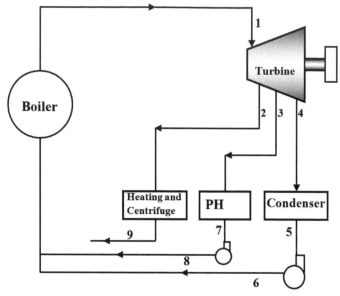

Fig. 8.21 Schematic diagram of condensing extraction turbine system.

Similarly, cogeneration efficiency can be evaluated using the Equation (8.17).

4.4 Biomass-derived products

In addition to energy generation, conversion, and storage methods, biomass-derived products like biochar, bio-oil, and syngas can be used for a variety of other purposes. These compounds can be used to clean up polluted environments by removing harmful pollutants like heavy metals from water and soil. By supplementing the soil with these elements, the water and soil quality is improved, which also aids in increasing plant and crop productivity. Biochar is an efficient material in methods employed for converting and storing energy because it offers high porosity, a vast surface area, availability of functional groups abundant in oxygen, high electrical conductivity, and affordable manufacture. Further, expanding its energy generation and storage technology capacity, doping biochar with heteroatoms, nanoparticles, and metal oxides can be done. Biochar-based materials have attracted attention for applications like fuel cell technology, supercapacitors, batteries, and oxygen electrocatalysis. Additionally, pyrolysis, energy production, syngas production, and trans-esterification can be done using biochar as a catalyst (Jaiswal et al., 2022; Ellabban et al., 2014; Shukla et al., 2017; Kumar et al., 2020). Employing biomass as a source of renewable energy in combination with highly effective heat and power cogeneration has been made possible with the help of bioenergy villages. Energy crops and liquid manure from neighboring farmers are used as feedstock for the biogas plants in these areas. The biogas is utilized in combined heat and power plants to produce heat and electricity, which are then fed into a heating system and made available to nearby families. With an energy efficiency above 80%, this decentralized utilization of waste heat from energy production can help lower the need for sources of primary power production. According to the scenario study, using liquid manure and locally cultivated energy crops as substrates also helps to lower greenhouse gas emissions. Because every kWh of power introduced into the grid and every kWh of heat absorbed by the homes in the heating network brings down the average greenhouse emissions, the prospective way to reduce emissions by this method is huge for larger biogas plant capacity and increased heat demand (Karschin and Geldermann, 2015; Andersen and Lund, 2007). Table 8.4 summarizes the application, benefits, and challenges of Biomass and Biogas Energy Technology.

Table 8.4 Applications of Biomass and Biogas energy along with the benefits and challenges in adopting Bioenergy technology (Energy Education, 2022; Jaiswal et al., 2022; Shukla et al., 2017; Dey et al., 2022).

Applications	
Biomass	Electricity generation Heat Fuel
Biogas	Generation of thermal energy Making sludge for fertilizer Catalyst
Benefits	
Biomass	Less strain on nature's resources. Less usage of pollution-causing fuels like charcoal and fuel wood. Decreases the possibility of drought and deforestation. Generation of employment and income prospects in remote areas.
Biogas	Significantly reducing the utilization of dangerous emissions-producing fuels such as liquefied petroleum gas, fuel wood, and charcoal. Less consumption of fertilizers and pesticides. Decreases the possibility of droughts and deforestation. More promising prospects for increasing agricultural output and revenue.
Challenges	
	Lower biofuel yield and competitive fossil fuel prices. Problems with emitted gaseous aerosols from inadequate thermal treatment procedures. Organic and inorganic pollutants, which are common in biomass materials, can be poisonous, and thermal processing of these materials may harm the environment. People who produce, transport, and use biomass-derived materials are prone to develop respiratory ailments due to the presence of ash in such materials.

5. Hydro Energy

Hydropower is the most commonly used renewable energy source in the world. With the least amount of greenhouse gas emissions into the environment, it is renewable and dependent on the hydrological cycle. By using a turbine, generator, and convertor, moving water can be converted to hydroelectric energy. The conversion station may have run-of-river installations on a relatively small scale or big-scale generation structures like dams. One or more turbines are turned by the water flowing through river channels and falling across dams (Pierce, 2012; Şen, 2018). Figure 8.22 explains the process of hydropower generation through a dam. A typical dam stores water behind it in form of a reservoir or lake. When water is let out of the dam, its flow rotates a turbine that is connected to a generator. The water flows back into the river on the dam's downstream edge.

Figure 8.23 depicts the flow chart for energy conversion through various available hydro resources (Şen, 2018). Kinetic energy is transformed into mechanical, then electrical energy by a generator when the water flow rotates the turbine. The energy obtained from tides' normal rise and fall is transformed into electrical energy through tidal power. The combined effects of the Moon's gravitational pull, Earth's rotation, and its revolution around the Sun are responsible for the generation of tides. The only drawback to tidal energy is that it can only be used close to coastal locations, still making it one of the emerging technologies with considerable potential. Coastal areas frequently have two high and two low tides per day. To generate electricity, the difference in water levels must be at least 5 meters. Tidal barrages, tidal gates, and tidal turbines (Fig. 8.24) are the technologies currently being employed to harness tidal energy for electricity production. Tidal barrages are among the most efficient tidal power sources. Tidal barrages are dams that harness the potential energy generated by the height difference between high and low tides. Electricity is produced when this

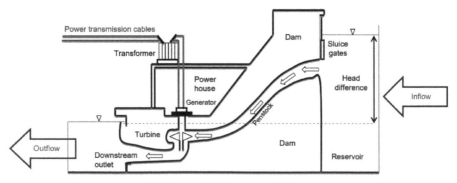

Fig. 8.22 Hydropower production (Şen, 2018).

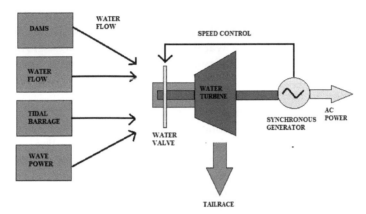

Fig. 8.23 Hydropower production through different sources (Şen, 2018).

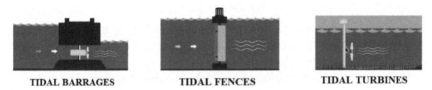

Fig. 8.24 Tidal technologies (Alfakih et al., 2019).

potential energy is used to turn a turbine by the flow of water. Tidal fences are turbines that function like enormous turnstiles, whereas tidal turbines are like submerged wind turbines. In both instances, electricity is produced when tidal currents' mechanical energy drives turbines that are linked to a generator. Compared to air, ocean water is 832 times denser. This allows water to exert more force on the turbines and generate considerably more energy than air currents (Energy Education, 2022; Ellabban, 2014; Şen, 2018). Figure 8.24 depicts the operation of tidal barrages, tidal fences, and tidal turbines as explained above in this section.

Let the water be captured and kept during high tide in a basin, and during low tide, let it pass through a turbine. The basin's surface area A is constant. At low tide, the water is still on top of it. The confined water is expected to release at low tide and has a mass of ϱAR at a centre of gravity of R/2 (Tiwari and Mishra, 2012). The formula for the highest potential energy per tide where water falls through R/2 is

$$Energy \ per \ tide = (\rho AR)g\tfrac{R}{2} \qquad (8.22)$$

In Equation (8.22), g stands for acceleration due to gravity. Taking into account the tidal period T, the average potential power equals

$$P = \frac{\rho AR_2 g}{2T} \quad \text{(in watts)} \qquad (8.23)$$

Wind-generated waves have the greatest energy density. Different forms of ocean waves are created as winds blow across the oceans; this energy transfer results in the water near the free surface serving as a natural energy reservoir for the wind. Wind waves have a low energy loss over thousands of kilometers, natural phenomena like refraction and reflection can make up for energy loss close to the shore, resulting in energy concentrations known as hot spots. Wave energy can be utilized by; (1) A partially submerged hollow structure constitutes the oscillating water column which is open to the seafloor below the water line. A 'Wells' turbine positioned beneath the device's roof creates a reciprocating flow by alternately pressuring and depressurizing the air inside the structure due to the sea surface's heaving action. No matter which way the airflow is moving through this sort of turbine, the direction of the revolution can be kept constant. (2) Wave-collection overtopping systems that use one or more low-head turbines to power the collected water from incident waves (Fig. 8.25).

(3) Floating or underwater heaving devices produce a heave motion that gets transformed into linear or rotational motion through the mechanical and/or hydraulic systems which are further used to power electrical generators (Fig. 8.26). (4) Pitching equipment consists of several floating bodies which are connected at their beams by hinges. The relative motion between floating bodies is utilized to pump high-pressure oil in hydraulic motors, which power the electric generators (Fig. 8.27).

(5) Surging instruments that use the wave's horizontal particle velocity to push a flexible bag toward the wavefront or push a deflector (Fig. 8.28) (Energy Education, 2022; Ellabban et al., 2014; Lemonis, 2004). Figures 25, 26, 27, and 28 demonstrate the mechanisms for overtopping, heaving, pitching, and surging, as described above in this section.

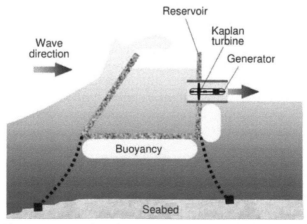

Fig. 8.25 Functioning of an overtopping device (floating type).

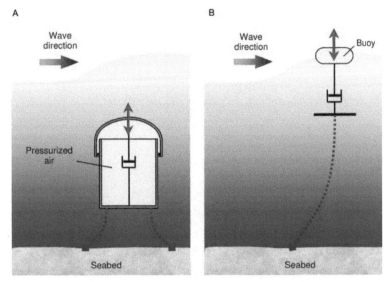

Fig. 8.26 Functioning of heaving devices via (A) Archimedes Wave. Swing concept, as suggested by Rademakers et al. (1998), and (B) floating buoy converter

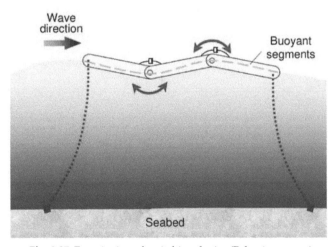

Fig. 8.27 Functioning of a pitching device (Pelamis concept).

Despite a step up in capacity growth, hydropower generation declined by 15 TWh (down 0.4%), falling to 4 327 TWh in 2021(BP, 2022; IEA, 2019). Droughts in many parts of the world were the reason for this decline. However, hydropower continues to be the most significant source of renewable electricity, superior to all other renewable technologies in terms of total output. In the Net Zero Emissions by 2050 Scenario, hydropower preserves a 2022–2030 average annual generation growth rate of about

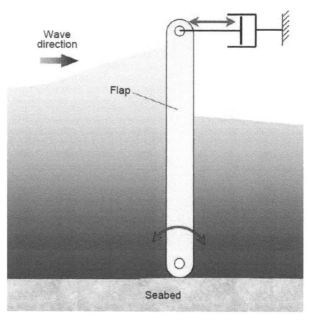

Fig. 8.28 Functioning of a surging device suggested by Yemm (1999) (Lemonis, 2004; Rademakers et al., 1998; Yem, 1999).

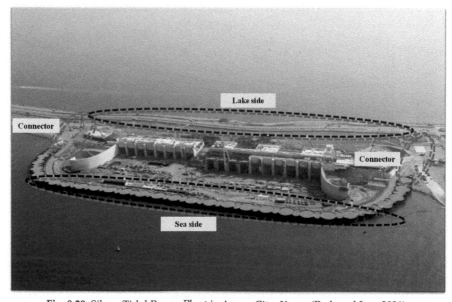

Fig. 8.29 Sihwa Tidal Power Plant in Ansan City, Korea (Park and Lee, 2021).

3% and is expected to supply roughly 5 700 TWh of electricity annually (IEA, 2019). Depending on their sizes, hydropower projects are divided into large and small hydro projects. The size standards or capacity used to categorize minor hydropower projects can differ by standards set by respective countries. In India, small hydropower plants are those that have a capacity of 25 MW or less. The Hydro and Renewable Energy Department evaluated the potential for small/mini hydropower projects, which is expected to be 21133 MW from 7,133 sites. About half of this potential is represented by the hilly states, primarily Arunachal Pradesh, Himachal Pradesh, Jammu & Kashmir, and Uttarakhand (Ministry of New & Renewable Energy–Government of India, 2022). With a 254 MW generation capacity, the Sihwa Tidal Power Plant in Korea is the world's largest and most costly tidal project. It comprises 10 generators, which have an annual energy output of more than 550 GWh (Park and Lee, 2021).

A summary of the applications, benefits, and challenges of hydro technologies in current use has been mentioned in Table 8.5 mentioned below.

Table 8.5 Applications of hydro energy along with the benefits and challenges in adopting technology (Energy Education, 2022; Ellabban et al., 2014; Shukla et al., 2017; Dey et al., 2022).

Applications	• Electricity generation • Agricultural processing • Tidal barrages are used in the protection of coasts during storms • Dams are used for water supply, irrigation and flood control • Wave energy can be used for commercial mariculture and fish farming
Benefits	• Dams are a fairly inexpensive way to produce electricity along with offering recreational pleasures like boating, fishing, etc. • Abundant, clean, and secure form of power. • Tidal and wave energy technology is relatively easy to build, maintain, and extract for towns near the coastline. • Tidal energy is a comparatively dependable energy source as opposed to other sources like wind, making it highly desirable for electrical grid management.
Challenges	• Dams have significant ecological effects on the nearby hydrology, can lead to erosion, and can upset the ecological balance, which can result in flooding of nearby communities and landscapes. • Only applicable in locations with abundant water supply. • Large wave or tidal power plants are too expensive to establish and maintain. • Tidal technologies have not evolved rapidly; the amount of power produced by tidal energy generation plants is extremely low and tide cycles do not usually match daily patterns of electricity use.

6. Geothermal Energy

The thermal energy that is produced and stored within the Earth is known as geothermal energy. A geothermal gradient is produced by the temperature difference between the Earth's core and surface, which means thermal energy is continuously transported to the surface. Geothermal resources are found where there is a high heat flow to the surface, making it possible to economically feasible to extract energy for power generation or for direct use in things like district heating or agriculture. The natural release of heat from the Earth's surface through hot springs, geysers, fumaroles, and disturbed ground is the most prominent example of a geothermal system. Typically, vapor-dominated and liquid-dominated geothermal systems are distinguishable. Fluid is present in liquid-dominated systems at temperatures ranging from mild to high, and they are found in tectonically active geological contexts. These systems' rock has permeable fractures that allow fluid to flow through them. Like the Pacific Ocean's Ring of Fire, most liquid-dominated systems are found in volcanic arcs. Wells produced from a system with a liquid-dominated reservoir typically have a reservoir temperature of above 200°C and can provide 2–10 MW of electrical power. Systems that are vapor-dominated (i.e., devoid of liquid water) are less prevalent (Ellabban et al., 2014; Geothermal Energy, 2020; Dey et al., 2022). Drilling wells between 3 and 10 kilometers down into the Earth's crust is required to produce geothermal energy. There are many ways to extract heat, but water and steam are used in most situations to take it from the Earth. Homes and buildings can be heated with hot water that is extracted from the earth. This can be accomplished by either pumping hot water directly through buildings or through a heat exchanger that transmits heat to the building. In a geothermal power plant, geothermal heat can also be used to generate electricity. Geothermal heat creates steam that turns turbines coupled with a generator to produce electricity. Only areas with particular geologic conditions can harness geothermal energy. Because of this, the world's most volcanically and tectonically active places are also where the majority of power generation occurs. For instance, heat and power plants can be found in Ecuador, Hawaii, California, New Zealand, Iceland, and Indonesia (Energy Education, 2022). Direct use and district heating systems, geothermal power plants, and geothermal heat pumps are the three primary categories of geothermal systems. For direct-use heating systems in the case of single buildings and district heating systems in the case of multiple buildings, hot water is supplied from springs or reservoirs that are close to the earth's surface. Pipes that transport hot water from the earth's surface are used to heat up buildings. Among the industrial uses of geothermal energy are food dehydration (drying), pasteurization of milk, and gold mining. Geothermal power production requires steam or water that is heated to high temperatures (between 300° and 700°F). Geothermal reservoirs are frequently found about 2 miles below the surface of the earth, and it is here that geothermal power plants are usually built. Geothermal heat pumps, which take advantage of constant surface temperatures, can be applied to buildings to heat and cool them. Wintertime heating is provided to buildings via geothermal heat pumps taking transferring heat through the soil or water. During the summer, the process gets reversed (Energy Explained –U.S. Energy Information Administration (EIA). Retrieved from website: https://www.eia.gov/energyexplained). Geothermal systems close to the surface draw heat from the topmost part of the earth's crust; most of the time, at a depth somewhere between 150 and 400 meters. Ground heat collectors, borehole heat exchangers, holes dug into the groundwater, and geothermal energy heaps are examples of such systems. The exploitation is indirect and needs further conversion using, e.g., heat pumps. It is currently being designed to use heat pipes directly at very low temperatures. Examples of common potential uses include road dicing and railroad switch heaters. Hydro-geothermal low-enthalpy systems, which exploit the heat retained in hot or warm water in deep aquifers, are known as deep geothermal systems. This type of heat reservoir is put to use through a heat exchanger, and occasionally even a heat pump. The generated thermal water can be utilized to heat spas, residential buildings, and greenhouses as well as can be supplied to the local and district heating grids. Roughly above 80°C, it is possible to convert heat to electrical energy using supplemental technologies like Organic Rankine Cycle or Kalina Cycle installations (Stober and Bucher, 2013]. Figure 8.30 depicts different types of shallow and deep geothermal systems along with variations in depth and temperature ranges at which energy is harnessed for heat production, storage, and power generation purposes. Figure 8.31 shows a flow chart for electricity generation through geothermal plants. Geothermal power stations generate steam by pumping fluids from subsurface reservoirs. The steam is then used to power turbines that produce electrical energy.

There have been 15.4 gigawatts (GW) of geothermal power installed globally as of 2019, of which 3.68 GW (23.86%) are in the United States. In 2020, a total of 202 MW of this power were further added to which Turkey (168 MW) and the United States (19 MW) were major contributors. By the end of 2020, the total geothermal power generation capacity reached 15,608 MW with this addition (Ritcher, 2021). A considerable portion of the electricity demand in nations like Iceland, El Salvador, Kenya, the Philippines, and New Zealand is met by geothermal energy, which also happens to meet more than 90% of the country's heating needs in Iceland (IRENA, 2019). In India's Himalayan region, there are many thermal springs available with temperatures around 100°C. The temperature differences in Jammu and Kashmir, Puga and Chumathang are greater than 100°C/km. In

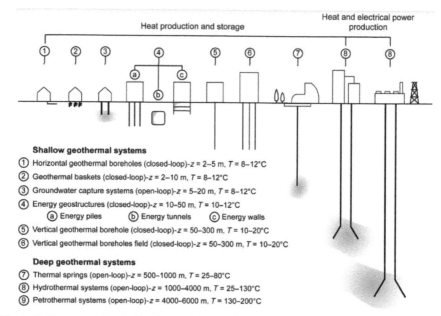

Shallow geothermal systems

① Horizontal geothermal boreholes (closed-loop)-z = 2–5 m, T = 8–12°C
② Geothermal baskets (closed-loop)-z = 2–10 m, T = 8–12°C
③ Groundwater capture systems (open-loop)-z = 5–20 m, T = 8–12°C
④ Energy geostructures (closed-loop)-z = 10–50 m, T = 10–12°C
    ⓐ Energy piles   ⓑ Energy tunnels   ⓒ Energy walls
⑤ Vertical geothermal borehole (closed-loop)-z = 50–300 m, T = 10–20°C
⑥ Vertical geothermal boreholes field (closed-loop)-z = 50–300 m, T = 10–20°C

Deep geothermal systems

⑦ Thermal springs (open-loop)-z = 500–1000 m, T = 25–80°C
⑧ Hydrothermal systems (open-loop)-z = 1000–4000 m, T = 25–130°C
⑨ Petrothermal systems (open-loop)-z = 4000–6000 m, T = 130–200°C

Fig. 8:30 Representation indicating various geothermal systems at different depths (Laloui et al., 2020).

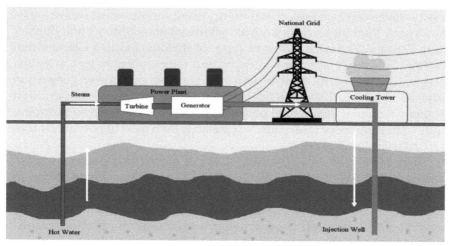

Fig. 8.31 A flow diagram representing the geothermal energy extraction process in a power plant (Islam et al., 2022).

northwest India, there are a total of 113 thermal springs, 112 of which are located in the Himalayan region (Tiwari and Mishra, 2012).

The Geysers, is the biggest geothermal field in the world, lying in the Mayacamas Mountains, about 75 miles north of San Francisco, California.

Fig. 8.32 The Geysers Geothermal Complex, California, USA (About Calpine. Retrieved from website: https://geysers.com/About-Calpine).

It consists of a network of 22 operational geothermal power facilities that collect steam from over 350 wells. In 2021, its net recorded generation was 5,519,169 megawatt-hours. A picture of this geothermal field is shown in Fig. 8.32 (Geysers By The Numbers. Retrieved from website: https://geysers.com/The-Geysers/Geysers-By-The-Numbers).

The seismicity concerns are one of the biggest obstacles in using geothermal energy. Various tools have been developed in recent years to help mitigate the dangers associated with land subsidence and earthquakes by predicting them as accurately as possible. It is expected that the development of artificial intelligence over the past 10 years will assist geothermal power plants in minimizing their harm. Risk management strategies to minimize environmental effects usually caused include: monitoring regional deformation and reservoir pressure, locating faults through research studies, installing strong barriers around sites, and using warning systems. Other measures consist of advancements in operational procedures and more efficient power plants, such as hybrid power plants (Soltani et al., 2021). Some emerging technologies include hybrid geothermal plants that are built on a binary cycle configuration or when the geothermal heat transported by brine is transferred to a secondary working fluid that drives the power cycle. They have shown significantly fewer emissions. Other hybrid technologies include incorporating solar technology, which require less area than a solar plant. This hybridization was appropriate for boosting power output on summer days and extending the useful life of deteriorating geothermal fields. It has been observed that solar heating can increase the geothermal plant's steam quality, giving it a thermodynamic lead over a stand-alone system. These hybrid technologies are a few latest advancements to utilize geothermal energy more efficiently (Zhou et al., 2013). Table 8.6 summarizes the application, benefits, and challenges faced in harnessing geothermal energy.

Table 8.6 Applications of geothermal energy along with the benefits and challenges in adopting technology (Energy Education, 2022; Ellabban et al., 2014; Shukla et al., 2017).

Applications	• Space Heating • Crop drying • Electricity generation • Industrial applications like food dehydration, gold mining, and milk pasteurizing
Benefits	• Technology is really beneficial to the environment because it emits minimal to no greenhouse gases. • Since it can be accessed all year round and is not dependent on the sun or wind, it is an incredibly reliable source of energy.
Challenges	• Since geothermal energy needs seismic sensing, drilling a test well, approval test, and other necessary preliminary research, the costs are generally high. • The process of evacuating groundwater for intermediate use may unintentionally release carbon dioxide, hydrogen sulphide and other harmful emissions into the atmosphere. • Plant construction can adversely affect land stability. • It is location specific or restricted.

7. Conclusion

According to data obtained for worldwide power generation, it is clear that the share of renewables has grown over the previous several years and has continued to rise due to the significant growth in solar and wind energy. In 2021, the contribution of renewable energy in power generation was approximately 13% (including hydro), exceeding the share of nuclear energy (9.8%). In 2021, coal's percentage in the electricity sector climbed marginally, from 35% to 36%. The amount of coal consumed increased by more than 6% in 2021 to 160 EJ, slightly higher than 2019 and the highest level since 2014. More than 70% increase in coal demand in 2021 was attributed to China and India, who saw their respective consumption rise by 3.7 and 2.7 EJ. This suggests even though the utilization of renewable energy resources has increased in the power sector, but there hasn't been much reduction in dependency on conventional fuels, which still dominate the power sector. In 2021, carbon dioxide emissions (measured in carbon dioxide equivalent) from energy usage, industrial processes, flaring, and methane increased by a combined 5.7%. To overcome environmental issues like climate change, global warming, and CO_2 emission, it is necessary to reduce this dependency on conventional fuels. Although many renewable energy sources are available today, none of them has evolved enough to compete against conventional fossil fuels. There are several challenges in harnessing renewable energy to its full potential; technical inefficiency, geographical barriers & economic policies are the main reasons behind them. Technical challenges include lower conversion efficiencies, and the inadequate technological ability to create, implement, run, administer, and support new energy services based on renewable energy. Resources like wind, hydro (tidal and wave), and geothermal are area-specific, limiting their global utilization. Also, uncertainties in climatic conditions (solar, wind) make them less reliable. For some countries, lack of incentives to attract the private sector, inconsistent policies, absence of subsidies, partial knowledge of market potential, lack of uniform technology, high initial costs, and insufficient government financial support are political and economic reasons behind lower employment of new renewable energy sources and technologies, compared to conventional fuels. To improve the role of renewable energy in meeting global energy demand, it's essential to overcome these challenges. A lot of work is being done to come up with new materials and transmission technologies to increase the conversion efficiency of these resources. There is a need for support and funding from both government and private organizations to promote research and development in the field of renewable energy. Perfect planning and implementation of policies and schemes by the government could play a vital role in creating awareness among the public to switch towards non-conventional resources. These small initiatives and changes in the renewable sector by every nation could yield tremendous results at a global scale and bring the world much closer to sustainability.

References

About Calpine. Retrieved from website: https://geysers.com/About-Calpine.

Alfakih, S.M., De, T., Jawad Ali Shah, S., Aneeq and Hayat, K. (2019). Simulation model of single structured tower hybrid wind and tidal energy cultivation based on Yemen's southwest coast. *E3S Web of Conferences*, 107: 01007. https://doi.org/10.1051/e3sconf/201910701007.

Andersen, A.N. and Lund, H. (2007). New CHP partnerships offering balancing of fluctuating renewable electricity productions. *Journal of Cleaner Production*, 15(3): 288–293. https://doi.org/10.1016/j.jclepro.2005.08.017.

Ang, T.-Z., Salem, M., Kamarol, M., Das, H.S., Nazari, M.A. and Prabaharan, N. (2022). A comprehensive study of renewable energy sources: Classifications, challenges and suggestions. *Energy Strategy Reviews*, 43: 100939. https://doi.org/10.1016/j.esr.2022.100939.

Bhadla Solar Park, Jodhpur District, Rajasthan, India. (2019). Retrieved from NS Energy website: https://www.nsenergybusiness.com/projects/bhadla-solar-park-rajasthan/

BP. (2022). Statistical Review of World Energy 2022. Retrieved from BP global website: https://www.bp.com/en/global/corporate/energy-economics/statistical-review-of-world-energy.html.

Broom, D. (2022). These 4 charts show the state of renewable energy in 2022. Retrieved from World Economic Forum website: https://www.weforum.org/agenda/2022/06/state-of-renewable-energy-2022/

Choudhary, P., Sachar, S., Khurana, T., Jain, U., Parvez, Y. and Soni, M. (2018). Energy Analysis of a Single Cylinder 4-Stroke Diesel Engine Using Diesel and Diesel-Biodiesel Blends. *International Journal of Applied Engineering Research*, 13: 10779–10788. https://www.ripublication.com/ijaer18/ijaerv13n12_95.pdf.

Contreras Montoya, L.T., Hayyani, M.Y., Issa, M., Ilinca, A., Ibrahim, H. and Rezkallah, M. (1 January 2021,). Chapter 8 –Wind power plant planning and modeling. *In*: (Ed.) E. Kabalci,, pp. 259–312. Retrieved from ScienceDirect website: https://www.sciencedirect.com/science/article/pii/B978012821724500012X.

Danes turn waste into raw material. Retrieved from Gaz Energie website: https://www.cng-mobility.ch/en/beitrag/danes-turn-waste-into-raw-material/

Daware, K. Solar Power System–How does it work? Retrieved from Electrical Easy website: https://bit.ly/3C5OGd7.

Dey, S., Sreenivasulu, A., Veerendra, G.T.N., Rao, K.V. and Babu, P.S.S.A. (2022). Renewable energy present status and future potentials in India: An overview. *Innovation and Green Development*, 1(1): 100006. https://doi.org/10.1016/j.igd.2022.100006.

Dobriyal, R., Negi, P., Sengar, N. and Singh, D.B. (2020). A brief review on solar flat plate collector by incorporating the effect of nanofluid. Materials Today: *Proceedings*, 21: 1653–1658. https://doi.org/10.1016/j.matpr.2019.11.294.

Ehtesham, M. and Jamil, M. (2020). Control techniques for enhancing performance of PV system under dynamic conditions. *International Journal of Energy Sector Management*, 15(1): 119–138. https://doi.org/10.1108/ijesm-02-2020-0017.

Ehtesham, Md. and Jamil, M. (2020). Control Techniques to Optimize PV System Performance for Smart Energy Applications. *Lecture Notes in Civil Engineering*, 91–105. https://doi.org/10.1007/978-981-15-2545-2_9.

Ellabban, O., Abu-Rub, H. and Blaabjerg, F. (2014). Renewable energy resources: Current status, future prospects and their enabling technology. *Renewable and Sustainable Energy Reviews*, 39: 748–764. https://doi.org/10.1016/j.rser.2014.07.113.

Energy Education. (2022). Retrieved from Student Energy website: https://studentenergy.org/energy-education/

Energy Explained–U.S. Energy Information Administration (EIA). Retrieved from website: https://www.eia.gov/energyexplained.

Geothermal Energy. (2020). Future Energy, 431–445. https://doi.org/10.1016/B978-0-08-102886-5.00020-7.

Geysers By The Numbers. Retrieved from website: https://geysers.com/The-Geysers/Geysers-By-The-Numbers.

Hall, D.O. (1991). Biomass energy. *Energy Policy*, 19(8): 711–737. https://doi.org/10.1016/0301-4215(91)90042-m.

Hodge, B.K. (2017). Alternative Energy Systems and Applications. Hoboken, Nj, Usa Wiley.

IEA. (2019). IEA –The global energy authority. Retrieved from Iea.org website: https://www.iea.org.

IRENA. (2019). International Renewable Energy Agency (IRENA). Retrieved from website: https://www.irena.org/

Islam, M.T., Nabi, M.N., Arefin, M.A., Mostakim, K., Rashid, F., Hassan, N.M.S., Muyeen, S.M. (2022). Trends and prospects of geothermal energy as an alternative source of power: A comprehensive review. Heliyon, 8(12): e11836. https://doi.org/10.1016/j.heliyon.2022.e11836.

Jaiswal, K.K., Chowdhury, C.R., Yadav, D., Verma, R., Dutta, S., Jaiswal, K.S. and Karuppasamy, K.S.K. (2022). Renewable and sustainable clean energy development and impact on social, economic, and environmental health. *Energy Nexus*, 7: 100118. https://doi.org/10.1016/j.nexus.2022.100118.

Kammen, D.M. (2004). Renewable Energy, Taxonomic Overview. *Encyclopedia of Energy*, 385–412. https://doi.org/10.1016/b0-12-176480-x/00308-9.

Kanoglu, M., Cengel, Y.A. and Cimbala, J.M. (2020). *Fundamentals and Applications of Renewable Energy*. McGraw-Hill Education.

Karschin, I. and Geldermann, J. (2015). Efficient cogeneration and district heating systems in bioenergy villages: An optimization approach. *Journal of Cleaner Production*, 104: 305–314. https://doi.org/10.1016/j.jclepro.2015.03.086.

Khare, V., Khare, C.J., Nema, S. and Baredar, P. (2022). Path towards sustainable energy development: Status of renewable energy in Indian subcontinent. *Cleaner Energy Systems*, 3: 100020. https://doi.org/10.1016/j.cles.2022.100020.

Khatiwada, D., Seabra, J., Silveira, S. and Walter, A. (2012). Power generation from sugarcane biomass—A complementary option to hydroelectricity in Nepal and Brazil. *Energy*, 48(1): 241–254. https://doi.org/10.1016/j.energy.2012.03.015.

Kumar, A., Bhattacharya, T., Mozammil Hasnain, S.M., Kumar Nayak, A. and Hasnain, M.S. (2020). Applications of biomass-derived materials for energy production, conversion, and storage. *Materials Science for Energy Technologies*, 3: 905–920. https://doi.org/10.1016/j.mset.2020.10.012.

Laloui, L. and Rotta Loria, A.F. (2020). Energy and geotechnologies. Analysis and Design of Energy Geostructures, 3–23. https://doi.org/10.1016/b978-0-12-816223-1.00001-1.

Lemonis, G. (2004). Wave and Tidal Energy Conversion. *Encyclopedia of Energy*, 385–396. https://doi.org/10.1016/b0-12-176480-x/00344-2.

Lennartsson, P.R., Erlandsson, P. and Taherzadeh, M.J. (2014). Integration of the first and second generation bioethanol processes and the importance of by-products. *Bioresource Technology*, 165: 3–8. https://doi.org/10.1016/j.biortech.2014.01.127.

Madvar, M.D., Ahmadi, F., Shirmohammadi, R. and Aslani, A. (2019). Forecasting of wind energy technology domains based on the technology life cycle approach. *Energy Reports*, 5: 1236–1248. https://doi.org/10.1016/j.egyr.2019.08.069.

Marr, B. (2022). The 3 Biggest Future Trends (and Challenges) in the Energy Sector. Retrieved from Forbes website: https://www.forbes.com/sites/bernardmarr/2022/02/11/the-3-biggest-future-trends-and-challenges-in-the-energy-sector/?sh=6ce991b227b.

Ministry of New & Renewable Energy–Government of India. (2022). Retrieved from website: https://mnre.gov.in/

Park, E.S. and Lee, T.S. (2021). The rebirth and eco-friendly energy production of an artificial lake: A case study on the tidal power in South Korea. *Energy Reports*, 7: 4681–4696. https://doi.org/10.1016/j.egyr.2021.07.006.

Parvez, Y. and Hasan, M.M. (2019). Exergy analysis and performance optimization of bagasse fired boiler. *IOP Conference Series: Materials Science and Engineering*, 691(1): 012089. https://doi.org/10.1088/1757-899x/691/1/012089.

Parvez, Yusuf. (2022). Exergy Based Performance Improvement of Cogeneration Plant of Sugar Mills. 10.36297/vw.jei.v4i2.63.

Pellegrini, L.F. and de Oliveira Junior, S. (2011). Combined production of sugar, ethanol and electricity: Thermoeconomic and environmental analysis and optimization. *Energy*, 36(6): 3704–3715. https://doi.org/10.1016/j.energy.2010.08.011.

Pierce, M.A. (2012). Encyclopedia of Energy. Salem Press.

Pina, E.A., Palacios-Bereche, R., Chavez-Rodriguez, M.F., Ensinas, A.V., Modesto, M. and Nebra, S.A. (2017). Reduction of process steam demand and water-usage through heat integration in sugar and ethanol production from sugarcane—Evaluation of different plant configurations. Energy, 138: 1263–1280. https://doi.org/10.1016/j.energy.2015.06.054.

Rademakers, L.W.M.M., van Schie, R.G., Schuttema, R.,Vriesema, B. and Gardner, F. (1998). Physical model testing for characterizing the AWS. Paper presented at the Third European Wave Energy Conference, Patras, Greece.

Raghu Ram, J. and Banerjee, R. (2003). Energy and cogeneration targeting for a sugar factory. *Applied Thermal Engineering*, 23(12): 1567–1575. https://doi.org/10.1016/s1359-4311(03)00101-7.

Rehman, S., Alhems, L.M., Alam, Md. M., Wang, L. and Toor, Z. (2023). A review of energy extraction from wind and ocean: Technologies, merits, efficiencies, and cost. *Ocean Engineering*, 267: 113192. https://doi.org/10.1016/j.oceaneng.2022.113192.

Renewable Energy in India. (2022). Retrieved from website: https://pib.gov.in/FeaturesDeatils.aspx?NoteId=151141&ModuleId=2.

Renewable Energy Projects | Solar Power Projects India, USA. (2020). Retrieved from Azure Power website: https://www.azurepower.com/project-overview?page=1.

Ritcher, A. (2021). Top 10 Geothermal Countries 2020 – Installed power generation capacity (MWe) | ThinkGeoEnergy–Geothermal Energy News. Retrieved from https://www. thinkgeoenergy.com/thinkgeoenergys-top-10-geothermal-countries-2020-installed-power-generation-capacity-mwe/

Saleem, M. (2022). Possibility of utilizing agriculture biomass as a renewable and sustainable future energy source. *Heliyon*, 8(2): e08905. https://doi.org/10.1016/j.heliyon.2022.e08905.

Sarwar, M. (2019). A First Course on Basics of Wind Power Plants. Retrieved from Medium website: https://medium.com/@sarwarel15/a-first-course-on-basics-of-wind-power-plants-11b26830c6d8.

Şen, Z. (2018). 3.7 Hydro Energy Production. *Comprehensive Energy Systems*, 304–334. https://doi.org/10.1016/b978-0-12-809597-3.00314-x.

Shukla, A.K., Sudhakar, K. and Baredar, P. (2017). Renewable energy resources in South Asian countries: Challenges, policy and recommendations. *Resource-Efficient Technologies*, 3(3): 342–346. https://doi.org/10.1016/j.reffit.2016.12.003.

Sihwa Tidal Power Plant | Tethys. Retrieved from Tethys Engineering website: https://tethys.pnnl.gov/project-sites/sihwa-tidal-power-plant.

Singh, B.P., Goyal, S.K. and Kumar, P. (2021). Solar PV cell materials and technologies: Analyzing the recent developments. *Materials Today: Proceedings*, 43: 2843–2849. https://doi.org/10.1016/j.matpr.2021.01.003.

Singh, D., Chaubey, H., Parvez, Y., Monga, A. and Srivastava, S. (2022). Performance improvement of solar PV module through hybrid cooling system with thermoelectric coolers and phase change material. *Solar Energy*, 241: 538–552. https://doi.org/10.1016/j.solener.2022.06.028.

Solar PV. Retrieved from website: https://studentenergy.org/conversion/solar-pv.

Solar thermal collectors –U.S. Energy Information Administration (EIA). (2016). Retrieved from website: https://www.eia.gov/energyexplained/solar/solar-thermal-collectors.php.

Soltani, M., Moradi Kashkooli, F., Souri, M., Rafiei, B., Jabarifar, M., Gharali, K. and Nathwani, J.S. (2021). Environmental, economic, and social impacts of geothermal energy systems. *Renewable and Sustainable Energy Reviews*, 140: 110750. https://doi.org/10.1016/j.rser.2021.110750.

Srivastava, S., Chaubey, H. and Parvez, Y. (2019). Performance evaluation of CI engine using diesel, diesel-biodiesel blends and diesel-kerosene blends through exergy analysis. *IOP Conference Series: Materials Science and Engineering*, 691: 012066. https://doi.org/10.1088/1757-899x/691/1/012066.

Stober, I. and Bucher, K. (2013). Applications of Geothermal Energy. *Geothermal Energy*, 35–60. https://doi.org/10.1007/978-3-642-13352-7_4.

Thareja, H. (2020). Utilization of Solar Thermal Energy. Retrieved from Solar Octa website: https://solarocta.com/utilization-of-solar-thermal-energy/

The World's Biggest Wind Farms. (2021, November). Retrieved from NES Fircroft website: https://www.nesfircroft.com/blog/2021/11/the-worlds-biggest-wind-farms?source=google.com.

Tiwari, G.N. and Mishra, R.K. (2012). Advanced Renewable Energy Sources. Cambridge: Royal Society of Chemistry.

U.S. Energy Information Administration. (2016). Photovoltaics and electricity–U.S. Energy Information Administration (EIA). Retrieved from website: https://www.eia.gov/energyexplained/solar/photovoltaics-and-electricity.php.

Verger, T., Azimov, U., and Adeniyi, O. (2022). Biomass-based fuel blends as an alternative for the future heavy-duty transport: A review. Renewable and Sustainable Energy Reviews, 161: 112391. https://doi.org/10.1016/j.rser.2022.112391.

Vertical-Axis Wind Turbines Could Reduce Offshore Wind Energy Costs. (October, 2018). Retrieved from Energy.gov website: https://www.energy.gov/eere/wind/articles/vertical-axis-wind-turbines-could-reduce-offshore-wind-energy-costs.

Wikipedia Contributors (December, 2018). Gansu Wind Farm. Retrieved from Wikipedia website: https://en.wikipedia.org/wiki/Gansu_Wind_Farm.

Wind Energy Technologies Office. How Do Wind Turbines Work? Retrieved from Energy.gov website: https://www.energy.gov/eere/wind/how-do-wind-turbines-work#:~:text=Wind%20turbines%20work%20on%20a.

Yahyaoui, I. (2018). Advances in Renewable Energies and Power Technologies Solar, Wind, Wave Energies and Fuel Cells. Elsevier Science Ltd.

Yemm, R. (1999). The history and status of the Pelamis Wave Energy Converter. Paper presented at IMECHE Seminar, Wave Power: Moving towards Commercial Viability, London.

Zhou, C., Doroodchi, E. and Moghtaderi, B. (2013). An in-depth assessment of hybrid solar–geothermal power generation. *Energy Conversion and Management*, 74: 88–101. https://doi.org/10.1016/j.enconman.2013.05.014.

9

Donar-π-Acceptor (D-π-A) Chromophores use as Smart Material for OLED

Md. Zafer Alam,[1] Md. Mohasin,[1] Mohd. Abdul Mujeeb,[2]
H. Aleem Basha[2] and Salman A. Khan[1],*

1. Introduction

1.1 OLED (Organic Light Emitting Diode)

An OLED is a device that emits light with the application of an external current of less than 5 V. It is a thin film (100 nm –500 nm) solid-state organic semiconductor light emitting diode, consisting of a thin flexible layer of organic electroluminescence substance (an organic compound) sandwiched between two electrodes (cathode and anode), one of which is transparent and on the passing of electricity to the thin layer of the organic compound they emit lights (Chang and Lu, 2013). OLED is also known as organic electroluminescent (organic EL) (Organic EL–R&D". Semiconductor Energy Laboratory. Retrieved 8 July 2019; Idemitsu, K. What is organic EL? Retrieved 8 July 2019). OLEDs have many beneficial properties such as fast response time, excellent luminescence efficiencies,

[1] Physical Sciences (Chemistry), School of Sciences, Maulana Azad National Urdu University, Hyderabad 500032, Telangana, India.
[2] Physical Sciences (Physics), School of Sciences, Maulana Azad National Urdu University, Hyderabad 500032, Telangana, India.
* Corresponding author: sahmad_phd@yahoo.co.in

very low energy consumption, lightweight and cost-effective, foldable or flat display fabrication, etc. (Zhu and Yang, 2013; Murawski et al., 2013; Schmidbauer et al., 2013; Xiao et al., 2011; Fan Yang, 2014; Sasabe and Kido, 2013; Chen and Ma, 2012). OLEDs are used in electronic display screens such as desktop display screens, mobile display screens, handheld game consoles, smartwatches, large computer displays, portable appliances, etc. (Raj et al., 2019). OLED is better than Light Emitting Diode, which is PN-diode-based semiconductor, and an OLED display is better than phosphorescent displays because of faster performance, less consumption of electricity, and more efficiency (Zou et al., 2020; Wang et al., 2013). The first electroluminescence organic material was observed in the early 1950s at Nancy University in France by Andre Bernanose when he applied a high AC voltage in the air to an acridine-orange dye, that happened due to the excitation of the dyes molecules that were dissolved in the cellophane thin films. Different researchers developed different advanced OLEDs at different times (Bernanose and Vouaux, 1953; Bernanose, 1955). OLEDs are generally classified as small molecules- and polymer-based OLEDs. The first small molecules (anthracene, naphthalene, benzene)-based OLEDs were synthesized by the Ching W. Tang group (Tang and Vanslyke, 1987). The first polymer-based LED was made in the UK by Roger Partridge at the National Physical Laboratory(NPL); he applied a poly-(N-vinyl carbazole) film up to 2.2 mm long between two charge injection electrodes. However, these LEDs had two limitations: low conductivity and difficulty in injecting electrons (Partridge, 1983).

OLEDs may be single-layer or multilayer depending on the use of organic electroluminescence substance between the electrodes (–ve and +ve electrodes). Typical OLEDs-devices are made up of different layers as shown in Fig. 9.1(a). In single-layer OLEDs, the organic EL substance present between the anode and cathode that is deposited on the substrate is shown in Fig. 9.1(b). Multilayer OLEDs are fabricated by inserting two/multilayers of the electron blocking layer (EBL) or hole-blocking layers (HBL) next to the EML "emissive layer", as this increases the efficiency of the device. Nowadays, OLED devices are constructed as cathode, emissive layer (EML), conducting layer (CDL), anode, and substrate (Li et al., 2015). Generally, the anode is made of indium tin oxide (ITO), as it facilitates the instillation of the holes into the highest occupied molecular orbitals of the organic layer, while Au or Pt is rarely employed as an anode. The CDL can be PEDOT: PSS (poly-(3,4-ethylene-dioxy-thiophene): polystyrene sulfonate), as it is favored to lower the energy barrier for hole injection; sometimes, organic materials in an OLED device act as a conductive layer/emissive layer. Generally, the cathode is made of Mg, Ca, and Ba, or their alloys, and is capped with Al or Ag. These materials help to inject electrons into the lowest unoccupied molecular orbital of the organic layer/emitter. When a DC-current is passed through the electrode, the holes are

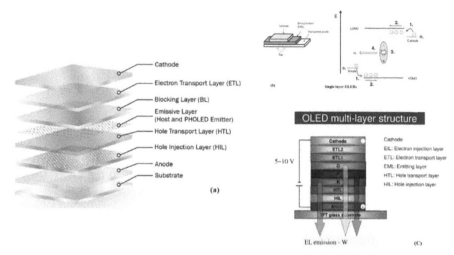

Fig. 9.1 (a), (b), (c): Different layers of ordinary OLEDs devices. (Copyright, Michael Eisenstein, Science Writer, Photonics spectra (2020); Alejandro Lorente, research gate (2017); H.W. Chen, slide serve (2010)).

transferred from the anode by HTL, and the electrons are transferred from the cathode by ETL or EML; then on the application of the DC bias, the electrons and holes recombine in the organic layer/emissive layer, and emit light. Their color depends on the organic substances or emissive substances while illumination/ brightness depends on the external DC current (Kukhta et al., 2006; Ding et al., 2008; Liu et al., 2008).

A typical OLED is constructed by six different layers as shown in Fig. 9.2(a). There is a protective layer present on the top seal and the bottom substrate, which is made of glass or plastic. Beneath the protective layers anode (+ve) and cathode (–ve) the electrode layers are present, between these electrodes there is the organic layer/emissive layer (light-emitting

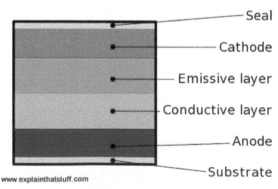

Fig. 9.2: Schematic structure of a bilayer OLED working principle (Copyright, Chris Woodford. Last updated: 1 September 2022).

layer), and the conductive layer which is an organic compound. The organic compound is electrically conductive, because of the delocalization of electrons due to its conjugation over the molecules. It works in the same way as the LED works, but instead of n-type and p-type semiconductors, the organic compound in the OLED device produces electrons and holes, as shown in Fig. 9.2(b). When electricity flows through the OLED, the cathode gains electrons while the anode receives holes from the power source. Due to the gain of electrons and holes, the EML becomes -vely charged, and the CDL becomes +vely charged/holes. Since the +ve holes are more mobile than the -ve electrons, the hole jumps from the CDL to the EML and meets the electrons, explodes, and releases energy (light/photon). This process is known as recombination. and continues until the electricity stops flowing. Generally, polymer OLEDs are made from a single EML. J.H. Burroughes et al. prepared the first light-emitting device composed of a single polymer organic layer of poly-(p-phenylene-vinylene). The first multilayered high-efficiency OLED was prepared by C.W. Tang et al. as reported (Tang and VanSlyke, 1987). However, many works have been done to improve the device efficiency, and one successful method achieved to improve the efficiency of OLEDs is the doping of fluorescent organic layer with high quantum yield (Choong et al., 2000; Kulkarni et al., 2004).

Based on the type of construction, there are two types of OLEDs: (1) passive and (2) active OLED. In passive OLED, the cathode and anode run perpendicular to each other, and the organic layer is present between them, which is shown in Fig. 9.3(a). Passive OLEDs are easier to manufacture; they consume more power, which is useful for smaller screens. In an active OLED, the organic layer is present between the layers of the cathode and anode, (Fig. 9.3(b)), the thin film transistor (TFT) resides on the anode. Iit requires less power than a passive OLED and is applicable for large screens (Copy Craig Freudenrich, Ph.D. How OLEDs Work. HowStuffWorks.com. <https://electronics.howstuffworks.com/oled.htm> 8 December 2022).

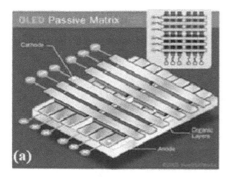 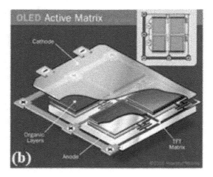

Fig: 9.3 (a), (b): The passive and active OLED (Copyright).

OLEDs do not require any backlight, as they emit light due to free electrons, which is why OLEDs are lighter and thinner than LCDs and they may be flexible and transparent enough to be used in foldable smartphones. Over the past two decades, it has been seen that the use of OLEDs has grown tremendously (Chang et al., 2012; Reineke et al., 2013; Jeon et al., 2011). The first low-voltage OLEDs were designed and developed by Kodiak, which is a simple bilayer structure. Kido was the first who designed and synthesized the first white organic light-emitting device by mixing emitters of different colors. In 2001, Thompson et al. prepared IR-based phosphors OLEDs. Lu et al. (2011) designed and synthesized a new light out-coupling OLED, which enhanced the device efficiency by a factor of 2.5 for any monochromatic OLED, with an excellent performance green OLED (290 lm/W) (Reineke et al., 2013). Over the past few decades, conventional LED (semiconductor) technology has improved incredibly, due to its energy conversion efficiency and minimal energy requirement. GaN-based blue LEDs were used in these conventional LEDs, but the problem arises for the formation of LEDs in the green-yellow color regions (green-yellow gap), which ultimately affects human eyes (photopic sensitivity). However, such types of constraints are not faced by OLEDs.

As OLEDs are best compared to ordinary phosphorescent displays because of their advanced features such as quantum efficiency (QE) of the emitter, current, and energy efficiency (CE), lifetime, brightness, angle of view, contrast, temperature range, low consumption of electricity, cheap to manufacture, etc. The organic materials used in LEDs give primary colors red, green, and blue (RGB) on display (Kukhta et al., 2006; Ding et al., 2008) as shown in Fig. 9.4. Many pi-conjugated molecules have been synthesized which exhibit semiconducting and conducting character and are applied as light-emitter substances and charge-transporter substance. An organic substance that transfers electrons must possess an electron affinity of higher than 3.0 eV, which enables the electrons injection from the metallic cathode. Over the last few years, several electron-withdrawing imides have been reported that are used as ETL material in the construction of OLEDs that

Fig. 9.4 Different colors of OLED display.

have excellent intramolecular charge transfer (ICT) properties (Kulkarni et al., 2004; Hughes and Bryce, 2005; Sasabe et al., 2007; Yoo et al., 2013). However, different organic substances have been reported that function as HTLs and only a few electron-deficient substances exhibited/ reported better electrons-transporter properties (Purvis et al., 2018; Usta et al., 2012; Hendsbee et al., 2014). Nowadays, tris-(8-hydroxy-quinoline)-aluminum (Alq₃) is widely used as an electron transporter substance in OLEDs (Kao, and Chiu, 2015; Bo et al., 2005; Li et al., 2015; Earmme et al., 2009). Besides this (Alq3), many other compounds such as derivatives of carbazole, derivatives of oxadiazole, derivatives of triazole, and derivatives of phenanthroline are used as electron transporters due to their high electron affinity (Kolosov et al., 2002). Many researchers are putting in efforts to design and synthesize a new organic substance for organic optoelectronics (Elbing and Bazan, 2008).

However, OLED has many advantages and disadvantages, such as OLEDs are flexible, consume less power, and exhibit singlet and triplet excitation radiation phenomenon because of the wide energy gap semiconductor, even if it is very thin, light in weight and safe for the environment, and colorful, etc. Therefore, researchers try to design foldable displays for android phones, portable game consoles, media players, wearable devices, etc. While OLED has disadvantages as it has a shorter lifetime as compared to other displays. The lifetime of red, green, and white OLED is about 5–25 years whereas blue OLED only lasts 1.6 years. An OLED is easily damaged by water and direct sunlight application, is more expensive, and its luminescence gradually degrades, etc. But researchers/ manufacturers are trying to upgrade the OLED, to increase the blue luminescence, roll-to-roll manufacturing, enhance the lifespan, and even manufacture it cheaper. Many OLEDs materials have been reported and several other researchers are trying to synthesize better organic materials that can be used in OLEDs. Similarly, chromophore materials are used in the fabrication of OLEDs devices due to the pi-conjugated system; the ETL materials for OLED fabrication must be electron-deficient-pi-systems with excellent chemical stability as well as thermal stability.

1.2 The chromophore

It is the part of a compound or molecule that is accountable for its color. The chromophore is a Greek word, "chroma + photos" which means "color bearer". It is an unsaturated group or region in molecules where two different molecular orbitals of the same molecule show their energy difference within the range of the visible spectrum. The compound which contains a chromophore is known as Chromogen. When visible light hits the chromophore, it absorbs the light and reflects at a specific angle after excitation, and gives different specific colors to the eyes. The quantum or moiety of the chromophore shows conformational changes in the molecules

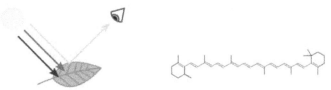

Fig 9.5 (a) absorption of visible light by chlorophyll; (b) conjugated -bonds in the β-carotene compound.

after an electron is excited from the ground state to the excited state. Chromophore groups may be ethylene, thio, nitroso, nitro, keto, and azo (Lipton, 2017). As seen in Fig. 9.5(a), the healthy leaf appears as greenish because the chlorophyll of the leaf absorbs only red and blue wavelengths of light whereas the green wavelengths of light are reflected by cell organelles of the plant (cell walls, etc.) or less absorbed by chlorophyll (Virtanen et al., 2020), while in Fig. 9.5(b), 11 conjugated-bonds are present in the β-carotene compounds, a chromophore, that is abundant in fungi (Lee et al., 2012).

The electrons in a—conjugation system capture photons of light and resonates within the specific distances of—molecular orbitals, like photons detected by a radio antenna within its length; this means that the longer the conjugated-system, the longer will be the wavelength of the captured photon. Generally, if the conjugated -system has less than eight conjugated double bonds, it appears colorless to the human eye because it absorbs ultraviolet region light, even though the compound absorbs blue or green light which depends not only on the conjugated double bonds but also on the functional group attached to the chromophore (Lipton, 2017). When a group is linked with a chromophore which is responsible for enhancing its potential to absorb visible light, it is known as an Auxochrome. It can be a saturated or an unsaturated functional group with non-bonding or one or more pairs of electrons attached to the chromophore and changes the absorption intensity and wavelength, e.g., OH, NH_2, NHR, COOH, Cl, CHO, SCH_3, etc. If the functional group was directly linked with the—conjugation system of the chromophore, it enhanced the wavelength at which the light was absorbed and the absorption peak was more intense (Auxochrome, 2021, 24 December). Benzene does not display any color but it exhibits pale yellow color due to the NO_2 group; similarly, p-hydroxynitrobenzene displays a deep yellow color, because of the OH group. An auxochrome is also known to produce a bathochromic shift, i.e., redshift because it enhances the wavelength of absorption, which is predicted by the Woodward-Fieser rules.

1.3 *Donor—Acceptor chromophore*

The complexes or coordination compounds are formed by the Lewis acid-base reaction of two or more different constituents or donor-acceptor mechanisms. Any neutral compound that can donate a pair of electrons acts

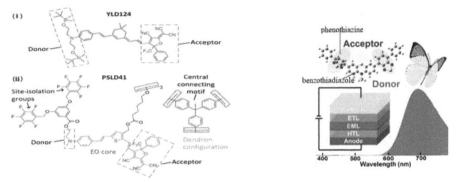

Fig. 9.6 (a), (b): Donor and Acceptor mechanisms.

as a donor while the constituent that accepts the pair of electrons is known as an acceptor, which may be a metal ion or an organic molecule. Many metal-free organic dyes are used in the manufacture of DSSC "dye-sensitized solar cells" (Shalini et al., 2016) and OLED "organic light emitting diode" devices (Song et al., 2020) because their efficiency is better than metal-based organic dyes such as ruthenium-based sensitizers (Shelke et al., 2013) and iridium-based OLEDs (Al-Attar et al., 2011), even though the heavy metals are more expensive and toxic (Wang et al., 2018). While metal-free organic synthesis is cheap, facile synthesis has excellent absorption energies, prolonged illumination, and stability at high temperatures. Metal-free dyes are divided into three parts: donor site, linker site, and acceptor site. The linker site is typically a conjugated-pi-system that links the es-donors (D) to es-acceptors (A) groups by a bridge, that is why it is also known as D-π-A.

In Fig. 9.6 (a) (i), the chromophore YLD124 has one donor group, one acceptor group, and a conjugated-system/bridge, while in Fig. 9.6 (a) (ii), the structure looks like a dendritic configuration which connecting three electro-optic substances, marked as green color, in each EO substance the core electro-optic site isolation group is marked as blue color. This site isolation group minimizes intermolecular interaction or dipole-dipole interactions of neighboring molecules (Sullivan et al., 2007), hence the head to the tail orientation of neighboring molecules is very low (Sullivan and Dalton, 2010; Palmer et al., 2014). Figure 9.6 (b) represents the D-A compound, within that phenothiazine acts as an electron donor and benzothiadiazole as an electron acceptor (Yao et al., 2014).

Various reaction methods have been reported for the synthesis of the donor-acceptor chromophore by different reaction methods for the fabrication of smart organic light-emitting diodes as shown in Fig. 9.7.

1.3.1 Suzuki coupling reaction

It is an important cross-coupling reaction used for the preparation of C-C compounds and polyolefins, styrenes, substituted biphenyls, etc. Suzuki

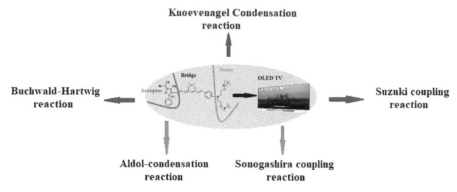

Fig. 9.7 Different reaction methods for the synthesis of OLEDs.

coupling is also known as the Suzuki-Miyaura reaction, in this reaction C-C compound is formed when organo-boronic acid reacts with organo-halide with a catalytic amount of "palladium (0) complex" (Miyaura and Suzuki, 1995; Suzuki, 1991).

$$R^1X + R^2BY_2 \xrightarrow[\text{base}]{\text{[Pd]catalyst.}} R^1\text{-}R_2$$

Suzuki-coupling reaction

Many reactions have been reported for the preparation of C-C derivatives by this reaction such as:V. Promarak et al. (2021) have synthesized a "pyrene-naphthalimide-based fluorophore by Suzuki-coupling reaction and catalyzed by palladium" for the fabrication of blue-green OLEDs devices. The pyrene-naphthalimide-based derivatives were synthesized by the following reaction. The boronic-acid-pinacol ester was prepared by borylation of alkylated **4-Bromo-1H,3H-benzo[de]isochromene-1,3-dione** with bis(pinacolato) diboron, with a catalytic amount of Pd(PPh3)Cl2/Pot. Acetate. The obtained intermediate further reacts with derivatives of Bromopyrene in the presence of Pd(PPh3), Na_2CO_3, and forms compounds 1, 2, and 3, (Scheme 1). The obtained compound was thermally stable at 347–447°C (with 5% wt. loss) due to the ICT mechanism. Among the three derivatives, compound 3 exhibits excellent performance due to the push-pull effect, excellent charge transfer, and correct HOMO and LUMO energy levels, which enhance and balance the hole-electron recombination in the device. The emitting solution for the device was stacked with the configuration ITO/PEDOT: PSS-Nafion (Nf)/1/2/3:10%CBP/TmPyPb/LiF/Al. The device with 3 showed maximum luminescence of 3389 cd m^{-2}, and a maximum EQE of 3.98%; similarly the luminance efficiency and turn-on voltage was found to be 3.22 cd A^{-1} and 3.2 V, respectively (Boonnab et al., 2021; Sneha and Dhanya, 2022).

The novel triphenylamino-naphthalimide (Scheme 2) was used to fabricate OLED devices because of their excellent thermal stability > 350°C

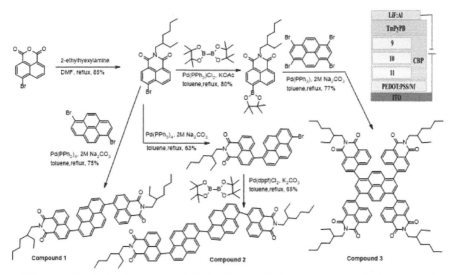

Scheme 1 Synthesis of pyrene-naphthalimide-based blue-green OLEDs device material.

with 10% weight loss, the electrochemical band-gap was estimated to be 2.48–2.54 eV. Compounds 1, 2, and 3 (synthesized by P. Rashatasakhon and co-worker) exhibited enhanced conjugation and packing patterns, restricting the molecular vibration in the solid thin-film state. Their absorption and emission peaks shifted towards longer and shorter wavelengths, respectively, and displayed blue emission compared to the solution state. Among the three derivatives of Scheme 2, compound 1 showed better performance in multilayer devices, while in single-layer devices compound 1 didn't display a satisfactory performance because of their electron transport barrier at EML and poor-film-forming ability and LiF/Al electrode. But after adding BCP with compound 1, it displayed better luminance. The solution for the emittance layer stacked in the configuration ITO/PEDOT: PSS/compound 1: CBP/BCP/LiF/Al. The device displayed excellent luminescence of 10404 cd m^{-2}, luminance efficiency of 3.77 cd A^{-1}, a current density of 410 mA m^{-2}, a quantum efficiency of 1.11% after application of turn-on voltage 5.8 V, and applied voltage 19 V, due to electron transfer from BCP to compound 1. The OLEDs device of compounds 2 and 3 displayed very poor performance because of the solubility problem, while the device with compound 1 displayed excellent yellowish-green luminescence with a CIE coordinate at (0.295, 0.600) (Sneha and Dhanya, 2022; Arunchai et al., 2015).

Zagranyarski et al. (2021) synthesized 3,4-dioxin-annulated 1,8-naphthalimide via the Suzuki-coupling reaction, which was used for the fabrication of a green emitter OLEDs device. Compounds 1 and 2 were synthesized by the following reaction as shown in Scheme 3. It was seen

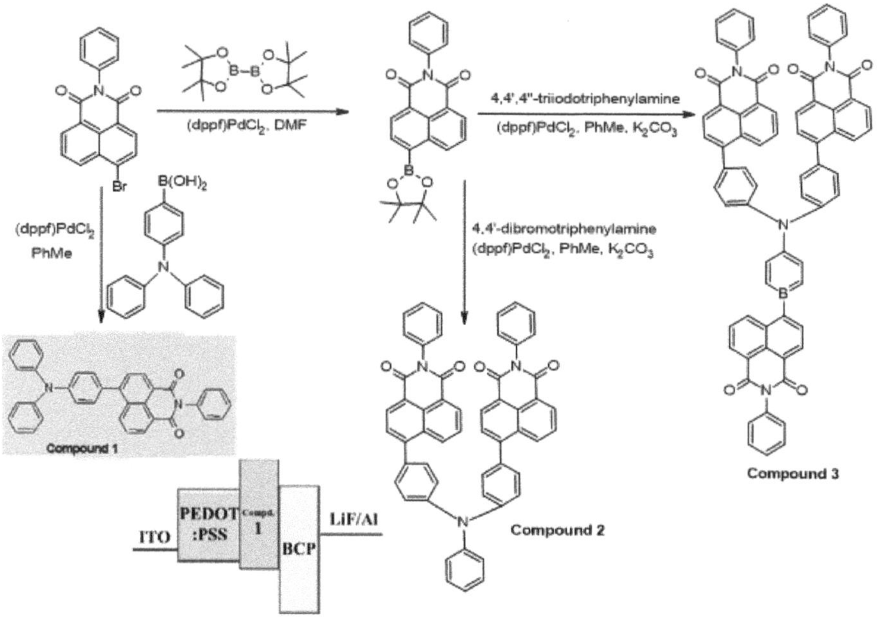

Scheme 2 Schematic diagram of derivatives of triphenylamine-phthalimide synthesis.

that upon adding the hexyl-oxy group to the dioxin moiety of 1 and 2, the FMO energy levels increased and the energy band gaps also enlarged. The performance of 1 and 2 was tested by using 1 and 2 as electroluminescence materials in an OLED device with configuration ITO|PVK: TPD|1/2|Al. The devices containing compound 1 as EML exhibited better performance and displayed cyan-greenish emission due to certain orientations of the aliphatic chain towards pi-conjugation, while the OLED device with compound 2 as EML did not displaygood performance because the longer-alkoxy group obstructed the recombination mechanism of the hole and electron. The OLED device with compounds 1 and 2 was thermally stable over 400°C. The turn-on voltage and applied voltage of the device with compound 1 was estimated to be 8 V and 19.6 V, respectively. It displayed a maximum luminescent intensity of 3031 cd m^{-2} and a current density of 6.9 cd A^{-1}, whereas the device with compound 2 displayed poor luminescence of 643 cd m^{-2} upon applying turn-on and applied voltage of 15 V and 26 V, respectively. Therefore, the current and power efficiency of OLED-1 was two-and-a- half times greater than OLEDs-2 (Zagranyarski et al., 2020).

The host material with excellent thermal stability and bipolar charge transporter naphthalimide-pyrene-based compound was synthesized by G. Bagdzinuas et al. (2018) for the manufacturing of an OLED device. The red-emitting host material (Scheme 4) was synthesized by treatment of 4-Bromo-N-(2-Ethylhexyl)-1,8-naphthalimide and pyrene-1-boronic

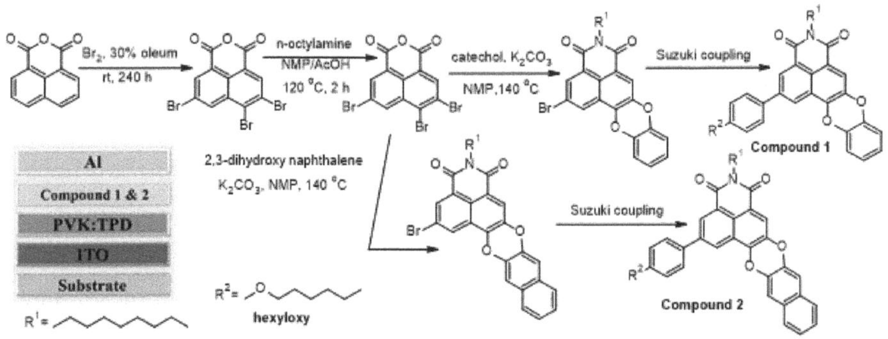

Scheme 3 Structure of green emitters derivatives of 1,8 naphthalimide for green OLEDs device.

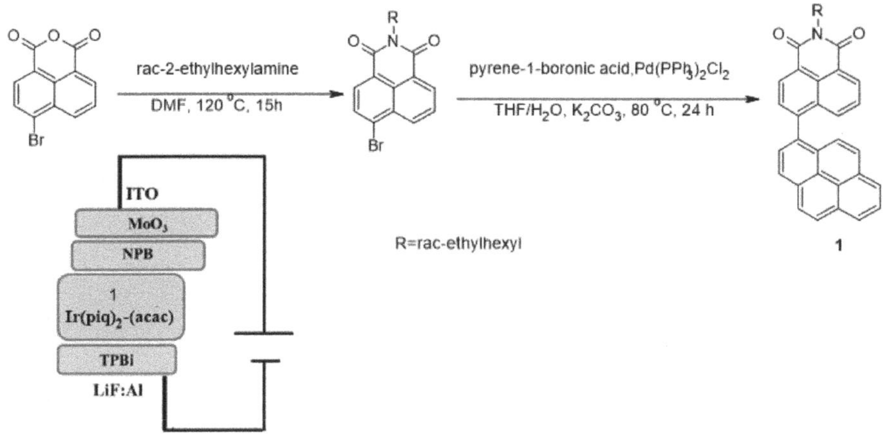

Scheme 4 Schematic diagram of the synthesis of host substance; stacks configuration for red OLEDs device.

acid with catalytic amounts of $Pd(PPh_3)_2Cl_2$ and K_2CO_3 in THF, the pyrene moiety attached at the C_4-position. Compound 1 was used as a dopant for the fabrication of different OLED devices and configured as $ITO/MoO_3/NPB/Ir(piq)_2(acac)$ (X%):1/TPBi/LiF/Al where X=10, 15, 25%, respectively. When 4–10 V external voltages were applied to the OLED device, it emitted red light with CIE coordinates (0.677, 0.319). The compound 1 and $Ir(piq)2(acac)$ based OLEDs device showed maximum current efficiency (CE), power efficiency (PE), and quantum efficiency (EQE) of 10.8 cd A^{-1}, 7 lm W^{-1}, 13.6%, respectively. These devices displayed a high luminescence of 15300 cd m^{-2}. But there are two non-doped devices with compound 1 with the configuration as $ITO/MoO3/NPB/TcTa/mCP/1/TSPO1/TPBi/Ca/Al$ and $ITO/MoO_3/mMTDATA/NPB/mCP/1/TSPO1/TPBi/LiF/Al$, which emit blue light (Bezvikonnyi et al., 2018).

Quinazoline-centered based bipolar host substances were reported by Z. Wang et al. (2015). These derivatives were highly thermal stable and exhibited high decomposition temperature and glass-transition temperature up to 447°C and 154°C, respectively. The desired host substance BITpQz, CzTpQz, BBIQz, and BCzQz were synthesized by the following reaction shown in (Scheme 5). The solution for the manufacturing of red Ph-OLEDs was stacked with the configuration of ITO/H04/TCTA/host: dopant 8%/ TPBi/Bphen/LiF/Al, where H04 as HIL/HTL, TCTA as EBL, Ir(mphmq)$_2$acac as EML, TPBi as ETL, and BPhen as the electron-injection and transporting layer. Photophysical properties of the desired compounds were analysed by UV-vis and PL, and maximum absorption peaks of BITpQz, CzTpQz, BBIQz, and BCzQz appeared at 337 nm, 365 nm, 327 nm, and 341 nm, respectively. The ICT strength and redshift bands increased due to the donor-acceptor ability of CzTpQz and BCzQz. The triplet energies of the host are 2.40, 2.40, 2.46, and 2.36 eV, respectively. The performance of BITpQz, CzTpQz, BBIQz, and BCzQz- based red Ph-OLEDs was analyzed and found that the turn-on voltage was 2.3 V, 2.5 V, 2.7 V, and 3.0 V, respectively. The host substance BITpQz displayed EQE of 19%, CE of 18.3 cd/A and PE of 21.7 lm/W; similarly the other host materials CzTpQz displayed 18.4%, 17.6 cd/A, 19.31 lm/W, for BBIpQz as 15.4%, 14.4 cd/A, 21.7 lm/W, for BCzQz as 18.4%, 16.7 cd/A, 15.7 lm/W, respectively. The performance of the host material was

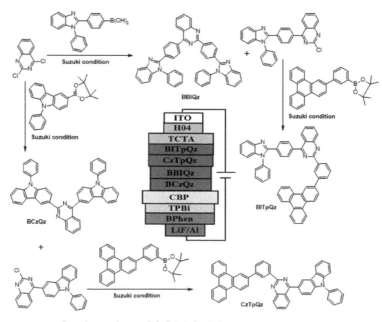

Suzuki condition: Pd(OAc)$_2$/PPh$_3$/K$_2$CO$_3$, 1,4-dioxane/H$_2$O

Scheme 5 Schematic diagram of the synthesis of Quinazoline-centred based bipolar host material; red Ph-OLED layer configuration.

much higher than the reference CBP-based OLED, the EQE was 10.3%, CE of 9.9 cd/A, 7.8 lm/W. The **BITpQz-based** red Ph-OLED showed the highest maximum efficiencies and low turn-on voltage of 2.3 V under low driving voltage, whereas **CzTpQz-based** red Ph-OLED displayed maximum luminescence of 20980 cd/ m² and weak roll-off character because of better recombination of electron-hole at high voltage (Zhang et al., 2016).

3.1.2 Knoevenagel condensation reaction

It is the nucleophilic addition reaction based on the synthesis of α, β-unsaturated ketone. It is a type of aldol condensation reaction which occurs between an active hydrogen compound and carbonyl group in the presence of weak base followed by dehydration in which a water molecule is released.

Knoevenagel Condensation reaction

R''-CH2-R''' or R''-CHR-R''' e.g., diethyl malonate, meldrum's acid, ethyl acetoacetate or malonic acid, or cyanoacetic acid or nitromethane; R''' and R'' must be powerful electron-withdrawing groups (Jones, 2004). There are many reactions that have been reported for the synthesis of the OLED host material, electron transport material, hole transport material, etc.

A highly thermally stable electroluminescent cyno-substituted phenylenevinylene (PV) monomeric and oligomeric chromophores were synthesized by F.E. Karasz et al. (2006). Compounds 1, 2, and 3 are synthesized by Knoevenagel-type condensation reaction as shown in Scheme 6. The emissive substance stacks with the configuration ITO/PEDOT: PSS/1-3/Ca/Al, where PEDOT: PSS is a hole injection layer. The thermal stability of compounds 1, 2 and 3 was analyzed by DSC, and it was found that all the compounds were stable up to 300£C with 5% weight loss. The optical properties of 1, 2 and 3 were analyzed by the absorption band gap and photoluminescence spectra; the absorption band gap of 1, 2 and 3 was displayed at 2.87, 2.85 and 2.83 eV, respectively. The fluorescence quantum yield was shown as 0.65, 0.46 & 0.68, respectively. The electroluminescence character of 1, 2 and 3 was analyzed in the PL spectra, and the turn-on voltage, maximum luminance, and luminance efficiency exhibited for 1 as 8 V, 21 cd/m², and 0.01 cd/A, for 2 as 8 V, 62 cd/m² and 0.05 cd/A, while for 3 it exhibited as 5 V, 25 cd/m² and 0.02 cd/A, respectively (Cirpan et al., 2006).

A series of multibranched isophorone-based dyes was structured and prepared by X.T. Tao and a co-worker (2006). Compounds 1, 2, and 3 shown in (Scheme 7) were synthesized by Knoevenagel condensation reaction according to the reported literature (Lemke, 1974). The solution for the

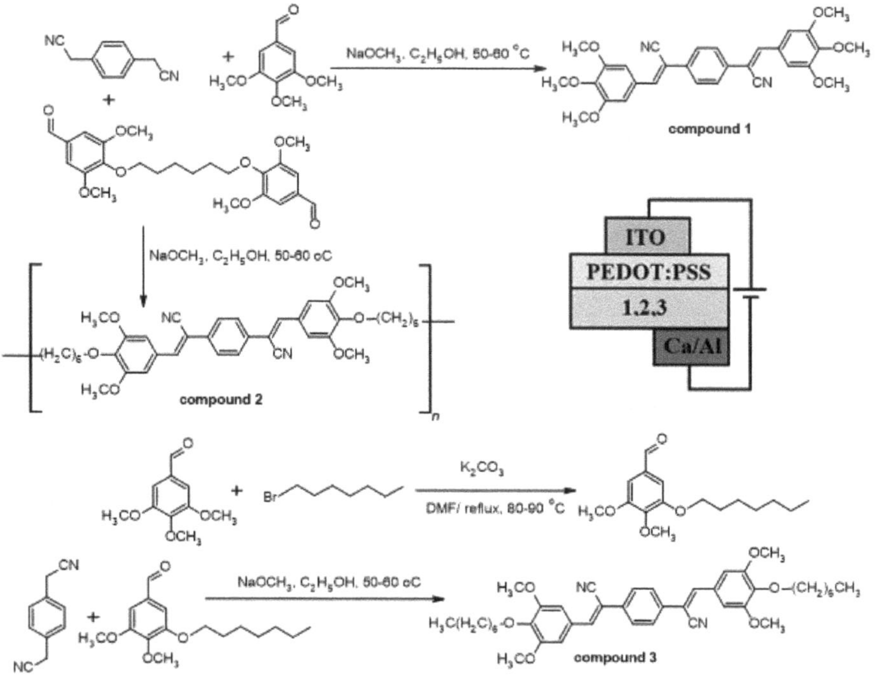

Scheme 6 Synthesis of cyno-substituted phenylenevinylene (PV) monomeric and oligomeric.

fabrication of red-OLED devices stacks in the configuration ITO/NPB/dyes 1, 2, and 3/Alq3/Mg: Ag/Ag: ITO as an anode, NPB as HTL, Alq_3 as ETL. Due to donor-pi-acceptor character and number of branches, these dyes exhibited excellent optical properties, two strong electronic absorption bands one near 350 nm due to pi-pi-transition and the second between 495–505 nm due to electron charge transfer transition. Although 1, 2 and 3 showed the same donor group and acceptor moiety, the electrochemistry character analysed by CV and HOMO/LUMO energy levels of 1, 2 and 3 were found to be –5.67–3.76, –5.68/–3.85 and –5.69/–3.88 eV, respectively. The performance of the dyes 1, 2 and 3 were displayed as, the turn-on voltage of 4.0V, 7.7V and 9.3V, the maximum luminescence of 1801 cd m^{-2}, 215 cd m^{-2} and 25.7 cd m^{-2}, respectively. the luminance efficiency/current efficiency was found to be 0.095/0.189, 0.0074/0.0232 and 0.00074/0.00248 lmW^{-1}/ cdA^{-1}, respectively. Compound 1 or single-branched red OLEDs displayed maximum luminescence at less turn-on voltage with CIE (0.672, 0.324) (Ju et al., 2006).

3.1.3 Buchwald-Hartwig amination reaction

It is the cross-coupling reaction used for the synthesis of carbon-nitrogen bonds when derivatives of amine and aryl halide are catalyzed by palladium (Li et al., 2010).

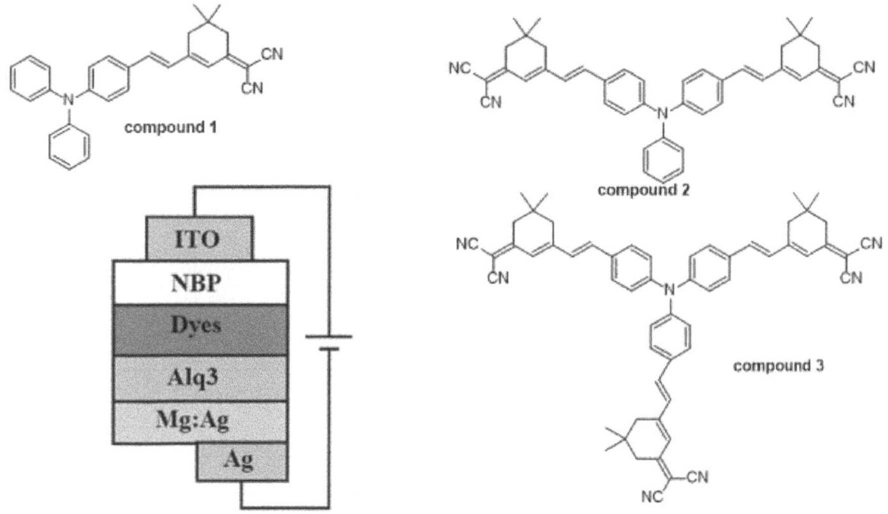

Scheme 7 Structure of isophorone-based dyes and stacks configuration of different layer solutions.

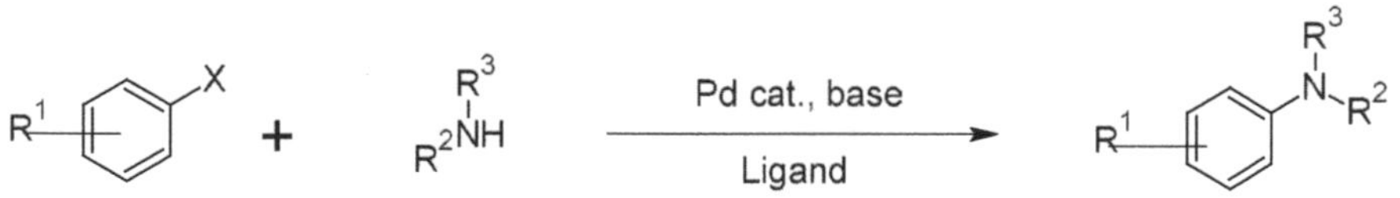

X= -Cl, -Br, -I, -OTf; R2= -alkyl, -aryl, -H; R3= -alkyl, -aryl

Buchwald-Hartwig amination reaction

There are many Buchwald-Hartwig amination reactions reported used for the synthesis of host materials for the OLED device such as:S.J. Su and a co-worker (2012) synthesized a host material 2,4,6-Tris(3-(9H-carbazol-9-yl)-phenyl)-1,3,5-triazine (TCPZ), having triazine core with three phenylcarbazole arms, that was used to fabricate the Ph-OLEDs. The TCPZ (Scheme 8) was synthesized when the mixture of TBrPZ, carbazole, lead-chloride, tri-tert-butylphosphine, and sod. ter-butoxide in anhyd. o-xylene stirred under N_2-atmosphere for 17 h at 120°C. Through column chromatography, pure TCPZ compounds were obtained with 79% yield. The HOMO and LUMO energy levels of TCPZ were analyzed and found to be 6.18 eV and 2.7 eV, respectively. For the fabrication of the red Ph-OLEDs device, the emitter solution was constructed with different configurations as of red emitter ITO/TPDPES: TBPAH/TAPC/TCPZ:4 wt.% Ir(piq)$_3$/TmPyBPZ/LiF/Al; of green emitter ITO/TPDPES: TBPAH/TAPC/TCPZ:8 wt.% Ir(PPy)$_3$/TmPyBPZ/LiF/Al; of blue emitter ITO/TPDPES: TBPAH/3DTAPBP/TCPZ:11 wt.% FIrpic/BP4mPy/LiF/Al. The lifetime and quantities data (t_1A_1, t_2A_2, t_3A_3) of the host material and dopant {TCPZ:11 wt.%(FIrpic), TCPZ:8 wt.%(Ir(PPy)$_3$), and TCPZ:4 wt.%(Ir(piq)$_3$)} is given in Table 9.1 (a), while the performance data (efficiency, intensity, current in V, etc.) of the co-deposited host material is given in Table 9.1 (b) (Su et al., 2012).

Table 9.1(a) The lifetime and quantities data of the host material and dopant.

Co-deposited	$_{pl}$(%)	Wavelength (nm)		t_1 0/ A$_1$	t_2 0/ A$_2$	t_3 0/ A$_3$
FIrpic	66 ± 1		350–650	0.45/ 0.50	1.92/ 0.49	6.58/ 0.10
Ir(PPy)$_3$	69 ± 1	b1	400–700	0.56/ 0.54	1.82/ 0.44	5.90/ 0.10
		b2	400–500	0.65/ 5.78	3.50/ 0.19	-
		b3	650–700	1.29/ 1.43	4.74/ 0.49	-
Ir(piq)$_3$	54 ± 1		520–820	1.15/ 1.10	-	-

Table 9.1 (b) The performance data of the co-deposited host material.

Host material	V$_{on}$ (V)	Maxm. Efficiency $_{p}/_{ext}$ (lmW^{-1}/%)	At 100 cd m^{-2} V/$_{p}/_{ext}$ (V/lmW^{-1}/%)	At 1000 cd m^{-2} V/$_{p}/_{ext}$ (V/lmW^{-1}/%)
FIrpic	2.3	20.5/19.1	3.02/15.8/17.9	4.25/8.85/14.1
Ir(PPy)$_3$	2.1	87.8/19.6	2.47/85.0/19.1	2.98/70.4/19.1
Ir(piq)$_3$	3.0	62.0/18.5	3.74/25.4/10.1	4.72/17.3/8.67
Ir(piq)3 & FIrpic	3.0	31.2/13.7	3.91/9.45/5.39	5.36/5.95/4.76

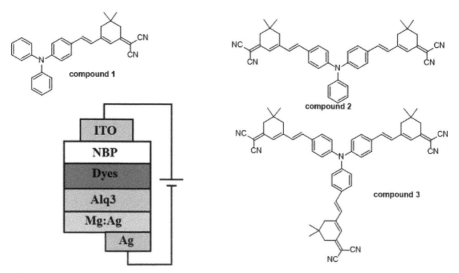

Scheme 8 Schematic structure of 2,4,6-tris(3-(carbazol-9-yl)-phenyl)-triazine (TCPZ) as EML for red, green, and blue OLEDs devices (Su et al., 2012).

Two orange-light-emitting substances or compounds "NBDI derivatives (1 and 2)" were synthesized by S.O. Jung et al. (2014). The compound was prepared via the Buchwald-Hartwig amination reaction or N-arylation. The 4-Bromo-1,8-naphthalic anhydride reacted with 1-naphthylamine, isoquinoline, and m-cresol giving the intermediate product, which further reacted with N,N-diphenyl, and N-naphthyl-N-phenyl in the presence of the catalytic amount of Pd complex to obtain NBID derivatives, which are shown in Scheme 9. The NBID derivatives were highly thermally stable due to donor-acceptor ability: NBID-1 was stable up to 368°C, NBID-2 was stable up to 407°C and their glass transition temperature was observed at 128°C and 148°C, respectively. In NBID derivatives, 1H-Benz[de]isoquinoline-1,3(2H)-dione, 2-(1-naphthalenyl) acts as an electron-acceptor while the benzene and naphthalene substituent acts as hole transporting/electron-donating site. The electrochemical properties of NBID-1 and 2 were analysed by CV, the electrochemical band gaps were observed as 2.42 eV. The HOMO and LUMO energy level of NBID-1 and 2 derivatives were estimated to be –5.48 eV and –5.46 eV and –3.06 eV and –3.04 eV, respectively. The maximum EL and the luminance efficiency of NBID-1 were exhibited at 574 nm and 6.6 cd/A, similarly the maximum EL and luminance efficiency of NBID-2 was exhibited at 588 nm and 5.9 cd/A, but the turn-on voltage of NBID-1 and 2 was found to be as 5.9 V and 6.3 V, respectively. The orange emission of NBID-1 and NBID-2 was displayed at 10 mA/cm² (CIE) coordinates of (0.46, 0.52) and (0.48, 0.52), respectively. The NBID-1 and 2 derivatives "orange emitting material" were highly efficient because of their bulky and asymmetrical structure and also their intermolecular dipole-dipole interaction (Jung et al., 2009).

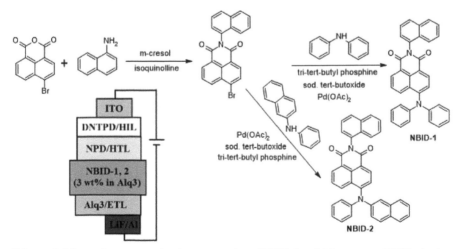

Scheme 9 Schematic structure for the preparation of NBID-1 and 2 for orange OLEDs device.

H.Y. Lu et al. (2019) constructed and synthesized new chiral TADF materials for the fabrication of orange-red OLEDs. They synthesized two enantiomeric compounds (–)-(R, R)-CAIDMAC and (+)-(S, S)-CAI-DMAC. First, they obtained (–)-(R, R)-1,2-diaminocyclohexane or (+)-(S,S)-1,2-diamino-cyclohexane by lactamizing of 4-Bromo-1,8-naphthalic-anhydride with 1,2-diaminocyclohexane, then further reacted it with 9,9-dimethyl-9,10-dihydroacridine in the presence of the catalytic amount of palladium for C-N coupling reaction, which is shown in Scheme 10. The thermal stability of both the enantiomers was analyzed by TGA and DSC; it was found that the enantiomers decomposed at high temperatures up to 405°C, displayed good electrochemical character, excellent thermally activated delayed fluorescence character with a small ΔE_{ST} value of 0.07 eV. The enantiomers displayed circularly polarized luminance (CPL) characters with $|g_{lum}|$ value of 9.2×10^{-4} and showed mirror-image circular dichroism (CD) characters. The orange-red OLEDs device was fabricated by dopping of CBP with 6 wt% of both the enantiomers or emitters: ITO/HAT-CN/

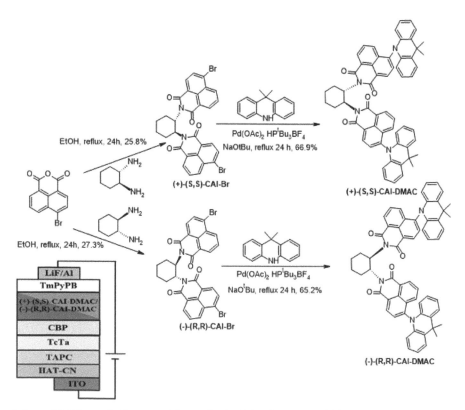

Scheme 10 Schematic diagram of the synthesis of chiral orange-red TADF emitters and device configuration.

TAPC/TcTa/CBP/(–)-(R,R)-CAIDMAC & (+)-(S,S)-CAI-DMAC/TmPyPB/LiF/Al; the orange-red emission band of the device with enantiomers was displayed at 592 nm. The external quantum efficiency (EQE) of (–)-(R,R)-CAIDMAC and (+)-(S,S)-CAI-DMAC were observed to be 12.4% and 12.3%, respectively. The maximum CE of (–)-(R,R)-CAIDMAC and (+)-(S,S)-CAI-DMAC were observed as 28.5 cd/A and 28.8 cd/A, and maximum PE at 28.8 lm/W and 26.6 lm/W, respectively (Wang et al., 2019).

3.1.4 Sonogashira cross-coupling reaction

This is a cross-coupling reaction used for the preparation of C-C bonds between alkyne and aryl/vinyl-halide in the presence of a catalyst such as palladium catalyst as well as copper co-catalyst with the mild base at room temperature in aqueous media (Sonogashira, 2002).

$$R^1-X \; + \; \overset{}{\underset{R^2}{\equiv}} \quad \xrightarrow[\text{base, r.t.}]{\text{[Pd] cat., [Cu] cat.}} \quad R^1\overset{}{\underset{R^2}{\equiv}}$$

R^1= aryl or vinyl, R^2= arbitrary, X= Cl, Br, I, or OTf

The Sonogashira coupling reaction

Sonogashira cross-coupling reactions have a wide-applications, such as donor-acceptor complex formation (Geenen et al., 2020), in pharmaceuticals, natural products, organic materials, and nanomaterials (Sonogashira, 2002), for the synthesis of tazarotene (used for the treatment of psoriasis, acne) (King and Yasuda, 2005), also used for the synthesis of SIB-1508Y (Altinicline) which is a nicotinic receptor agonist (King and Yasuda, 2004).

V. Ervithayasuporn and co-worker (2010) synthesized two new classes of host substances via the Sonogashira coupling reaction or 'click-chemistry' catalyzed by palladium. The synthesis of mono and oligo-(p-phenylene-ethynylene) "host substance" is shown in (Scheme 11). It was synthesized according to the reported literature (Ervithayasuporn et al., 2010) and grafted with polyhedral oligomeric silsesquioxanes (POSS). The host substance exhibits excellent thermal stability in the air of over 330°C with 5% weight loss and a glass transition temperature of 80°C. It exhibits maximum photoluminescence in the blue emission range with a quantum yield of up to 80%. For the fabrication of an OLED device, the solution was configured as ITO/PEDOT: PSS/PVK/dopant(1–13% w/w): PBD: PVK/MPT/LiF/Al. Owing to the expanded -conjugation system, oligo-pPEs POSS molecules retain the maximum photoluminescence QE in the solid state compared to mono-pPEs POSS molecules. Among other derivatives, the pyrene-based dopant material shows a turn-on voltage of 5.9 V–6.5 V with a very low luminance of 2

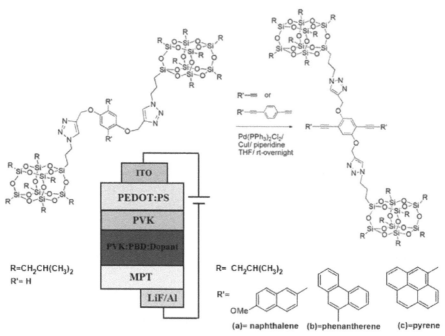

Scheme 11 Synthesis of mono- oligo-pPEs POSS host material; device architect with dopped materials.

cd m⁻², while maximum EQE and luminance efficiencies of 1.50% and 4.54 cd A⁻¹, respectively. But after 5% of dopant, the output light was enhanced by 100 cd m⁻² at 8.0 V and 2.34 mA cm⁻² of current density (Ervithayasuporn et al., 2010).

3.1.5 Aldol-condensation reaction

It is a condensation reaction in which two carbonyl compounds, either aldehyde or ketone, react with each other to give β-hydroxy aldehyde or β-hydroxy ketone. In this reaction, the nucleophile or enolate ion attack the carbonyl compound and forms derivatives of ketone or aldehyde of β-hydroxy, then undergo dehydration to form a conjugated enone. It is an important organic-chemical reaction that facilitates the formation of carbon-carbon bonds easily.

Aldol-condensation reaction

The novel donor-acceptor bipolar host-substance (indolo-(3,2,1-jk)-carbazole-pyrimidine) was effective for 4CzIPN. 4CzIPN was used as a green thermally activated delayed fluorescence emitter (TADF) and applied to synthesize the green OLEDs devices. The host substance (ICzPyr) has an electron donor site as indolo-(3,2,1-jk)-carbazole, while the electron-withdrawing site as pyrimidine. The ICzPyr was synthesized by Y. Hiraga et al. (2021) by Suzuki coupling reaction followed by an aldol condensation reaction which is shown in Scheme 12. The host substance was thermally and

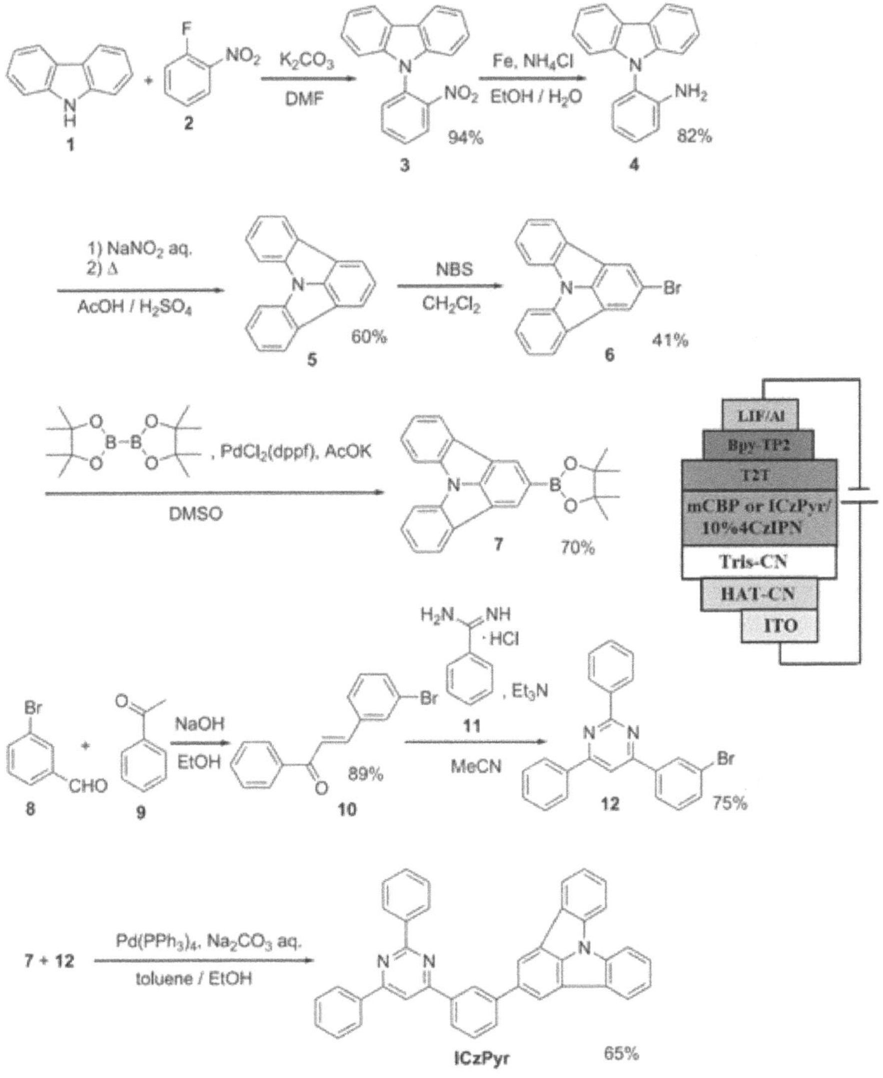

Scheme 12 Synthesis of ICzPyr, a host substance for the fabrication of green-organic LEDs.

morphologically most stable and has a high thermal stability up to 398°C with 5% weight loss, and the glass transition temperature was calculated as 110°C, which was analyzed by DSC and TGA. The bipolar electrochemical properties of the host substance were analyzed by CV and DPV, and it was found that the host substance showed a reversible oxidation potential at 0.96 V and a reduction potential at –2.33 V. The performance of the host substance was better than dopped substance because the current efficiency of mCBP (dopped substance) and ICzPyr were observed to be 35.1 cd/A and 47.8 cd/A, the luminous efficiency was observed as 13.4 lm W^{-1} and 30.1 lm W^{-1} and EQE were observed as 10.9% and 10.0%, respectively. The lifetime of mCBP and ICzPyr was determined at an initial luminescence of 10000 cd m^{-2} and found to be 1.0/h and 5.5/h, respectively, which revealed that ICzPyr-based devices are five times longer operated than mCBP-based devices (Hiraga et al., 2021).

Conclusion

In this chapter, we have demonstrated the synthesis of the host material, electron transporter, hole transporter, and hole blocker for OLEDs through different methodologies, i.e., Suzuki cross-coupling reaction, Buchwald-Hartwig amination reaction, Sonogashira cross-coupling reaction, and Aldol-condensation reaction. The obtained compound exhibits high luminescence and better mobility due to planer geometry, and expanded pi-conjugation is an important feature for OLED materials. The donor-acceptor potential of the chromophore moieties enhanced the HOMO and LUMO energy level separation for the fabrication of OLEDs. Even the ICT mechanism contributes immensely to luminescence emission. There are many dopants or emissive substances that have been reported which give full solid color red, blue, green, and even white colors with a low power input of 4.5 V. Due to high thermal stability and glass transition temperature, ICT mechanism, excellent CIE coordinate, low turn-on, and applied voltage, enhanced HOMO and LUMO energy levels as well as triple energy gap the OLED devices displayed superior performance, maximum luminescence, long lifetime, excellent quantum efficiencies, power efficiency, luminous efficiency, etc.

References

Al-Attar, H.A., Griffiths, G.C., Moore, T.N., Tavasli, M., Fox, M.A., Bryce, M.R. and Monkman, A.P. (2011). Highly efficient, solution-processed, single-layer, electrophosphorescent diodes and the effect of molecular dipole moment. *Adv. Funct. Mater.*, 21: 2376–2382. https://doi.org/10.1002/adfm.201100324.

Arunchai, R., Sudyoadsuk, T., Prachumrak, N., Namuangruk, S., Promarak, V., Sukwattanasinitt, M. and Rashatasakhon, P. (2015). Synthesis and characterization of new triphenylamino-1,8-naphthalimides for organic light-emitting diode applications. *New Journal of Chemistry*, 39(4): 2807–2814. doi:10.1039/c4nj01785d.

Auxochrome. (2021, 24 December). In Wikipedia. https://en.wikipedia.org/wiki/Auxochrome.
Bernanose, A. (1955). The mechanism of organic electroluminescence. *J. Chim. Phys.*, 52: 396. doi:10.1051/JCP/1955520396.
Bernanose, A. and Vouaux, P. (1953). Organic electroluminescence type of emission. *J. Chim. Phys.*, 50: 261. doi:10.1051/JCP/1953500261.
Bezvikonnyi, O. Gudeika, D., Volyniuk, D., Grazulevicius, J.V., and Bagdziunas, G. (2018). Pyrenyl substituted 1,8-naphthalimide as a new material for weak efficiency-roll-off red OLEDs: a theoretical and experimental study. *New Journal of Chemistry*, 42(15): 12492–12502. doi:10.1039/c8nj01866a.
Bo, C.L., Cheng, C.P., You, Z.Q. and Hsu, C.P. (2005). Charge transport properties of tris(8-hydroxyquinolinato)aluminum(III): why it is an electron transporter. *J. Am. Chem. Soc.*, 127: 66–67. https://doi.org/10.1021/ja045087t.
Boonnab, S., Chaiwai, C., Nalaoh, P., Manyum, T.,Namuangruk, S., Chitpakdee, C. … and Promarak, V. (2021). Synthesis, Characterization, and Physical Properties of Pyrene-Naphthalimide Derivatives as Emissive Materials for Electroluminescent Devices. *European Journal of Organic Chemistry*, 2021(17):, 2402–2410. doi:10.1002/ejoc.202100134.
Chang, Song, Y., Wang, Z., Helander, Qiu, J., Chai, L., … Lu, Z. (2012). Highly efficient warm white organic light-emitting diodes by triplet exciton conversion. *Adv. Funct. Mater.*, 23 (6): 705–712, 2013. doi:10.1002/adfm.201201858.
Chang and Lu, Z.-H. (2013). White Organic Light-Emitting Diodes for Solid-State Lighting. *Journal of Display Technology*, 9(6): 459–468. doi: 10.1109/jdt.2013.2248698.
Chen, Y. and Ma, D. (2012). Organic semiconductor heterojunctions as charge generation layers and their application in tandem organic light-emitting diodes for high power efficiency. *J. Mater. Chem.*, 22: 18718–18734. https://doi.org/10. 1039/c2jm32246c.
Choong, V.-E. Shen, J., Curless, J., Shi, S., Yang, J. and So, F. (2000). "Efficient and durable organic alloys for electroluminescent displays". *Journal of Physics D: Applied Physics*, 33(7): 760–763. doi:10.1088/0022-3727/33/7/302.
Cirpan, A., Rathnayake, H.P., Gunbas, G., Lahti, P.M. and Karasz, F.E. (2006). New conjugated materials containing cyano substituents for light-emitting diodes. Synthetic Metals, 156(2–4): 282–286. doi:10.1016/j.synthmet.2005.12.008.
Copy Craig Freudenrich, Ph.D. How OLEDs Work. HowStuffWorks.com. <https://electronics. howstuffworks.com/oled.htm> 8 December 2022.
Ding, G., Xu, Z., Zhong, G., Jing, S., Li, F. and Zhu, W. (2008). Synthesis, photophysical and electroluminescent properties of novel naphthalimide derivatives containing an electron-transporting unit. *Res. Chem. Intermed.*, 34: 299–308. https://d oi.o rg/10.1163/156856708783623401.
Ding, G., Xu, Z., Zhong, G., Jing, S., Li, F. and Zhu, W. (2008). Synthesis, photophysical and electroluminescent properties of novel naphthalimide derivatives containing an electron-transporting unit. *Res. Chem. Intermed.*, 34: 299–308. https://d oi.o rg/10.1163/156856708783623401.
Earmme, T., Ahmed, E., and Jenekhe, S.A. (2009). Highly efficient phosphorescent light-emitting diodes by using an electron-transport material with high electron affinity. *J. Phys. Chem. C.*, 113: 18448–18450. https://doi.org/10.1021/jp907913d.
Elbing, M. and Bazan, G.C. (2008). A new design strategy for organic optoelectronic materials by lateral boryl substitution". Angew Chemie – Int. Ed., 47: 834–838. https://doi.org/ 10.1002/anie.200703722.
Ervithayasuporn, V., Abe, J., Wang, X., Matsushima, T., Murata, H., and Kawakami, Y. (2010). Synthesis, characterization, and OLED application of oligo(p-phenylene ethynylene)s with polyhedral oligomeric silsesquioxanes (POSS) as pendant groups. *Tetrahedron*, 66(48): 9348–9355. doi:10.1016/j.tet.2010.10.009.
Fan, Yang, C. (2014). Yellow/orange emissive heavy-metal complexes as phosphors in monochromatic and white organic light-emitting devices. *Chem. Soc. Rev.*, 43: 6439–6469. https://doi.org/10.1039/c4cs00110a.

Geenen, S.R., T. Schumann, T. and Mueller, T.J.J. (2020). Fluorescent Donor-Acceptor-Psoralen Cruciforms by Consecutive Suzuki-Suzuki and Sonogashira-Sonogashira One-pot Syntheses. *The Journal of Organic Chemistry*. doi:10.1021/acs.joc.0c01059.

Hendsbee, A.D., Sun, J.P., Rutledge, L.R., Hill, L.G. and Welch, G.C. (2014). Electron deficient diketopyrrolopyrrole dyes for organic electronics: Synthesis by direct arylation, optoelectronic characterization, and charge carrier mobility. *J. Mater. Chem.*, A 2: 4198–4207. https://doi.org/10.1039/c3ta 14414c.

Hiraga, Y., Kuwahara, R. and Hatta, T. (2021). Novel indolo[3,2,1-jk]carbazole-based bipolar host material for highly efficient thermally activated delayed-fluorescence organic light-emitting diodes. *Tetrahedron*, 94: 132317. doi:10.1016/j.tet.2021.132317.

Hughes and Bryce, M.R. (2005). Electron-transporting materials for organic electroluminescent and electrophosphorescent devices. *J. Mater. Chem.*, 15: 94–107. https://doi.org/ 10.1039/b413249c.

Jang, J. (2006). Displays develop new flexibility. *Mater. Today*, 9: 46–52. https://doi.org/10.1016/S1369-7021(06)71 447-X.

Jeon, Jang, Son and Lee, (2011). External Quantum Efficiency Above 20% in Deep Blue Phosphorescent Organic Light-Emitting Diodes. *Advanced Materials*, 23(12): 1436–1441. doi:10.1002/adma.201004372.

Jones, G. (2004). The Knoevenagel Condensation. *Organic Reactions*. 204–599. doi:10.1002/0471264180.or015.02.

Ju, H.D. Tao, X.T., Wan, Y. Shi, J.H., Yang, J.X., Xin, Q., ... and Jiang, M.H. (2006). Structure and properties of multibranched isophorone-based materials for organic light-emitting diodes. *Chemical Physics Letters*, 432(1–3): 321–325. doi:10.1016/j.cplett.2006.09.066.

Jung, S.O. Yuan, W., Ju, J.U., Zhang, S., Kim, Y.H., Je, J.T. and Kwon, S.K. (2009). A New Orange-Light-Emitting Materials Based on (N-naphthyl)-1,8-naphthalimide for OLED Applications. *Molecular Crystals and Liquid Crystals*, 514(1): 45/[375]–54/[384]. doi:10.1080/15421400903217751.

K. What is organic EL? Retrieved 8 July 2019.

Kao, P.C. and Chiu, C.T. (2015). MoO3 as a p-type dopant for Alq3- based p-i-n homojunction organic light-emitting diodes. *Org. Electron.*, 26: 443–450. https://doi.org/10.1016/j.orgel.2 015.08.018.

King, A.O. and Yasuda, N. (2005). A practical and efficient process for the preparation of Tazarotene. *Org. Process Res. Dev.*, 9(5): 646–650. doi:10.1021/op050080x.

King, A.O. and Yasuda, N. (2004). Palladium-Catalyzed Cross-Coupling Reactions in the Synthesis of Pharmaceuticals Organometallics in Process Chemistry, Top. *Organomet. Chem.*, 6: 205–245. doi:10.1007/b94551.

Kolosov, D., Adamovich, V., Djurovich, P. and Thompson, M.E. (2002). 1,8-Naphthalimides in phosphorescent organic LEDs: The interplay between dopant, exciplex, and host emission. *J. Am. Chem. Soc.*, 124: 9945–9954. https://doi.org/ 10.1021/ja0263588.

Kukhta, A., Kolesnik, E., Grabchev, I. and Sali, S. (2006). Spectral and luminescent properties and electroluminescence of polyvinyl carbazole with 1,8-naphthalimide in the side chain. *J. Fluoresc.*, 16: 375–378. https://doi.org/10.1007/s10 895-005-0064-6.

Kukhta, A., Kolesnik, E., Grabchev, I. and Sali, S. (2006). Spectral and luminescent properties and electroluminescence of polyvinyl carbazole with 1,8-naphthalimide in the side chain. *J. Fluoresc.*, 16: 375–378. https://doi.org/10.1007/s10 895-005-0064-6.

Kulkarni, , Tonzola, Babel, A. and Jenekhe, S.-A. (2004). Electron Transport Materials for Organic Light-Emitting Diodes. *Chemistry of Materials* 16(23): 4556–4573. doi:10.1021/cm049473l.

Kulkarni, A.P., Tonzola, C.J. and Babel, A. (2004). Electron transport materials for organic light-emitting diodes. *Chem. Mater.*, 14: 4556–4573. https://doi.org/10.1021/cm049473l.

Lee, S.C., Ristaino, J.B. and Heitman, J. (2012). Parallels in Intercellular Communication in Oomycete and Fungal Pathogens of Plants and Humans. *PLOS Pathogens*, 8(12): e1003028. doi:10.1371/journal.ppat.1003028. PMC 3521652. PMID 23271965.

Lemke, R. (1974). Knoevenagel-Kondensationen in Dimethylformamid*. *Synthesis*, 1974(05): 359–361. doi:10.1055/s-1974-23322.

Li, C., Li, T., Li, A., Cui, G., Zhang, R. and Liu, S. (2010). Performance enhanced OLEDs using a Li3N doped tris(8-hydroxyquinoline) aluminum(Alq3) thin film as electron-injecting and transporting layer. Symp Photonics Optoelectron SOPO 2010 – *Proc.*: 8–11. https:// doi.org/10.1109/ SOPO.2010.5504074.

Li, G. Yu, X., Liu, D., Liu, X., Li, F. and Cui, H. (2015). Label-free electrochemiluminescence aptasensor for 2,4,6-trinitrotoluene based on bilayer structure of luminescence functionalized graphene hybrids. *Anal. Chem.*, 87: 10976–10981. https://doi.org/10.1021/ acs.analchem.5b02913.

Lipton, M. (2017). Chapter 1. Electronic Structure and Chemical Bonding. Purdue: Chem 36505: Organic Chemistry I (Lipton) (LibreTexts ed.). Purdue University.

Liu, J., Cao, J., Shao, S., Xie, Z., Cheng, Y., Geng, Y. Wang, Y., Jing, X. and Wang, F. (2008). Blue electroluminescent polymers with dopant-host systems and molecular dispersion features: Polyfluorene as the deep blue host and 1,8-naphthalimide derivative units as the light blue dopants. *J. Mater. Chem.*, 18: 1659–1666. https://doi.org/10.1039/b716234k.

Miyaura, N. and Suzuki, A. (1995). Palladium-Catalyzed Cross-Coupling Reactions of Organoboron Compounds. *Chemical Reviews*, 95300(7): 2457–2483. CiteSeerX 10.1.1.735.7660. doi:10.1021/cr00039a007.

Murawski, , Leo, K. and Gather, (2013). Efficiency roll-off in organic light-emitting diodes. *Adv. Mater.*, 25: 6801–6827. https://doi.org/10.1002/adma.201301603.

Palmer, R., Koeber, S., Elder, D.L., Woessner, M., Heni, W., Korn, D. ... and Koos, C. (2014). High-Speed, Low Drive-Voltage Silicon-Organic Hybrid Modulator Based on a Binary-Chromophore Electro-Optic Material. *Journal of Lightwave Technology*, 32(16): 2726–2734. doi:10.1109/jlt.2014.2321498.

Partridge, R. (1983). Electroluminescence from polyvinylcarbazole films: 1. Carbazole cations. *Polymer.*, 24(6): 733–738. doi:10.1016/0032-3861(83)90012-5.

Purvis, L.J., Gu, X., Ghosh, S., Zhang, Z., Cramer, C.J. and Douglas, C.J. (2018). Synthesis and characterization of electron-deficient asymmetrically substituted diarylindenotetracenes. *J. Org. Chem.*, 83: 1828–1841. https://doi.org/10.1021/acs.joc. 7b02756.

"Organic EL –R&D". Semiconductor Energy Laboratory. Retrieved 8 July 2019.

Raj, A., Gupta, M. and Suman, D. (2019). Simulation of multilayer energy-efficient OLEDs for flexible electronics applications. *Procedia Comput. Sci.*, 152: 301–308. https://doi.org/ 10.1016/j.procs.2019.05.013.

Reineke, , Thomschke, M., Lüssem, B. and Leo, K. (2013). White organic light-emitting diodes: Status and perspective. *Reviews of Modern Physics*, 85(3): 1245–1293. doi:10.1103/ revmodphys.85.1245.

Sasabe, H. and Kido, J. (2013). Development of high-performance OLEDs for general lighting. *J. Mater. Chem.*, C 1: 1699–1707. https://doi.org/10.1039/c2tc00584k.

Sasabe, H. Gonmori, E., Chiba, T. et al. (2008). Wide-energy gap electron-transport materials containing 3,5-dipyridylphenyl moieties for an ultra-high efficiency blue organic light-emitting device. *Chem. Mater.*, 20: 5951–5953. https://d oi.org/10.1021/cm801727d.

Schmidbauer, Hohenleutner, A. and König, B. (2013). Chemical degradation in organic light-emitting devices: Mechanisms and implications for the design of new materials. *Adv. Mater.*, 25: 2114–2129. https://doi.org/10.1002/adma. 201205022.

Shalini, S., Balasundaraprabhu, R., Kumar, Prabavathy, N., Senthilarasu, S. and Prasanna, S. (2016). Status and outlook of sensitizers/dyes used in dye-sensitized solar cells (DSSC): A review. *International Journal of Energy Research*, 40(10): 1303–1320. doi:10.1002/er.3538.

Shelke, R.S., Thombre, S.B. and Patrikar, S.R. (2013). Status and perspectives of dyes used in dye-sensitized solar cells. *International Journal of Renewable Energy Research*, 3: 12–19.

Sneha, K. and Dhanya, S. (2022). A systematic review on 1,8-naphthalimide derivatives as emissive materials in organic light-emitting diodes. *J. Mater. Sci.*, 57: 105–139.

Song, Shao, W., Jung, J., Yoon, S.J. and Kim, J. (2020). Organic Light-Emitting Diode Employing Metal-Free Organic Phosphor. *ACS Applied Materials & Interfaces.* doi:10.1021/ acsami.9b20181.

Sonogashira, K. (2002). Development of Pd-Cu catalyzed cross-coupling of terminal acetylenes with sp²-carbon halides. *J. Organomet. Chem.*, 653(1–2): 46–49, doi:10.1016/s0022-328x(02)01158-0.

Su, S.J., Cai, C., Takamatsu, J. and Kido, J. (2012). A host material with a small singlet–triplet exchange energy for phosphorescent organic light-emitting diodes: Guest, host, and exciplex emission. *Organic Electronics*, 13(10): 1937–1947. doi:10.1016/j.orgel.2012.06.009.

Sullivan, P.A. and Dalton, L.R. (2010). Theory-inspired development of organic electro-optic materials." *Acc. Chem. Res.*, 43(1): 10–18.

Sullivan, P.A., Rommel, H., Liao, Y., Olbricht, Akelaitis, Firestone, Kang, J.W., Luo, J., Davies, J.A., Choi, , Eichinger, , Reid, , Chen, A., Jen, , Robinson and Dalton, L.R. (2007). Theory-guided design and synthesis of multichromophore dendrimers: An analysis of the electro-optic effect. *J. Amer. Chem. Soc.*, 129(24): 7523–7530.

Suzuki, A. (1991). Synthetic Studies via the cross-coupling reaction of organoboron derivatives with organic halides. *Pure Appl. Chem.*, 63(3): 419–422. doi:10.1351/pac199163030419.

Tang and VanSlyke, S.A. (1987). Organic electroluminescent diodes. *Applied Physics Letters*, 51(12): 913–915. doi:10.1063/1.98799.

Tang, C.W. and Vanslyke, S.A. (1987). Organic electroluminescent diodes. *Applied Physics Letters*, 51(12): 913. Bibcode:1987ApPhL..51..913T. doi:10.1063/1.98799.

Usta, H., Kim, C., Wang, Z., Lu, S., Huang, H., Facchetti, A. and Marks, T.J. (2012). Anthracenedicarboximide-based semiconductors for air-stable, n-channel organic thin-film transistors: materials design, synthesis, and structural characterization. *J. Mater. Chem.*, 22: 4459–4472. https://doi. org/10.1039/c1jm14713g.

Virtanen, O., Constantinidou, E. and ,E. (2020). Chlorophyll does not reflect green light – how to correct a misconception. *Journal of Biological Education*: 1–8. doi:10.1080/00219266.2020 .1858930.

Wang, J., Zhang, F., Zhang, J. et al. (2013). Key issues and recent progress of highly efficient organic light-emitting diodes. *J. Photochem. Photobiol. C. Photochem. Rev.*, 17: 69–104. https://doi.org/10.1016/j.jphotochemrev.2013.0 8.001.

Wang, J., Zhao, Z., Zhou, S., Zhang, X. and Bo, H. (2018). The antitumor effect and toxicity of a ruthenium (II) complex *in vivo*. *Inorganic Chemistry Communications*, 87: 49,–52. doi: 10.1016/j.inoche.2017.12.003.

Wang, Y.F., Lu, H.Y., Chen, C., Li, M., and Chen, C.F. (2019). 1,8-Naphthalimide-based circularly polarized TADF enantiomers as the emitters for efficient orange-red OLEDs. *Organic Electronics*. doi:10.1016/j.orgel.2019.03.020.

Xiao, L., Chen, Z., Qu, B., Luo, J., Kong, S., Gong, Q. and Kido, J. (2011). Recent progress on materials for electrophosphorescent organic light-emitting devices. *Adv. Mater.*, 23: 926–952. https://doi.org/10.1002/adma.201003128.

Yang, Z. Song, J., Zeng, H. and Wang, M. (2019). Organic composition tailored perovskite solar cells and light-emitting diodes: Perspectives and advances. *Mater. Today Energy*, 14: 100338. https://doi.org/10.1016/j.mtener.2019.06.013.

Yao, L., Zhang, S., Wang R., Li, W., Shen, F., Yang, B. and Ma, Y. (2014). Highly Efficient Near-Infrared Organic Light-Emitting Diode Based on a Butterfly-Shaped Donor-Acceptor Chromophore with Strong Solid-State Fluorescence and a Large Proportion of Radiative Excitons. *Angewandte Chemie International Edition*, 53(8): 2119–2123. doi:10.1002/ anie.201308486.

Yoo, S.J. Yun, H.J. Kang, I. Thangaraju, K. Kwon, S.K. and Kim,Y.H. (2013). A new electron transporting material for effective hole-blocking and improved charge balance in highly efficient phosphorescent organic light-emitting diodes. *J. Mater. Chem.*, C 1: 2217–2223. https://doi.org/10.1039/c3tc00801k.

Zagranyarski, Y. Mutovska, M., Petrova, P., Tomova, R., Ivanov, P. and Stoyanov, S. (2020). Dioxin-annulated 1,8-naphthalimides—Synthesis, spectral and electrochemical properties, and application in OLED. *Dyes and Pigments*: 108585. doi:10.1016/j. dyepig.2020.108585.

Zhang, Z. Xie, J., Wang, H., Shen, B., Zhang, J., Hao, J. ... and Wang, Z. (2016). Synthesis, photophysical and optoelectronic properties of quinazoline-centered dyes and their applications in organic light-emitting diodes. *Dyes and Pigments*, 125: 299–308. doi:10.1016/j.dyepig.2015.10.042.

Zhu, M. and Yang, C. (2013). Blue fluorescent emitters: Design tactics and applications in organic light-emitting diodes. *Chem. Soc. Rev.*, 42: 4963–4976. https://doi.org/10.1039/c3c s35440g.

Zou, S.J., Shen, Y., Xie, F.M., Chen, J.D., Li, Y.Q. and Tang, X.J. (2020). Recent advances in organic light-emitting diodes: Toward smart lighting and displays. *Mater. Chem. Front.*, 4: 788–820. https://doi.org/10.1039/c9qm00716d.

10

Recent Advancements in Smart Textiles and Their Applications

Priyanka Gupta#,* and *Ankur Shukla#,**

1. Introduction

Textiles are essential to our lives since they provide us with clothes, safety, and aesthetics. However, as technology has advanced and consumer needs have changed, there is a growing global need for smart materials and intelligent textiles. In other words, the textile industry has also been dominated by technology (Çelikel, 2020). The term "technical or engineering textiles" refers to goods, components, and fibers employed more for functionality than aesthetic appeal. Textile items that can operate differently than standard cloth and are usually capable of performing a specific function are considered smart textiles. Due to superior performance and functionality, numerous important industries used technical textiles in various ways, such as aerospace, packaging, hazard protection, shipping, sports, agriculture, defense, healthcare, and construction. Figure 10.1 shows the application areas of smart textiles. The development of intelligent and ultra-intelligent clothing has a significant impact with the potential to revolutionize the world of fashion, sport, vehicles, interior design, medicine, and food processing.

Department of Textile and Fibre Engineering, Indian Institute of Technology Delhi, India.
# Equal contribution authors
* Corresponding authors: priyankaparul00@gmail.com; ankurshukla1893@gmail.com

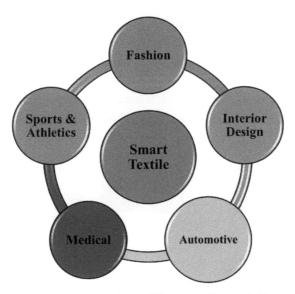

Fig. 10.1 Smart textiles in different application fields.

The intelligent textile may detect environmental changes or stimuli from mechanical, thermal, chemical, electrical, magnetic, or other sources and respond accordingly. Smart textiles must contain three specific elements: controlling units, actuators, and sensors (Bendkowska et al., 2005). Textile products such as fibers and filaments, yarns together with woven, knitted, or nonwoven structures, which can interact with the environment, are smart textiles. The end use and functionality of any textile substrate depend on various factors, shown by a hierarchical relation among all the textile substrates in Fig. 10.2. One can alter any material properties to modify the textile as a responsive-intelligent textile substrate. The creation of completely functional smart clothing is made feasible by developing new fibre and textile materials and miniaturized electronic components. These smart clothes are worn like regular apparel and assist in various scenarios based on the intended applications (Syduzzaman et al., 2015).

Based on their performance, smart textiles are categorized into three types: passive smart textiles, active smart textiles, and ultra-smart textiles (Fig. 10.3). Passive smart textiles can only sense their surroundings since they are merely sensors. The most common types of passive smart textiles include UV-protecting clothes, conductive fibers, plasma-treated clothing, and waterproof materials. Active smart textiles can detect and respond to environmental stimuli; in addition to the sensor function, they feature an actuator function. Active smart textile applications include phase change materials, shape-memory materials, and heat-sensitive dyes. An ultra-smart textile is composed of a unit that functions similarly to a brain,

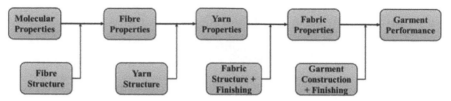

Fig. 10.2 Hierarchical relation of fibre, yarn, fabric, and garment.

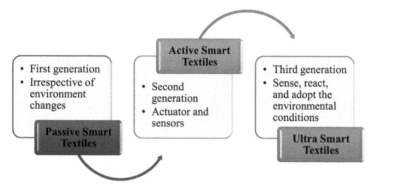

Fig. 10.3 Classification of smart textiles.

with cognition, reasoning, and activation capabilities to environmental conditions/stimuli. Spacesuits, musical jackets, and wearable computers are examples of ultra-smart textiles (Çelikel, 2020; Syduzzaman et al., 2015).

Our surrounding materials are becoming 'intellectualized' as they can interact, communicate, and sense. Various materials can introduce intellectual properties to the textiles, such as metal fibers, optical fibers, conductive ink, shape-memory materials, nano-particles coating, chromic materials, phase change materials, etc. Based on their unique properties, all the such materials provide some specific functions in the textile. Based on the integrated material and their functional properties, end applications of smart textiles can be decided in different sectors. This chapter includes the development of smart or intelligent textiles and involves collaboration with different research disciplines, such as nanotechnology, materials science, design, electronics, and computer engineering, among others.

2. Sportech

Textile materials are used in virtually every sport, from exercising to camping to football. High-performance textile fibers and fabrics are used in uniforms, equipment, and sports facilities. The characteristics and requirements of different types of sportswear are very different from each other (Table 10.1). The specifications for active sportswear fall into two

Table 10.1 Functional requirements for textiles in the sports sector (Kothari and Sanyal, 2003).

Textiles for Sports	Functional Requirements
Tents, Climbing ropes, Parachutes, Muscle support	Protection
Sleeping bag, Artificial turf	Insulation, Compressibility
Rackets, Fishing rods, Bicycle frames	Reinforcement
Skiwear, Windbreakers, Rainwear, Tracksuits	Moisture vapor permeability, Low fluid resistance, Waterproofing, Sunlight absorption, and thermal retention
Swimwear, Leotards, Skating costume, Cycling costume	Stretchability, Opacity, Low fluid resistance for water and air
Snowboard wear, Baseball uniform, Football uniform, Athletic tracksuits, Shirts for tennis, volleyball, rugby, golf (+slacks)	High tenacity, Resistance to abrasion, Stretchability, Sweat absorption, Fast drying, Cooling

categories: (1) functional (lightweight, low fluid resistance, high tenacity, stretchability, thermal regulation, UV protection, vapor permeability, and sweat absorption and release); and (2) aesthetic (softness, surface texture, handle, luster, color, and comfort) (Chaudhari et al., 2004).

The four comfort characteristics: (i) thermo-physiological comfort, (ii) sensorial/tactile comfort, (iii) mobility, and (iv) psychological comfort, should all be met by a well-designed sports textile. This might be done by properly managing air permeability, moisture vapor, and moisture vapor while maintaining the necessary thermal insulation. The expertise of polymer science and textile science has been used in developing an engineered athletic textile. Because athletes sweat a lot while exercising, the microclimate's temperature and humidity rise, encouraging the growth of microorganisms. Consequently, finishing a textile with an antimicrobial agent is also crucial. Using high-density fabric, polymeric coating, and film lamination, waterproof breathable textile has been created to regulate temperature and moisture vapor permeability. Like natural fibre, synthetic fiber may be produced to order by converting its circular cross-section to a Y or C form. Different fibers have been used to create single to multilayer fabrics for the fabric's outside and interior layers (Chowdhury et al., 2014).

2.1 Nanotechnology

The use of nanotechnology in textile materials such as nanofibers, nanocomposite fibers, and nano-finished textiles imparts multiple properties, making them suitable for use in the sports clothing sector. Numerous textile businesses have used nanotechnology to create a variety of sports gear. For extreme cold weather sports like mountaineering and skiing, a Swiss company (Scholler) has created a nano-based technology to develop garments with the ideal balance of comfort, air permeability, wind and water resistance, and self-cleaning properties. The manufactured apparel also benefits from water- and snow-repellent characteristics (Sawhney et al., 2008). Figure 10.4 depicts some of the most important nano-sport apparel and footwear characteristics.

The latest ski jacket from the Nano-tex Company includes a Nano-tex covering that makes the two-layer laminates scuff-resistant as well as windproof, waterproof, and breathable. While preserving the textiles' softness and breathability, it offers a more durable finish than the conventional repellent coating (Chowdhury et al., 2014).

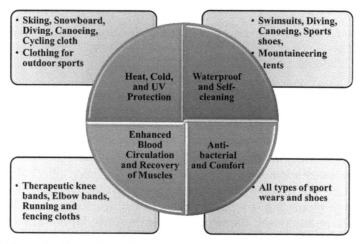

Fig. 10.4 Nanotechnology-enhanced sportswear and footwear properties (Harifi and Montazer, 2017).

2.2 Auxetic materials

Auxetic materials have a negative Poisson's ratio (NPR), meaning they expand laterally in one or more perpendicular directions when extended axially. Advancements in production methods for auxetic textiles and composites lead to the feasibility of protective and form-fitting auxetic garments. Early adoption of new technologies and quick product development, launch, and replacement cycles are traits of the sporting goods industry. Auxetic textiles can potentially improve the mechanical qualities of textiles used in sports, such as clothing, equipment, and the prevention and treatment of injuries. Due to two major sports shoe lines (the Under Armour Architech sports shoe range and the Nike Free RN Flyknit sports shoe) that use auxetic structures, this industry is among the first to see goods based on auxetic materials (Duncan et al., 2018). D3O offers the Trust Helmet Pad System, which includes pads with re-entrant auxetic geometry that provide a better fit to the head and less acceleration during a collision. Moreover, two-fold curvature and multi-axial expansion of auxetic materials may enhance athletic apparel's comfort, fit, and durability and personal protective equipment (PPE). A multi-axial expansion could also be advantageous in filtration applications, while the increased bending stiffness of gradient structures could lower the mass of skis, snowboards, and tennis rackets.

2.3 Smart materials

Globally, an increasing number of smart apparels including socks, gloves, jeans, shirts, and bras, are being used to track sporting activity. This is a fantastic opportunity to observe athletes as they practice and compete, maybe leading to more efficient training and improved safety standards. In this context, optical fibers can be embedded in textiles to transfer data signals and light or detect information about stress and strain inside the fabric. Sportswear has lately been produced using smart coatings, including phase change materials (PCM), moisture-responsive shape memory polymer, and conductive coatings, which can perceive and react to environmental stimuli. While retaining the fabrics' breathable and tactile qualities, smart coating with nanoparticles can provide attributes like anti-bacterial, water-repellence, UV protection, and self-cleaning. Smart membranes are coatings that provide protection. Fabrics are protected from water by fluoroethylene membrane (PFTE) and Gore-Tex, whereas waterproof breathable fabrics employ polyurethane membranes, such as Porelle Dry and Dorminaz NX. Cycling, sprinting, and swimming are just a few activities where smart coatings might help reduce drag. The ability of the garment to absorb and release heat is provided by the PCM microcapsule, which can be inserted or infused into textiles and fibres. The postural characteristics, i.e., movement and posture, can be provided by strain sensors built into a piece of clothing. Extreme sports have only recently begun to embrace smart clothes. Pressure sensors integrated into apparel can be adapted to measure muscle activity, while strain sensors can be used to estimate muscle fatigue. Lastly, smart clothing is the best option for this kind of research. Smart clothing offers an environmentally friendly method for tracking physiological information about players' performances and serves as a safety measure in competitive sports by alerting the team if an issue emerges (Scataglini et al., 2020).

3. Protech

Traditional clothing is lighter, less physically demanding, more breathable, and more comfortable but offers less protection from pathogens and outside environmental risks. Textiles with additional functionality can be created due to technological innovation, including self-cleaning textiles, flame retardant textiles, anti-bacterial textiles, UV protecting textiles, super hydrophobic textiles, electro-spun nanofiber textiles, thermo-regulating textiles, fire retardant textiles, medicinal textiles, and industrial textiles, etc. (Abdelhameed, el-deib, et al., 2018; Abdelhameed, Rehan, et al., 2018; Emam, Abdelhamid, et al., 2018; Emam, Darwesh et al., 2018; Shakeri et al., 2019; Shukla et al., 2017). Protective clothing serves as a barrier between the wearer and potential hazards, but it shouldn't abandon the wearer's comfort (Bhattacharjee et al., 2019). There are numerous varieties of protective clothing described in the literature, including biological protective clothing, chemical protective clothing, fire retardant clothing, UV protective clothing, cold weather protective clothing, and microbiological protective clothing (Bach et al., 2019; Bhuiyan et al., 2019; Dhineshbabu and Bose, 2019; Ghaffari et al., 2019; Su et al., 2019; Yin et al., 2019).

In the context of fire protective clothing, due to the utilized flame retardant's aqueous solubility, conventional flame-retardant treatments frequently exhibit a lack of resistance to soaking in water or to home washing processes. Microencapsulation may be beneficial to shield the active flame retardants from wet treatment effects, extending the duration of the effect and allowing treated fabrics to pass flammability tests that require a pre-soaking in water before testing. Additionally, the adherence of the microencapsulated flame retardant to the textile might be tailored by selecting the right polymer coating to retain the resilience to leaching tests or several wash cycles using bleach-activated detergents. It may be conceivable to offer a polymer coating for microencapsulation that serves as a protective covering for the intumescent flame retardant mixture, as well as providing enough adhesion and contributing to the carbonic component of the finish by carefully choosing the polymer coating.

Protective and military clothing serves smart textiles' most rapidly expanding and key growth segments. In the military and in its applications, textiles are commonly and extensively used for, such as uniforms, protective equipment, socks, gloves, sheets, sweaters, and sandbags, and other things (Steffens et al., 2019). Textiles for military apparel encounter numerous problems that must be overcome for better performance with the understanding of four forms of Weapons of Mass Destruction (WMD) named as biological, chemical, radiological, and nuclear (Cirincione et al., 2005). Bulletproof vests are military artifacts that shield soldiers from missiles and other military debris. Table 10.2 indicates the percentage of accidents caused by various factors, most of which occur on the battlefield. Ballistic armor is an example of protective textile, and its bullet-proof protection's effectiveness depends on the kind of projectile kind and its velocity. High-performance fibers such as Kevlar®, Spectra®, Twaron®, Zylon®, and Dyneema® are commonly used in ballistic armor. In contrast, selecting a suitable textile structure, yarn types, coatings, and composites is critical to achieving greater penetrating resistance (Scott, 2005).

Table 10.2 Reasons for ballistic casualties in a general battle (Scott, 2005).

Cause of Casualty	Percentage (%)
Fragments	59
Bullets	19
Others	22

In personal protection equipment, smart textiles are used for integrated position location, physiological monitoring, cooling and heating systems, and communication (Dolez and Mlynarek, 2016). UV radiation protection of fabric is determined by a variety of parameters such as fiber and yarn type, weave, color, fabric construction factor, finishing type, presence of UV absorber in textile materials, etc. ZnO, TiO_2, and Carbon nanotube are some of the most usable nanoparticles with the textile substrate to demonstrate excellent UV blocking property (Mondal, 2022).

Apart from the UV and fire protective clothing, antimicrobial protection is one of the most important requirements as textiles provide a good medium for the growth of microorganisms with essential conditions like nutrients (sweat and soil), moisture, oxygen, suitable temperature, and large surface area. Therefore, there is a need to suppress microbiological growth on fabrics used in both industrial and garment applications. As a result, new quality standards for antimicrobial agents include using an environmentally friendly production process and maintaining the agent's inherent functionality while assuring long product service life and the retention of desired functional and wear attributes (Purwar and Joshi, 2004).

4. Meditech

Medical textiles are a subset of technical textiles that have been specifically manufactured to satisfy the demands of medical and surgical applications that call for a combination of strength, flexibility, and, occasionally, moisture and air permeability. Figure 10.5 depicts the required characteristics of textiles for medical applications. Medical textiles are an emerging sector of the technical textiles industry, combining textile technology with medical science. Its growth is filled due to the constraint improvements in healthcare and innovations in the textile field. Some textile-based medical and healthcare products are shown in Fig. 10.6.

Bandages are designed to perform various specific functions depending upon the final medical requirements. Light support bandages are ideal for managing mild sprains, strains, and soft tissue injuries. Compression bandages are mainly used to treat venous leg ulcers and varicose veins. The basic function of these types of bandages is compression, retention, and support. The regulation of blood flow and prevention swelling is closely interlinked with this property. Compression hosiery can be used as

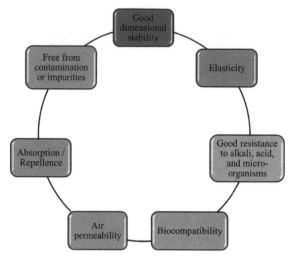

Fig. 10.5 Requirement of textile materials for medical applications.

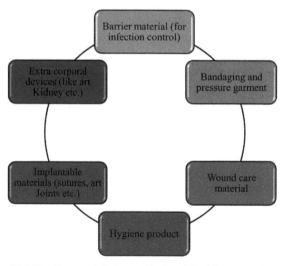

Fig. 10.6 Textile materials for medical and healthcare products.

an alternative to compression bandages to treat active ulcers. Orthopaedic bandages are used in plaster to repair broken limbs and fitting of prosthetic limbs. Another important product is pressure garments, which play a vital role in properly healing wounds and reducing the scarring effects. These garments prevent and control hypertrophic scar formation by applying counter pressure to the wounded area. Table 10.3 summarizes use of textile substrate in various medical products.

Table 10.3 Use of textile substrate in various medical products (Rajendran and Anand, 2002).

Applications	Fiber Used	Fabric Type
Gauges	Alginate, Chitosan, Cotton, Glass Fiber	Nonwoven, Woven, Knitted
High support bandage	Cotton, Elastomeric Fiber Yarn, Viscose	Nonwoven, Woven, Knitted
Compression bandage	Cotton, Elastomeric Fiber Yarn, Viscose	Nonwoven, Woven, Knitted
Orthopaedic bandage	Cotton, PP, PET, Lyocell, Polyurethane Fiber	Woven, Knitted
Wound contact layer	Alginate, Chitosan, Cotton	Nonwoven, Knitted, Woven
Base layer	Cotton, Plastic Film	Nonwoven, Woven
Absorbent pad	Cotton, Viscose, Lyocell	Nonwoven
Protective eye pads	Cotton, Viscose	Woven, Nonwoven
Wadding	Cotton, Viscose, Linter, Wood Pulp	Nonwoven

Sutures are the most popular implanted materials used in various surgical applications such as wound closure, ligation of blood vessels during wound healing, and tissue apposition (Rohani Shirvan and Nouri, 2020). The key qualities influencing the suture design are knotting strength, absorbability, softness, stiffness, biocompatibility, tensile strength, ease of sterilization, and easy handling. Some surgical sutures may also be regarded as intelligent fibers, as shown in Fig. 10.7. These sutures are significantly absorbed into the body system until the wound has healed sufficiently. For this purpose, shape memory materials are used to make smart sutures with required biodegradable and biocompatible properties.

Numerous types of medical textiles have the potential to microencapsulate antimicrobial compounds inside their fibers, yarns, and fabrics to prevent bacterial and fungal infections. The possibility of a significant reduction in post-operative infections following surgery is made possible by the controlled release of antibiotics from textiles in contact with the skin (Kennedy and Mehmood, 2007). Smart textiles can deliver heat to the body to cure ailments such as arthritis, muscle discomfort and spasm, joint stiffness, and Raynaud syndrome. Cardiovascular disorders are the major cause of death worldwide, and maximum research on wearable and portable e-health technologies has decided to focus on them (Coyle and Diamond, 2016). The LifeShirt®, produced by Vivometrics, was one of the

Fig. 10.7 Schematic for the tightening of sutures on the application of the stimulus.

first commercially accessible smart clothes (Heilman and Porges, 2007). The shirt has respiratory function sensors woven into it around the patient's chest and belly, as well as a three-lead, single-channel ECG recorder and a three-axis accelerometer to monitor the subject's posture and activity level.

In the crucial initial few days following surgery, surgical sutures with microencapsulated antibiotics may provide a controlled release of antibiotics near the site of the surgical wound, accelerating patient healing and reducing post-operative infection (Holme, 2007). The invention of medication microencapsulation allows a controlled, delayed release of the active component to be absorbed through the skin. This could be a potential field for developing nano capsules or tiny microcapsules for use in specific applications, such as medical dressings (Nelson, 2001). Overall, the discipline of biomedical engineering has several potential uses for the relatively new research field of smart fabrics and interactive textiles.

5. Hometech

We need several devices in our homes for day-to-day life. Often it becomes difficult to track all the electronic equipment, especially when you are equipped with some important job. Every day, an increasing demand for electronics has escalated the market competition for new and innovative devices, which has led to the development of smart electronics. Now the devices could be connected to each other or internet of things (IoT) or cloud storage. Thus, it is quite convenient to access the devices via a handheld device like a mobile phone (Bremner, 2015). With increasing market demand for smart devices, textiles offer many more advantages than any other devices in terms of mobility. If any devices could relate to smart textiles having embedded sensors and electronics for accessing the home electronic devices, it could lead to a revolutionary change in our lives. Electronics integration into textiles often requires flexible circuits, which are achieved using printed circuits over flexible substrates. Polyimide is one of the preferred flexible substrates as it offers high flexibility combined with robustness and lightweight characteristics. Polyimide shows highly stable chemical and electrical behavior across a wide temperature range. However, polyimide film itself should not be used for building large circuit sizes since they easily get distorted and, therefore, the printed circuit may also deform (Arevalo et al., 2015; Francioso, De Pascali, Bartali et al., 2013; Francioso, De Pascali, Siciliano, et al., 2013; Francioso et al., 2011; Khaleel, 2014; Li et al., 2012; Moon et al., 2010; Plovie et al., 2015; Shibata et al., 2014). However, small-size polyimide circuits could be easily used and may be integrated into the yarn core. The PlaceIt project (van den Brand et al., 2014) focused on developing an organic field effect transistor. On top of a metallic fiber lies a polyimide layer, which serves as the structure's gate.

This multilayer method, which is also applied to the drain and source of the transistor, is supposed to preserve the pliability and softness of the yarn. The fabrication of intelligent fabrics utilizing traditional weaving and knitting techniques is an intriguing trend. This renders the intelligent cloth elastic and flexible, enabling this novel ergonomic electronic interface to conform to the body and enhance mobility. A textile-based antenna is required to provide sensing and localization in order for smart textile data to contribute to a wireless sensor network (WSN) (Rogier, 2015). The antenna topology chosen for a smart textile depends on the surface area, form, and function of the structure it will cover. More antennae can be used to compensate for body contours or parts where the fabric is to be worn. Such functionality could be achieved using RFID (Radio Frequency Identification) tags (Koski et al., 2015).

Textile-based pressure sensors are often used for sensing the different pressure areas over a surface, such as car seats or chairs, for posture monitoring. Also known as eCushion, it is a smart cushion for monitoring and analyzing body posture. Different algorithms and calibration studies have been performed, which could increase body posture recognition to up to 85%. Although, the challenges remain in eliminating wire crosstalk and the rotation behavior of the yarns. Another issue with textiles is the 'scaling'. The variation of conductivity in textiles is higher compared to any metal. Therefore, the size of the object affects the change in resistance while sensing, which has to be taken into account for accurate sensing (Xu et al., 2013). For real-time monitoring of body posture, microcontrollers are generally used with sensing fabrics. Long-term use of pressure-sensing fabric based on silver or copper-coated conductive yarns would result in permanent deformation of the yarns. To prevent this, shape memory alloy is twisted with regular yarns and then used for weaving or knitting. Ni-Ti alloy wire was covered with polyurethane to prevent short-circuiting of the yarns. Shape memory yarns could retain the shape even after 100 cycles, as shown in Fig. 10.8. The presence of polyurethane coating imparts high wash durability to the fabric. Gupta et al. extensively studied the effect of shape memory polymer in knitted textile fabrics. It was observed that, unlike shape memory metal, the shape memory polymers could expand or contract upon applying the thermal stimulus, thus opening a new world for skin-friendly shape memory applications (Gupta et al., 2022; Gupta, Garg, et al., 2020; Gupta, Narayana, et al., 2020). Integrating a shape memory fiber in a textile is far more beneficial since it offers higher flexibility. Therefore, it can easily conform to different loops in knitting or weaving in a woven fabric.

Not just for furnishing but smart textiles could be used even for toys for psychotherapy purposes. In a study, three dolls embedded with a haptic feedback device were developed to communicate non-verbally with the

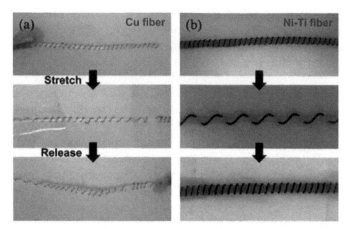

Fig. 10.8 Comparison of (a) copper-coated yarn and (b) polyurethane-coated shape memory Ni-Ti alloy wire after 100 cycles (Xu et al., 2013).

children. It is easy to shape the toys as per the requirement with embedded electronics compared to integration with wearable clothing (Righetto et al., 2012).

One of the most exciting applications is the interior design for hometech. Interior design involves using smart textiles in furnishings and surfaces to connect them to the cloud. These intelligent textile surfaces can be engaged passively or actively, such as by sitting or lying down or touching, tapping, or gesturing. Active involvement will only occur if the user desires it. Using intelligent textiles to exchange or collect data is likely more accurate (Xia et al., 2015). The linked textile would engage in passive interaction if its main purpose was to capture data during interactions, which would only include the person's ID, the time of the encounter, and the location (Kim et al., 2015). With the help of shape memory materials, the fabric has the ability to adapt the changes with respect to external temperature, and the changes depend on the fabric structure and specific programming method (Stylios and Wan, 2007). Figure 10.9 shows the possibility of intelligent curtains that can be produced using shape memory polymer with textiles.

6. Clothtech

Smart clothing is a new garment feature that can provide interactive reactions by sensing signals, processing information, and actuating responses (Tao, 2001). For this sort of clothing, terms such as interactive clothing, intelligent clothing, smart garments, and smart apparel are used interchangeably. It is possible to comprehend the idea of smart clothing within the context of functional clothes. Wearing functional clothes satisfies certain functional requirements, such as providing protection from harsh situations or

Fig. 10.9 Intelligent curtains for interior decoration.

achieving high-tech performance. Not only functional requirements but the human body comfort in terms of thermal comfort, free movement of the body, tactile comfort, etc., are desirable properties that smart clothing must fulfil (Fig. 10.10).

The integration of technology defines how smart the fabric is since higher integration with technology imparts higher functionality to the fabric. Based on technology integration with clothing, the smart textile can be categorized as shown in Fig. 10.11.

The most widely used shape memory materials can be integrated into the textile substrate to make it functional and smart. Various researchers focused on fabric designs to enhance and control the functional and aesthetic aspects of the fabrics integrated with shape memory materials (Berzowska and Coelho, 2005; Stylios and Wan, 2007). Huang et al. looked at how to form memory supercapacitors and regular fabrics that may be combined to create intelligent textiles that could store energy and have a shape memory effect. Additionally, they suggested that these shape-memory fabrics could be used to make wearable and portable electronic devices provided they had pre-designed shapes that allow for other functionalities beyond energy storage (Huang et al., 2015; 2016). Figure 10.12 shows textiles' new sculptural fashion artifacts integrated with memory materials, such as two electronically animated dresses and Oricalco shirts with programmed sleeves.

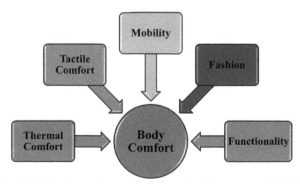

Fig. 10.10 Body comfort properties required from a smart clothing.

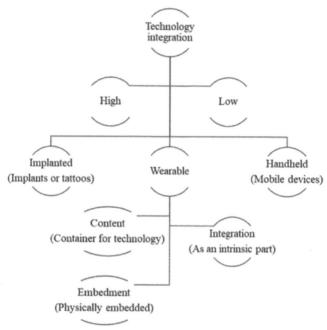

Fig. 10.11 Degree of integration of technology with clothing (Reproduced from Suh et al., 2010).

Fig. 10.12 New fashion approach using shape memory materials: (a) rising skirt and closing flower (b) Orcalco shirt.

7. Conductive Textiles

The convergence of textiles and electronics (e-textiles) can be relevant for developing smart materials capable of accomplishing a wide spectrum of functions found in rigid and non-flexible electronic products nowadays. Creating lightweight and flexible components and fibrous structures with a high electrical conductivity that can survive the pressures involved with wearing and caring for the textile is a significant barrier to the success of wearable e-textile technology. Utilization of conventional metallic conductors frequently leads to rigid and inflexible fabrics that cannot retain their functionality when subjected to harsh environmental conditions or after performing fabric care procedures such as washing. Flexible, deformable, elastic, and long-lasting conductive threads are crucial for constructing smart fabrics that record and transmit data and enable computing while supporting the human body's drape and mobility. Using the adaptability and versatility of textile structures, as well as advancements in the field of particle and fibrous materials, researchers have pursued a variety of techniques to address this issue over the past few decades. To create textile electrodes, flexible conductive yarns, entirely metal yarns, or natural/synthetic yarns combined with conductive fibres have been stitched into clothing (Catrysse et al., 2003).

Conductive textiles can be prepared either by polymer coating over the textiles or by metal integration with textiles. Often electroless plating of metals is done over the textile fabrics (Gan et al., 2007). Though the process results in a highly conducting metal coating over the fabric, the time required for conductive fabric production is higher than the fabrication of conductive polymer-coated fabrics. Shukla et al. (2022) proposed a fast method for producing a highly conducting polypyrrole-coated polyester fabric where a conducting fabric can be obtained within 3 minutes of the polymerization process. The conductivity of polymer-based textile fabrics is highly dependent on the dopants or electrolytes used during the polymerization process, especially during electrochemical polymerization (Shukla et al., 2022).

Conductive fabrics are required for many applications, such as antennas and flexible sensors. Different applications require a different range of conductivities (Krifa, 2021). Some textile-based sensors follow the electrochemical principle for sensing the analyte. However, the conductivity also depends on relative humidity, enabling their use as humidity sensors (Devaux et al., 2011; Schreuder-Gibson et al., 2003). A few sensors that measure the change in electrical conductivity when exposed to volatile organic compounds (VOCs) and certain biological compounds, such as urea or uric acid, are known as conductometric sensors (Qi et al., 2014; Sharma et al., 2002). Not just the chemical stimulus, textile-based sensors also respond to physical stimuli such as pressure, motion, and mechanical strain (Melnykowycz et al., 2014; Meyer et al., 2006; Wang et al., 2016). The basic mechanism of sensing involves detecting a change in resistance or capacitance. Thus, the sensors are also categorized as resistive or capacitive (Nur et al., 2018; Wang et al., 2017). Recent developments have also been exploring interdigital electrode-type textile-based sensors. These coplanar capacitive sensors work on the same principle as parallel plate capacitive sensors. Still, they are more practical from the application point of view since they do not require another surface where a dielectric medium is to be sandwiched (Teodorescu and Teodorescu, 2020).

8. Summary

A lot of innovation has been done and is still going on to develop smart and responsive textile materials by integrating technology into textiles. Such textiles are often termed technical textiles with numerous applications in the fashion, medical, sports, and automotive industries. In sports, the major requirements from textiles are resistance to UV, moisture management, and aesthetics. For such properties, several finishes may render the clothing properties such as enhanced antimicrobial, thermal insulation, moisture management, and stretchability so that body movement is not hindered. These properties are further improved when nanotechnology is involved in sports garments. Such highly durable finishes impart multiple properties, such as scuff resistance, higher skin breathability, and improved moisture management. It also preserves the softness and flexibility of the clothing. Often, they are attached with technologies such as therapeutic knee bands and compression stockings for improved blood flow and to prevent any injury. But mechanical properties of the textile material deteriorate over time, leaving them unusable. Such situations can be avoided by improving the mechanical properties of the textiles by using either auxetic materials or shape memory materials such as shape memory alloy (Ni-Ti) or shape memory polymers, which easily conform to the fabric construction, whether it is knitted or woven. There are various applications for smart fabrics, but the base for such applications requires the use of electrically conductive fabrics. Electrically conducting fabrics may be metal-coated, polymer-coated, or carbon-based fabrics that conduct electricity. They are often used for providing conducting media or as a sensor depending on the type of application. Thus, multiple developments and applications require smart textiles since they are closest to the human body. This requires continuous research and development to utilize their untapped potential.

References

Abdelhameed, R.M., el-deib, H.R., El-Dars, F.M.S.E., Ahmed, H.B. and Emam, H.E. (2018). Applicable Strategy for Removing Liquid Fuel Nitrogenated Contaminants Using MIL-53-NH2@Natural Fabric Composites. *Industrial & Engineering Chemistry Research*, 57(44): 15054–15065. https://doi.org/10.1021/acs.iecr.8b03936.

Abdelhameed, R.M., Rehan,M. and Emam, H.E. (2018). Figuration of Zr-based MOF@cotton fabric composite for potential kidney application. *Carbohydrate Polymers*, 195: 460–467. https://doi.org/10.1016/j.carbpol.2018.04.122.

Arevalo, A., Byas, E., Conchouso, D., Castro, D., Ilyas, S. and Foulds, I.G. (2015). A versatile multi-user polyimide surface micromachinning process for MEMS applications. *10th IEEE International Conference on Nano/Micro Engineered and Molecular Systems*, pp. 561–565. https://doi.org/10.1109/NEMS.2015.7147492.

Bach, A. J.E., Maley, M.J., Minett, G.M., Zietek, S.A., Stewart, K.L. and Stewart, I.B. (2019). An Evaluation of Personal Cooling Systems for Reducing Thermal Strain Whilst Working in Chemical/Biological Protective Clothing. *Frontiers in Physiology*, 10. https://www.frontiersin.org/articles/10.3389/fphys.2019.00424.

Bendkowska, W., Tysiak, J., Grabowski, L. and Blejzyk, A. (2005). Determining temperature regulating factor for apparel fabrics containing phase change material. *International Journal of Clothing Science and Technology*, 17(3/4): 209–214. https://doi.org/10.1108/09556220510590902.

Bhattacharjee, S., Joshi, R., Chughtai, A.A. and Macintyre, C.R. (2019). Graphene Modified Multifunctional Personal Protective Clothing. *Advanced Materials Interfaces*, 6(21): 1900622. https://doi.org/10.1002/admi.201900622.

Bhuiyan, M.A.R., Wang, L., Shaid, A., Shanks, R.A. and Ding, J. (2019). Advances and applications of chemical protective clothing system. *Journal of Industrial Textiles*, 49(1): 97–138. https://doi.org/10.1177/1528083718779426.

Bremner, D. (2015). The IoT Tree of Life [Research Reports or Papers]. *Knowledge Transfer Network, Sensors & Instrumentation Leadership Committee*. http://www.gambica.org.uk/IoTWhitePaper.

Catrysse, M., Puers, R., Hertleer, C., Van Langenhove, L., van Egmond, H. and Matthys, D. (2003). Fabric sensors for the measurement of physiological parameters. TRANSDUCERS '03. *12th International Conference on Solid-State Sensors, Actuators and Microsystems. Digest of Technical Papers* (Cat. No.03TH8664), 2: 1758–1761 Vol. 2. https://doi.org/10.1109/SENSOR.2003.1217126.

Çelikel, D.C. (2020). Smart E-Textile Materials. *In*: Advanced Functional Materials. IntechOpen. https://doi.org/10.5772/intechopen.92439.

Chaudhari, S.S., Chitnis, R.S. and Ramkrishnan, D.R. (2004). Waterproof Breathable Active Sports Wear Fabrics. 17.

Chowdhury, P., Samanta, K.K. and Basak, S. (2014). Recent Development in Textile for Sportswear Application. *International Journal of Engineering Research*, 3(5): 6.

Cirincione, J., Wolfsthal, J.B. and Rajkumar, M. (2005). Deadly Arsenals: Nuclear, Biological, and Chemical Threats. Carnegie Endowment.

Coyle, S. and Diamond, D. (2016). Medical applications of smart textiles. *In: Advances in Smart Medical Textiles, Elsevier*, pp. 215–237. https://doi.org/10.1016/B978-1-78242-379-9.00010-4.

Devaux, E., Aubry, C., Campagne, C. and Rochery, M. (2011). PLA/Carbon Nanotubes Multifilament Yarns for Relative Humidity Textile Sensor. *Journal of Engineered Fibers and Fabrics*, 6(3): 155892501100600300. https://doi.org/10.1177/155892501100600302.

Dhineshbabu, N.R. and Bose, S. (2019). UV resistant and fire retardant properties in fabrics coated with polymer based nanocomposites derived from sustainable and natural resources for protective clothing application. *Composites Part B: Engineering*, 172: 555–563. https://doi.org/10.1016/j.compositesb.2019.05.013.

Dolez, P.I. and Mlynarek, J. (2016). 22 - Smart materials for personal protective equipment: Tendencies and recent developments. *In*: V. Koncar (Ed.). *Smart Textiles and Their*

Applications, Cambridge, UK: Woodhead Publishing, pp. 497–517. https://doi.org/10.1016/B978-0-08-100574-3.00022-9.

Duncan, O., Shepherd, T., Moroney, C., Foster, L., Venkatraman, P.D., Winwood, K., Allen, T. and Alderson, A. (2018). Review of Auxetic Materials for Sports Applications: Expanding Options in Comfort and Protection. *Applied Sciences*, 8(6): Article 6. https://doi.org/10.3390/app8060941.

Emam, H.E., Abdelhamid, H.N. and Abdelhameed, R.M. (2018). Self-cleaned photoluminescent viscose fabric incorporated lanthanide-organic framework (Ln-MOF). Dyes and Pigments, 159: 491–498. https://doi.org/10.1016/j.dyepig.2018.07.026.

Emam, H.E., Darwesh, O.M. and Abdelhameed, R.M. (2018). In-growth metal organic framework/synthetic hybrids as antimicrobial fabrics and its toxicity. Colloids and Surfaces B: Biointerfaces, 165:, 219–228. https://doi.org/10.1016/j.colsurfb.2018.02.028.

Francioso, L., De Pascali, C., Bartali, R., Morganti, E., Lorenzelli, L., Siciliano, P. and Laidani, N. (2013). PDMS/Kapton Interface Plasma Treatment Effects on the Polymeric Package for a Wearable Thermoelectric Generator. *ACS Applied Materials & Interfaces*, 5(14): 6586–6590. https://doi.org/10.1021/am401222p.

Francioso, L., De Pascali, C., Farella, I., Martucci, C., Cretì, P., Siciliano, P. and Perrone, A. (2011). Flexible thermoelectric generator for ambient assisted living wearable biometric sensors. *Journal of Power Sources*, 196(6): 3239–3243. https://doi.org/10.1016/j.jpowsour.2010.11.081.

Francioso, L., De Pascali, C., Siciliano, P., De Risi, A., Bartali, R., Morganti, E. and Lorenzelli, L. (2013). Structural reliability and thermal insulation performance of flexible thermoelectric generator for wearable sensors. 2013 *IEEE SENSORS*: 1–4. https://doi.org/10.1109/ICSENS.2013.6688438.

Gan, X., Wu, Y., Liu, L., Shen, B. and Hu, W. (2007). Electroless copper plating on PET fabrics using hypophosphite as reducing agent. *Surface and Coatings Technology*, 201(16): 7018–7023. https://doi.org/10.1016/j.surfcoat.2007.01.006.

Ghaffari, S., Yousefzadeh, M. and Mousazadegan, F. (2019). Investigation of thermal comfort in nanofibrous three-layer fabric for cold weather protective clothing. *Polymer Engineering & Science*, 59(10): 2032–2040. https://doi.org/10.1002/pen.25203.

Gupta, P., Garg, H., Mohanty, J. and Kumar, B. (2020). Excellent memory performance of poly (1,6-hexanediol adipate) based shape memory polyurethane filament over a range of thermo-mechanical parameters. *Journal of Polymer Research*, 27(12): 382. https://doi.org/10.1007/s10965-020-02345-5.

Gupta, P., Mohanty, J., Garg, H. and Kumar, B. (2022). Memory behaviour of polyester knitted fabric integrated with temperature-responsive shape memory polymer filament. *Journal of Industrial Textiles*, 51(4_suppl.): 5952S–5972S. https://doi.org/10.1177/15280837211073752.

Gupta, P., Narayana, H., Shankr, P., Kumar, B. and Pan, N. (2020). Chapter 8—Shape memory polymers for design of smart stocking. In: A. Gefen (Ed.), Innovations and Emerging Technologies in Wound Care, Cambridge, Mass., *USA: Academic Press*, pp. 141–154. https://doi.org/10.1016/B978-0-12-815028-3.00008-0.

Harifi, T. and Montazer, M. (2017). Application of nanotechnology in sports clothing and flooring for enhanced sport activities, performance, efficiency and comfort: A review. *Journal of Industrial Textiles*, 46(5): 1147–1169. https://doi.org/10.1177/1528083715601512.

Heilman, K.J. and Porges, S.W. (2007). Accuracy of the LifeShirt® (Vivometrics) in the detection of cardiac rhythms. *Biological Psychology*, 75(3): 300–305. https://doi.org/10.1016/j.biopsycho.2007.04.001.

Holme, I. (2007). Innovative technologies for high performance textiles. *Coloration Technology*, 123(2): 59–73. https://doi.org/10.1111/j.1478-4408.2007.00064.x.

Huang, Y., Hu, H., Huang, Y., Zhu, M., Meng, W., Liu, C., Pei, Z., Hao, C., Wang, Z. and Zhi, C. (2015). From Industrially Weavable and Knittable Highly Conductive Yarns to Large Wearable Energy Storage Textiles. *ACS Nano*, 9(5): 4766–4775. https://doi.org/10.1021/acsnano.5b00860.

Huang, Y., Zhu, M., Pei, Z., Xue, Q., Huang, Y. and Zhi, C. (2016). A shape memory supercapacitor and its application in smart energy storage textiles. *Journal of Materials Chemistry A*, 4(4): 1290–1297. https://doi.org/10.1039/C5TA09473A,

Kennedy, J.F. and Mehmood, A.G. (2007). Medical Textiles and Biomaterials for Healthcare. *In*: S.C. Anand, J.F. Kennedy, M. Miraftab and S. Rajendran (Eds.). *Woodhead, Cambridge.* ISBN: 1-85573-683-7.

Khaleel, H.R. (2014). Design and Fabrication of Compact Inkjet Printed Antennas for Integration Within Flexible and Wearable Electronics. *IEEE Transactions on Components, Packaging and Manufacturing Technology*, 4(10): 1722–1728. https://doi.org/10.1109/TCPMT.2014.2352254.

Kim, H., Kim, I. and Kim, J. (2015). Designing the Smart Foot Mat and Its Applications: As a User Identification Sensor for Smart Home Scenarios. 1–5. https://doi.org/10.14257/astl.2015.87.01.

Koski, K., Moradi, E., Hasani, M., Virkki, J., Björninen, T., Ukkonen, L. and Rahmat-Samii, Y. (2015). Electro-textiles: The enabling technology for wearable antennas in wireless body-centric systems. *In: 2015 IEEE International Symposium on Antennas and Propagation & USNC/URSI National Radio Science Meeting*, pp. 1203–1204. https://doi.org/10.1109/APS.2015.7304990.

Kothari, V.K. and Sanyal, P. (2003). Fibres and fabrics for active sportswear. *Asian Textile Journal-Bombay*, 12(3): 55–61.

Krifa, M. (2021). *Electrically Conductive Textile Materials:Application in Flexible Sensors and Antennas. Textiles*, 1(2): Article 2. https://doi.org/10.3390/textiles1020012.

Li, Y., Torah, R., Beeby, S. and Tudor, J. (2012). An all-inkjet printed flexible capacitor on a textile using a new poly(4-vinylphenol) dielectric ink for wearable applications. 2012 IEEE SENSORS, pp.1–4. https://doi.org/10.1109/ICSENS.2012.6411117.

Melnykowycz, M., Koll, B., Scharf, D. and Clemens, F. (2014). Comparison of Piezoresistive Monofilament Polymer Sensors. *Sensors*, 14(1): Article 1. https://doi.org/10.3390/s140101278.

Meyer, J., Lukowicz, P. and Troster, G. (2006). Textile Pressure Sensor for Muscle Activity and Motion Detection. 2006 10th IEEE International Symposium on Wearable Computers, pp. 69–72. https://doi.org/10.1109/ISWC.2006.286346.

–Mondal, S. (2022). Nanomaterials for UV protective textiles. Journal of Industrial Textiles, 51(4_suppl.), 5592S-5621S. https://doi.org/10.1177/1528083721988949.

Moon, J.-H., Baek, D.H., Choi, Y.Y., Lee, K.H., Kim, H.C. and Lee, S.-H. (2010). Wearable polyimide: PDMS electrodes for intrabody communication. *Journal of Micromechanics and Microengineering*, 20(2): 025032. https://doi.org/10.1088/0960-1317/20/2/025032.

Nelson, G. (2001). Microencapsulation in textile finishing. *Review of Progress in Coloration and Related Topics*, 31(1): 57–64. https://doi.org/10.1111/j.1478-4408.2001.tb00138.x.

Nur, R., Matsuhisa, N., Jiang, Z., Nayeem, M.O.G., Yokota, T. and Someya, T. (2018). A Highly Sensitive Capacitive-type Strain Sensor Using Wrinkled Ultrathin Gold Films. *Nano Letters*, 18(9): 5610–5617. https://doi.org/10.1021/acs.nanolett.8b02088.

Plovie, B., Dunphy, S., Dhaenens, K., Van Put, S., Vandecasteele, B., Bossuyt, F. and Vanfleteren, J. (2015). 2.5D Smart Objects Using Thermoplastic Stretchable Interconnects. International Symposium on Microelectronics, 2015(1): 000868–000873. https://doi.org/10.4071/isom-2015-THP51.

Purwar, R. and Joshi, M. (2004). Recent Developments in Antimicrobial Finishing of Textiles — A Review. *AATCC Review*, 4(3): 22–26. https://search.ebscohost.com/login.aspx?direct=true&db=teh&AN=12825458&site=ehost-live.

Qi, J., Xu, X., Liu, X. and Lau, K.T. (2014). *Fabrication of textile based conductometric polyaniline gas sensor. Sensors and Actuators B: Chemical*, 202: 732–740. https://doi.org/10.1016/j.snb.2014.05.138.

Rajendran, S. and Anand, S.C. (2002). Developments in Medical Textiles. *Textile Progress*, 32(4): 1–42. https://doi.org/10.1080/00405160208688956.

Righetto, M., Smith, G.C. and Tabor, P. (2012, February). Aura: Wearable Devices for Non-verbal Communication between Expectant Parents. https://citeseerx.ist.psu.edu/document?repid=rep1&type=pdf&doi=10632a2ff18348c971a01a33ea2905d401e0ff66.

Rogier, H. (2015). Textile Antenna Systems: Design, Fabrication, and Characterization. In: X. Tao (Ed.). Handbook of Smart Textiles. *Springer*, pp. 433–458. https://doi.org/10.1007/978-981-4451-45-1_38.

Rohani Shirvan, A. and Nouri, A. (2020). Medical textiles. *In: Advances in Functional and Protective Textiles Elsevier*, pp. 291–333. https://doi.org/10.1016/B978-0-12-820257-9.00013-8.

Sawhney, A.P.S., Condon, B., Singh, K.V., Pang, S.S., Li, G. and Hui, D. (2008). Modern Applications of Nanotechnology in Textiles. *Textile Research Journal*, 78(8): 731–739. https://journals.sagepub.com/doi/abs/10.1177/0040517508091066.

Scataglini, S., Moorhead, A.P. and Feletti, F. (2020). A Systematic Review of Smart Clothing in Sports: Possible Applications to Extreme Sports. *Muscles, Ligaments & Tendons Journal (MLTJ)*: 10(2).

Schreuder-Gibson, H.L., Truong, Q., Walker, J.E., Owens, J.R., Wander, J.D. and Jones, W.E. (2003). Chemical and Biological Protection and Detection in Fabrics for Protective Clothing. *MRS Bulletin*, 28(8): 574–578. https://doi.org/10.1557/mrs2003.168.

Scott, R.A. (2005). Textiles for Protection. Elsevier.

Shakeri, M., Hemmatinejad, N. and Bashari, A. (2019). Synthesis of peptide nanotubes for fabricating a new type of bio-functional textiles. *Journal of Industrial Textiles*, 48(8): 1257–1273. https://doi.org/10.1177/1528083718760803.

Sharma, S., Nirkhe, C., Pethkar, S. and Athawale, A.A. (2002). Chloroform vapour sensor based on copper/polyaniline nanocomposite. *Sensors and Actuators B: Chemical*, 85(1): 131–136. https://doi.org/10.1016/S0925-4005(02)00064-3.

Shibata, S., Niimi, Y. and Shikida, M. (2014). Flexible thermal MEMS flow sensor based on Cu on polyimide substrate. *2014 IEEE Sensors Journal*, 424–427. https://doi.org/10.1109/ICSENS.2014.6985025.

Shukla, A., Basak, S., Ali, S.W. and Chattopadhyay, R. (2017). Development of fire retardant sisal yarn. *Cellulose* 24(1): 423–434. https://doi.org/10.1007/s10570-016-1115-7.

Shukla, A., Das, D. and Sen, K. (2022). Electrically-assisted chemical vapor polymerization: A novel method for *in situ* polymerization of pyrrole. *Journal of Applied Polymer Science*, e53443. https://doi.org/10.1002/app.53443.

Shukla, A., Sen, K. and Das, D. (2022). Studies on electrolytic and doping behavior of different compounds and their combination on the electrical resistance of polypyrrole film via electrochemical polymerization. *Journal of Applied Polymer Science*, 139(33): e52793. https://doi.org/10.1002/app.52793.

Steffens, F., Gralha, S.E., Ferreira, I.L.S. and Oliveira, F.R. (2019). Military Textiles:An Overview of New Developments. *Key Engineering Materials*, 812: 120–126. https://doi.org/10.4028/www.scientific.net/KEM.812.120.

Stylios, G.K. and Wan, T. (2007). Shape memory training for smart fabrics. *Transactions of the Institute of Measurement and Control*, 29(3–4): 321–336. https://doi.org/10.1177/0142331207069479.

Su, Y., Yang, J., Li, R., Song, G. and Li, J. (2019). Effect of compression on thermal protection of firefighting protective clothing under flame exposure. *Fire and Materials*, 43(7): 802–810. https://doi.org/10.1002/fam.2739.

Suh, M., Carroll, K.E. and Cassill, N.L. (2010). Critical review on smart clothing product development. *Journal of Textile and Apparel, Technology and Management*, 6(4).

Syduzzaman, M., Patwary, S., Farhana, K. and Ahmed, S. (2015). Smart Textiles and Nanotechnology: A General Overview. *Journal of Textile Science & Engineering*, 5. https://doi.org/10.4172/2165-8064.1000181.

Tao, X. (2001). (Ed.). Smart Fibers, Fabrics and Clothing, North America: CRC Press LLC.

Teodorescu, M. and Teodorescu, H.-N. (2020). Capacitive Interdigital Sensors for Flexible Enclosures and Wearables. 2020 International Conference on Applied Electronics (AE), pp. 1–6. https://doi.org/10.23919/AE49394.2020.9232783.

van den Brand, J., de Kok, M., Sridhar, A., Cauwe, M., Verplancke, R., Bossuyt, F., de Baets, J. and Vanfleteren, J. (2014). Flexible and stretchable electronics for wearable healthcare. 2014 44th European Solid State Device Research Conference (ESSDERC), pp. 206–209. https://doi.org/10.1109/ESSDERC.2014.6948796.

Wang, C., Li, X., Gao, E., Jian, M., Xia, K., Wang, Q., Xu, Z., Ren, T. and Zhang, Y. (2016). Carbonized Silk Fabric for Ultrastretchable, Highly Sensitive, and Wearable Strain Sensors. Advanced Materials, 28(31): 6640–6648. https://doi.org/10.1002/adma.201601572.

Wang, C., Zhang, M., Xia, K., Gong, X., Wang, H., Yin, Z., Guan, B. and Zhang, Y. (2017). Intrinsically Stretchable and Conductive Textile by a Scalable Process for Elastic Wearable Electronics. *ACS Applied Materials & Interfaces* 9(15): 13331–13338. https://doi.org/10.1021/acsami.7b02985.

Xia, H., Grossman, T. and Fitzmaurice, G. (2015). NanoStylus: Enhancing Input on Ultra-Small Displays with a Finger-Mounted Stylus. *Proceedings of the 28th Annual ACM Symposium on User Interface Software & Technology*, pp. 447–456. https://doi.org/10.1145/2807442.2807500.

Xu, W., Huang, M.-C., Amini, N., He, L. and Sarrafzadeh, M. (2013). eCushion: A Textile Pressure Sensor Array Design and Calibration for Sitting Posture Analysis. *IEEE Sensors Journal*, 13(10): 3926–3934. https://doi.org/10.1109/JSEN.2013.2259589.

Yin, W., Wang, Y., Liu, L. and He, J. (2019). Biofilms: The Microbial "Protective Clothing" in Extreme Environments. *International Journal of Molecular Sciences*, 20(14): Article 14. https://doi.org/10.3390/ijms20143423.

11

Recent Trends in UV Protection Materials for Textile Functionalization Applications

Tahsin Gulzar,[1] *Shumaila Kiran,*[1,*] *Tahir Farooq,*[1]
Sadia Javed,[2] *Nosheen Aslam,*[2] *Atizaz Rasool*[1] and
Iqra Bismillah

1. Introduction

The issues that people face when dealing with unsafe UV radiation are not hidden from the rest of the world; the prevalence of skin disease has increased in recent years (Gentile et al., 2021), with white people bearing the brunt of the consequences. However, because of the protective effects of melanin, the side effects of UV radiation are very low in South Asians (Rana, 2021). As a result, different preservative strategies for proper UV protective clothing must be considered (Franco et al., 2021). The substantial reason for skin cancer is a longtime acquaintance with solar ultraviolet radiation (UVR). The prevalence of various skin diseases such as aging, acne, and skin cancer is increasing as the ozone layer's density decreases (Roberts et al., 2021) than another part of sunlight, UV radiation, has side effects

[1] Department of Applied Chemistry, Government College University Faisalabad, Pakistan.
[2] Department of Biohemistry, Government College University Faisalabad, Pakistan.
[*] Corresponding author: shumaila.asimch@gmail.com

on the skin and eyesight, but it is also considered an immune exploitive (Souak et al., 2021). Ultraviolet radiation ranges from 100–400 nm, and it is divided into three bands of different wavelengths (Kidile, 2019). Fabric manufacturing and fabric auxiliaries finishing are two factors that protect against ultraviolet radiation. Different dyes also provide good opposition to UVR, and protection against it increases with dye concentration. Soft colors reproduce solar energy more competently than vivid colors, letting incident radiation enter the cloth and being reinforced by shimmering actions (Benli et al., 2021).

UV protective clothing may be defined as something that individuals keep to themselves and risk if they sojourn in the territory of menace with an abridged hazard of damage. To protect against the dangerous effects of ultraviolet radiation, there are three ways, i.e., a minimum acquaintance with sunlight, using protective clothes, and using sunscreens (Gabros et al., 2019). Aside from sunblock creams, textile ingredients and fixtures are principally used for UV fortification. To protect from ultraviolet radiation, textile items include different items like caps, canopies, shoes, and different textile materials (Alam et al., 2021). The ability of textile fabric to block ultraviolet radiation is dependent on the additives and their chemical structures. Ultraviolet protection (UPF) is also affected by the wearing conditions, fabric materials, humidity, and different additives used in fiber processing (Kibria et al., 2022).

Natural fibres like silk, cotton, and wool have lower ultraviolet protection than synthetic fibres like polyethylene terephthalate (PET). Cotton fibers, in their natural gray, state have the highest UPF because they contain pectin, waxes, and natural colors or pigments, whereas washed-out fibers have high UV pellucidity. Natural fibers like linen and hemp have a UPF factor of 20% and 10–15%, respectively, and are considered effective UV protective agents (Sankaran et al., 2021). Protein fibres also have diverse effects on permitting UV radiation. The uncolored or washed-out fabrics did not show better UV protection values than the colored fabrics. Wool soaks up energy in the region of 280–400 nm and even beyond 400 nm. Acquaintance with sunlight compensates for the quality of the silk's color, strength, and resilience in both dry and wet situations (Sankaran et al., 2021).

UVR has been linked to sunburn, sun allergy, rapid skin aging, eye damage, and skin cancer (Bernhard et al., 2022). The World Health Organization has advised minimizing sun exposure to lessen the likelihood of getting such illnesses (Alfredsson et al., 2020). To do this, the most typical form of protection is to cover any exposed body areas with sunscreen. However, in recent years, textile constructions have gained traction as a type of defense against the sun's harmful ultraviolet (UV) radiation (Kocic et al., 2019). Prevention from sun damage can be made easier and more successful if the proper textile structure is used, such as clothes or seeking shade (a parasol, etc.) (Bernhard et al., 2022). UV-protective textiles can offer incredibly excellent UV protection. Depending on the basis, they can provide a UPF of up to 80, allowing you to spend the entire day outside (Beslay et al., 2020).

Due to the unique characteristics of this group of substances, which include porous structure, enhanced performance, and large surface power that guarantee higher communication with UV materials to enhance the reliability and sensory attributes, nanomaterials could provide greater performance and sturdiness to UV protective substances (Baig et al., 2021). TiO$_2$ nanomaterial for UV protective fabrics, silver nanoparticles for antioxidant fabrics, and silver-titanium nanoparticles for chemical or biologically protected fabrics are a few of the specialized uses of nanomaterial for UV protective materials, etc. (Mondal, 2022).

2. Effects of UV Radiation on Human Health

Regarding the influence on human health, there are big variances among UVA, UVB, and UVC. UVA is also known as "crystal transmission", while communal glass can obstruct 90% of the radiation below 300 nm and permit radiation of about 350 nm. UVA radiation is thought to cause premature ageing and crumpling of skin and body parts, even though it eliminates vitamin A in the skin and has a significant impact on the body's collagen fibers. It intensely pierces into the skin causing tanning of the skin, but it does not cause skin burns. A brown color pigment known as melanin absorbs the UVR and disperses the energy into less harmful heat and protects the skin from tissue damage. But nowadays it is considered that UVA radiation engenders highly reactive chemical intermediates that can attack and damage the DNA leading to skin cancer. Immune suppression is mainly caused by the UVA radiation that leads to the diversity of contagious sicknesses like malaria, measles, chicken pock, herpes, and fungal diseases etc. rather than UVB (Ali et al., 2021).

UVB is also recognized as skin burn and is also considered as the main cause of skin cancer, sunburn, and cataracts (Ali et al., 2021). It shows adverse effects on the collagen fibers and destroys the DNA and vitamin A in the skin. UVB radiation upsurges the melanin manufacture as a way of defense which leads to a long-lasting tan with a 2-day lag phase after radiation. It is also known as sunburn and can damage the DNA at the molecular level and permanently destroy the genomic makeup (Premi et al., 2019). UVB also has adverse effects on vision i.e., it damages the retina and lenses of the eye. UVB radiation is also the constituent that helps to synthesize the vitamin D made in the skin (Bikle et al., 2020). So, in this way, it is good for human health as vitamin D is very essential for the normal working of our nervous system bone development etc. UVC radiation is known as bacterial region and is very dangerous for human health as it has the highest energy and destroys the DNA directly (Soundharaj et al., 2022). A brief overview of the effects of UVR on humans is given in Fig. 11.1.

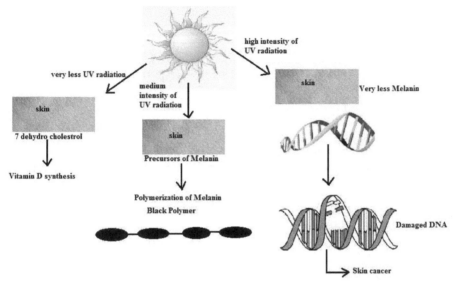

Fig. 11.1 Brief overview of the effects of UV radiation on human.

3. Ultraviolet Absorbers for Textiles

Long-term contact with short-wavelength electromagnetic radiation can damage the surface of many materials. Among these short wavelengths, UVR can damage and fade the color of textile fabrics, and it may also cause sunburn on humans and other animals' bodies (Butola, 2018). The degree of ultraviolet defense provided by textiles is determined by the type and chemical configuration of the materials as well as the product's envisioned determination. Because fabric engineering includes a diversity of process arrangements, natural UV absorbers may be detached from the fabrics during this stage. In terms of the process of synthesis and way of application on various textiles, reports on imparting ultraviolet-blocking properties through various kinds of UV absorbers have varied over the last three decades. The field's recent use of nanomaterials has drawn attention to the effectiveness and long-standing toughness of UV defense agents with multifunctional competencies (Sankaran et al., 2021). Figure 11.2 depicts the display of UVR in the case of defenseless and protected UV-protected fabrics.

3.1 Various types of UV protectors

Ultraviolet absorbers are those substances that are chemically organic or inorganic and that absorb UVB and ultraviolet radiation and replenish

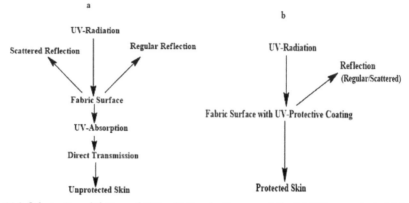

Fig. 11.2 Schematic exhibition of UV radiation in the case of the (a) UV-unprotected fabrics and (b) UV-protected fabrics.

the energy of the environment. The absorbing molecules imprison the high energy of UVR and transfers it into vibrating molecules and the environment. An effective ultraviolet absorber must have the following properties:

1. It should absorb the maximum amount of UVR and then dispel the absorbed energy into the environment to avert the color of the fabric or fabric damage.
2. The UV range 290–340 nm has the maximum absorption, though the visible zone has none.
3. Heat constancy is obligatory, as is congruity with other additions in the final composition.
4. It should be innocuous and non-poisonous to the skin.

The chemical structures of UV absorbers for textiles are given in Fig. 11.3.

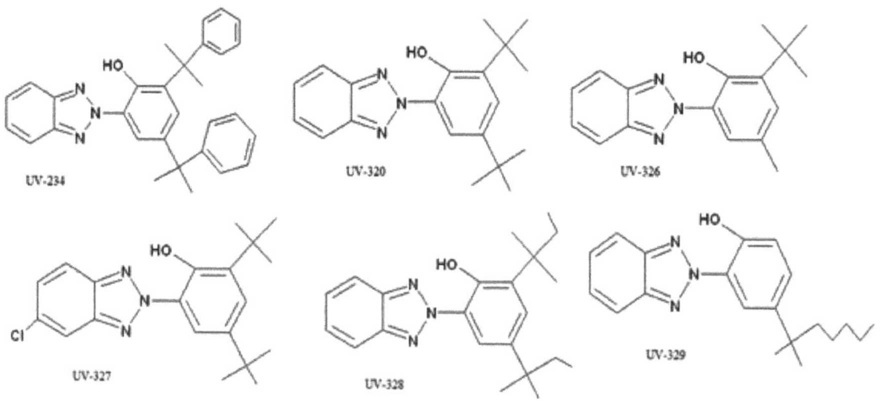

Fig. 11.3 Chemical structures of UV absorbers for textiles (Kawakami et al., 2018).

3.1.1 Development of ultraviolet protective fabric with natural herbs

Many experiments and studies were carried out using synthetic dyes from petrochemicals based on perilous chemical processes that not only threaten the environment but also affect human health (Islam et al., 2021). The basic reason for increased attention to the use of natural sources is their high compatibility with the environment and due to the accessibility of natural colours via plants, fungi, animals, and minerals (Zyczynski et al., 2019). Herbal finishing, which is named due to the use of herbs in textile finishing, comprises leaves, flowers, grass, fruits, roots, bark, etc. The researchers have tried to use some ecologically friendly natural elements as a substitute for chemically improved ultraviolet inhibitors (Kidile, 2019). It was previously described that some fabrics dyed with natural dye have antibacterial and ultraviolet protection capabilities (Islam et al., 2021). The ultraviolet protection is assigned by the fabric evaluation system, which specifies an ultraviolet protection factor value, which is a time factor for the Caucasian skin shield associated with no protection measures. For example, if the person shows visible erythema (sunburn), fabric with a UPF of 50 extends that time to five minutes times the protection factor, i.e., 250 minutes, or approximately four hours (Nisar et al., 2021). Tulsi, habitually known as the Queen of Herbs, has UV absorption competencies, making the finished item cost-effective and globally valuable (Ghosh, 2021).

By applying a dye-fixing agent, the color fastness of the walnut tree wood may also be enhanced. Dye fixing agents form a protective layer on colored silk materials. The color fastness of silk textiles is increased by applying a protective coating to the fiber, which decreases its water solubility (Wang et al., 2022). Different research is now looking for natural items to apply in a diversity of applications, particularly in medicinal engineering. Consequently, Psidium guajava leaf extract was employed to give versatile competencies to cotton materials, like antibacterial and antioxidant features, besides ultraviolet defense, due to the enrichment of phytoconstituents such as flavonoids and phenols, as well as terpenoids, tannins, saponins, etc. (Zayed et al., 2022). Rather et al. (2021) used various biomordants like Quercus infectoria bark, Punica granatum peels, Terminalia arjuna cover, Sapium sebiferum, and Cinnamomum camphora leaf extracts to modify the colorimetric and useful finishing properties of painted cloth samples using ferrous sulphate (Fe^{2+}) and alum (Al^{3+}) as reference mordants. A controlled amount of peanut cover extract applied to wool material demonstrated improved antibacterial and UV-protective properties. Pre-mordanting improved the antibacterial capabilities, but post-mordanting improved the UV-protective qualities.

Mothilal et al. (2021) discovered that the methanol extract of Musa acuminate greenery significantly improves the prominence of shielding properties such as antibacterial and ultraviolet defense. The researchers applied S. cumini (L.) leaf extract as a UV protective and antibacterial finish to bamboo textiles and found that it amplified UV protective and antibacterial qualities to a higher degree, resulting in good defense. This unique plant source with antibacterial and UV-protecting properties may be employed to make curative clothes as well as ordinary dresses. This investigation also delivers a new supply of natural plant materials that may be combined with developing skills like microencapsulation and nanotechnology to produce active and enduring fabric materials (Rachel and Keerthana, 2018).

3.1.2 Ultraviolet protection by fabric engineering

In several countries, the rising understanding of the side impacts of ultraviolet rays and consistent, active protection are the real themes among common people. In specialized journals, daily papers, and on the internet, a lot of dissimilar contributions can be observed where dermatologists, meteorologists, biologists, and other specialists make us cautious about UVR and ozone reduction and give us some suggestions or advice for active defense (Neale et al., 2021). The problem of UVR is multidisciplinary, with textile scientists being the primary focus. Although clothing made of appropriate fabrics provides more UVR protection, not all types of clothing provide the same level of protection (Khan et al., 2020). Stankovic et al. (2022) measured and established the positive effect of yarn folding on the UV protection properties of hemp-woven fabrics. The collapsing activity was used to deliver crossbreed hemp yarns from which woven textures with high UV protection limits were made in the review presented by Stankovic et al. (2017). Both these investigations submitted a few suggestions of how bright, defensive hemp-based material textures can be made in a cleaner way by utilizing altogether advantageous mechanical tasks.

Hopefully, the fabric structure is the most important and predictable aspect of all. Closer and tighter structures, as expected, provide better UV protection. As a result, the bulk of past research has focused on knit structures, which are less permeable and provide good UV protection. The capacity to get woven materials with an adequate UV assurance factor, then again, is critical, as sewn textures are greater for sports as well as easy-going summer-style clothing. The current assortment of information on the UV assurance element of sewn materials is meager, and it applies to a great extent to textures made on machines with rather huge measures (Louri et al., 2018). The amount of fabric coating increases the UPF and width of the cloth while decreasing its air penetrability. Also, switching from an angle interlock to an orthogonal interlock with the same base knit had no visible effect on UPF. Though switching from polyester to polypropylene (PP) has a noteworthy influence on UPF, to get actual results, at least three coats of cotton or polyester fabric devoid of any UV-resistant coating are essential.

As for their long-term resistance to UV contact, cotton and polyester fabrics are especially suitable for open-air usage (Nasreen et al., 2018).

3.1.3 *Ultraviolet protection provided by woven fabrics made with cellulose fibers*

It becomes perilous to defend humans from the hazardous effects of UVR coming from the sun, especially in those areas where the ozone layer has depleted, leading to skin cancer. Textiles are thought to be the most effective materials for UV protection, so it is critical to improve the protective factors. Summer clothing is mostly made of natural and renewed cellulose. The prevalence of fiber-built textile materials in summer is related to their outstanding sterile properties. Nevertheless, these materials do not frequently deliver good UV protection (Krysiak et al., 2022). This drawback or failure of natural cellulosic fibers can be minimized by proper engineering approaches and construction parameters. An investigation has been performed on the use of anti-allergenic, safe, and ecological natural colors in dyeing cellulose textile materials due to the increase in the reputation of a natural lifestyle. The public has been stimulated nowadays to use natural and biodegradable goods (Newman et al., 2020). *Gossypium arboreum*, often known as "tree cotton", is a plant that is widely cultured in Pakistan and India and used to make cotton textiles. Cotton textiles are especially famous due to their special properties like smoothness, ease of washing, and flexibility. However, one disadvantage of cotton material is that it provides only a weak defense against UVR, allowing radiation to pass through easily due to its low mass and high absorbency. The human body is very sensitive to environmental changes, especially radiation coming from the sun, which causes serious problems like skin cancer, eye problems, skin burns, etc. (Sahar and Ali, 2018).

Many natural extracts were utilized in the dyeing and the improvement towards UV protection from cellulose textile materials, such as eucalyptus leaf extract, tea extracts, and many extracts from Mediterranean flora (da Silva et al., 2018), as shown in Fig. 11.4a. The results support the use of some plant extracts (M. azedarach) to raise the UV protection potential of woven fabrics such as cotton and cotton/polyester by changing the printing style. The results showed a remarkable decrease in air passage in all printed samples. Furthermore, the color fastness abilities in the case of screen-printed samples, i.e., washing, perspiration, light, and abrasion, were found to range from good to brilliant for all the above-mentioned data (Hassan et al., 2020). These plant extracts were used in the staining and expansion of ultraviolet protection from fiber textile materials (Fig. 11.4a). Away from herbal extracts, certain agrarian dispensation waste was efficiently used to recover the ultraviolet resistance of cotton garments. It has been discovered

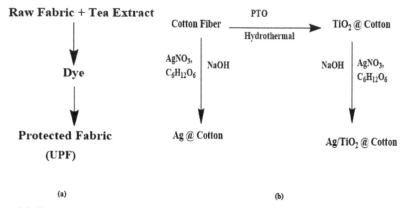

Fig. 11.4(a) Tea extract in preparing ultraviolet-protected cotton fabric: (a) Scheme for the synthesis of TiO$_2$ NPs and Ag NPs coated onto cotton fabric. Abbreviations: NP = nanoparticles; PTO = potassium titanium oxalate.

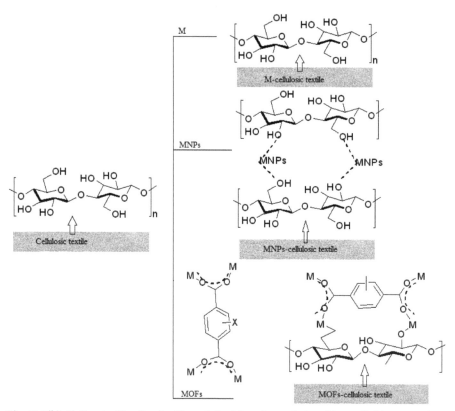

Fig. 11.4(b) Cotton textiles fixed with cyclodextrin using a citric acid crosslinking procedure for improved UV protection.

that using natural dyes after pre-treatment with conventional mordants can improve the UV protection abilities of cotton fabric while posing no environmental hazards (Kocic et al., 2019).

In another study, cotton textiles were fixed with cyclodextrin using a citric acid crosslinking procedure in the presence of sodium hypophosphite, then painted using a softened extract of the broadleaf holly leaf, which was used as a natural ultraviolet absorbent (Fig. 11.4b). The UV protection aspect was used to define the anti-ultraviolet property of fabrics tinted with a macerated extract. Cotton textiles attached with cyclodextrin show better-quality anti-ultraviolet and wrinkle recovery competencies when associated with unchanged samples, as well as outstanding toughness in contradiction of 30 washing series but with a loss of ductile strength (Liu et al., 2020).

3.1.4 UV protective finishes

To sum up, the material designing way to deal with further developing the UV assurance properties of material textures by planning the yarn structure through suitable fiber and turning strategy choice is a basic, cleaner, and financially savvy approach that gives another idea of the creation of innovated material with high UV insurance, lower ecological effect, and more noteworthy maintainability. This engineering idea delivers a truthful method to make new functionalized textile products exploiting normal mechanical industrial events throughout the interweaving stage(s), evading the usage of any extra mechanical or chemical wet treatments. Polyvinyl butyral (PVB) can be used as a binder matrix to contain several UV-range radiation absorbers. These coating agents can successfully be applied to textile surfaces to provide UV absorption. The absorptive behavior of the substrate is only marginally altered by the covering polymer. Organic UV-absorbers, inorganic particles, and metal pigments can all be successfully combined in a PVB coating agent for use on textile surfaces. Organic UV-absorbers can be successfully incorporated into a PVB coating agent for use on textile surfaces (Kocic et al., 2019).

Using TiO_2 particles, an excellent absorption characteristic below 400 nm may be produced. Metal particles, on the other hand, do not have such high absorption levels. This could be because metal pigments are small, isolated particles with plenty of interspaces for radiation to travel through it. Titanium oxide or zinc oxide particles are substantially smaller, allowing the coating to cover a larger cross area. Absorptive molecules are undoubtedly the best choice for this method. The molecular absorbers' absorption bands appear to be modestly influenced by their chemical surroundings (Grethe et al., 2018). During the commercial washing procedure, several water-soluble pale fiber-sensitive ultraviolet absorbers byproducts of regular triazine were used as additives. Experimentations presented that through the distinctive washing process at 40–60°C, a covalent link formed between the applied absorbers and the cellulose fiber. This approach has resulted in UV-protective textiles that have lasted for a long period (Baji et al., 2020).

UV absorbers should be unaffected by UVR to be perfect UV filters. A UV filter has been investigated chiefly to confirm a long-term protection function over the finest amount and composition. Some cinnamic acids, e.g., , are known to be endangered due to their resistance to breakdown by the cyclodextrin unit. Cinnamic acids' photostability recovers when the size of the cinnamic acids matches the size of the cyclodextrin cavity. As a result, innovative UV absorbers are likely to provide better UVA and UVB protection than the most widely used UV absorbers (Thompson et al., 2021). Some studies have been made to use gold and silver NPs to dye materials with colors while also including functionality such as antimicrobial activity and/or UV protection. According to the literature, three dissimilar actions are used to incorporate gold and silver NPs into fabric or fiber materials: saturation of fabrics in the synthesized metal nanoparticle solution, making metal nanoparticles in situ in fabrics, and manufacturing polymer-nanoparticles composites followed by a spinning process to form dyed fibers (Abou Elmaaty et al., 2018). Singh et al. (2019) investigated the effects of regulated UV contact on the thermal and mechanical properties of electro-spun polymer fibers made of poly-vinyl cinnamate (PVCN), a copolymer of vinyl cinnamate and vinyl alcohol. PVCN was chosen because it rapidly produces smooth fibers, is UV-sensitive, and conducts photo-dimerization without the need for a photo-initiator. PVCN forms a cyclobutane ring through a random cross-linking photo-addition between a UV-excited cinnamoyl group on a polymer chain and a comparable unexcited group on the same or different chain, as illustrated in Fig. 11.5.

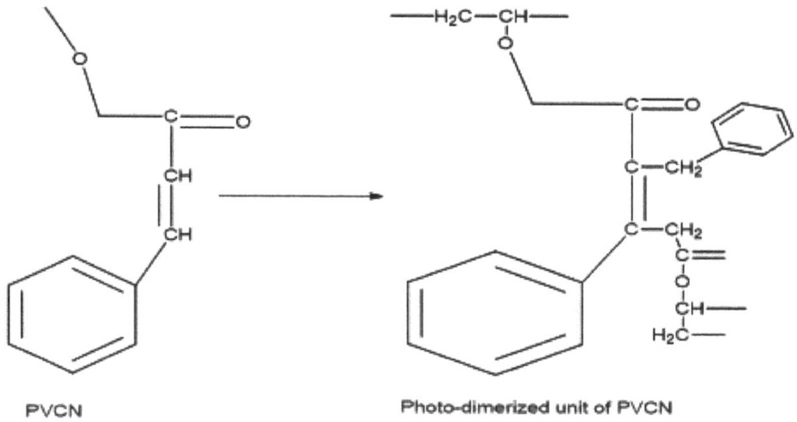

PVCN   Photo-dimerized unit of PVCN

Fig. 11.5 Photo-dimerization of PVCN in the presence of UV light to yield the cyclobutane ring.

The majority of antimicrobial products act by leaching or transferring from a surface onto a microorganism, poisoning it, disrupting a life process, or triggering a lethal mutation. The antibacterial agent dose is crucial for effectiveness. When too little of the substance is applied, the microorganism becomes uncontrollable and adaptable and if too much of it is used, it can hurt other living creatures as well. When used in clothes, this type of substance has limited durability and the potential to cause a range of other issues. Users may have rashes and other skin irritations because of the chemical, which may alter natural skin bacteria, cross the skin barrier, and/ or cause rashes and other skin irritations. Another class of antimicrobials that bind to textile materials has an entirely different mode of action. When this substance comes into close contact with a microorganism, it makes the substrate's surface antimicrobials active by rupturing the microorganism's cell membrane. These provide textiles with long-lasting antibacterial properties (Shah et al., 2021).

4. Conclusions and Future Perspectives

Numerous organic and synthetic materials' lifespans and efficacy are significantly shortened by the damaging UV rays. As a result, before these items are sold, they must first be filtered. Textile fabrics, both naturally derived and synthetic, that are used for a variety of purposes, including the preservation of a country's past, are one of these delicate materials. Recent advancements in environmentally friendly textile material coverings for the defense of diverse textile fibers were covered in this chapter. Nanomaterials were also investigated for their capacity to filter damaging UV radiation and their detrimental effects. The primary UV protection measures were also suggested. From all the above discussion, it is concluded that UV rays are harmful to the human body and should be avoided in direct contact with the body. As cloth covers the whole body, we can use fabric as a UV-protective layer. The method of knitting in fabric can be modified for this purpose. We can also modify the finishing of the fabric. When herbal extracts are used for this purpose, UV rays find it difficult to harm the body. However, because UV-protective clothing is a relatively new invention, few studies have been conducted to investigate its effectiveness. Sun protection may be best obtained by using a combination of UV textiles and sunscreen, with the materials trying to cover a significant chunk of the body while the latter safeguarding the residual shapes revealed to UVR. This is in line with the regularly updated transition from employing UPF measures to employing GPF (garment protection factor) levels, which also considering BSA (body surface area) when determining the photoprotective amount. The best method for sun protection with the least amount of danger is probably a mix of wearing photo-protective clothing and using sunscreen on visible body parts. To learn more about the subject, however, further research is required.

References

Abou Elmaaty, T., El-Nagare, K., Raouf, S., Abdelfattah, K., El-Kadi, S. and Abdelaziz, E. (2018). One-step green approach for functional printing and finishing of textiles using silver and gold NPs. *Royal Society of Chemistry Advances*, 8(45): 25546–25557.

Alam, I.K., Moury, N.N. and Islam, M. T. (2021). Synthetic and Natural UV Protective Agents for Textile Finishing. *Sustainable Practices in the Textile Industry*, 207–235.

Alfredsson, L., Armstrong, B.K., Butterfield, D.A., Chowdhury, R., de Gruijl, F.R., Feelisch, M., ... and Young, A.R. (2020). Insufficient sun exposure has become a real public health problem. International Journal of Environmental Research and Public Health, 17(14): 5014.

Ali, Z., Jatoi, M.A., Al-Wraikat, M., Ahmed, N. and Li, J. (2021). Time to enhance immunity via functional foods and supplements: Hope for SARS-CoV-2 outbreak. Altern. Ther. Health Med, 27: 30–44.

Baig, N., Kammakakam, I. and Falath, W. (2021). Nanomaterials: A review of synthesis methods, properties, recent progress, and challenges. *Materials Advances*, 2(6): 1821–1871.

Baji, A., Agarwal, K., and Oopath, S.V. (2020). Emerging developments in the use of electro-spun fibers and membranes for protective clothing applications. *Polymers*, 12(2): 492.

Benli, H. and Bahtiyari, M.İ. (2021). Testing Acorn and Oak Leaves for the UV protection of wool fabrics by dyeing. *Journal of Natural* Fibers, 1–14.

Bernhard, A., Caven, B., Wright, T., Burtscher, E. and Bechtold, T. (2022). Improving the ultraviolet protection factor of textiles through mechanical surface modification using calendering. *Textile Research Journal*, 92(9–10): 1405–1414.

Beslay, M., Srour, B., Méjean, C., Allès, B., Fiolet, T., Debras, C., ... and Touvier, M. (2020). Ultra-processed food intake in association with BMI change and risk of overweight and obesity: A prospective analysis of the French NutriNet-Santé cohort. *PLoS Medicine*, 17(8): e1003256.

Bikle, D. and Christakos, S. (2020). New aspects of vitamin D metabolism and action Addressing the skin as source and target. *Nature Reviews Endocrinology*, 16(4): 234–252.

Butola, B.S. (Ed.). (2018). The Impact and Prospects of Green Chemistry for Textile Technology. Woodhead Publishing.

da Silva, M.G., de Barros, M.A.S., de Almeida, R.T.R., Pilau, E.J., Pinto, E., Soares, G. and Santos, J.G. (2018). Cleaner production of antimicrobial and anti-UV cotton materials through dyeing with eucalyptus leaves extract. *Journal of Cleaner Production*, 199: 807–816.

Franco, M., Shani, A. and Poria, Y. (2021). Always the sun: The Uniqueness of sun exposure in tourism. *Tourism Review International*, 25(1): 19–30.

Gabros, S., Nessel, T.A. and Zito, P.M. (2019). Sunscreens and photoprotection.

Gentile, P. and Garcovich, S. (2021). Adipose-Derived Mesenchymal Stem Cells (AD-MSCs) against Ultraviolet (UV) Radiation Effects and the Skin Photoaging. *Biomedicines*, 9(5): 532.

Ghosh, J. (2021). Development of UV Protective Finished Fabric using Herbal Synthesized Colloidal Solution of Silver Nanoparticles. Journal of the Institution of Engineers (India): Series E, 102(2): 359–368.

Grethe, T., Schwarz-Pfeiffer, A., Grassmann, C., Engelhardt, E., Feld, S., Guo, F., ... and Mahltig, B. (2018). Polyvinylbutyral (PVB) coatings for optical modification of textile substrates. In: *Polymer Research: Communicating Current Advances, Contributions, Applications and Educational Aspects Formatex Research Center*, pp. 36–45.

Islam, R. and Sherman, B. (Eds.). (2021). Cover Crops and Sustainable Agriculture. CRC Press.

Kawakami, T., Isama, K. and Ikarashi, Y. (2018). Determination of benzotriazole UV absorbers in textile products made of polyurethane fibers by high-performance liquid chromatography with a photo diode array detector. *Journal of Liquid Chromatography & Related Technologies*, 41(13-14): 831–838.

Khan, A., Nazir, A., Rehman, A., Naveed, M., Ashraf, M., Iqbal, K., ... and Maqsood, H.S. (2020). A review of UV radiation protection on humans by textiles and clothing. *International Journal of Clothing Science and Technology*, 32(6): 869–890.

Kibria, G., Repon, M., Hossain, M., Islam, T., Jalil, M.A., Aljabri, M.D. and Rahman, M.M. (2022). UV-blocking cotton fabric design for comfortable summer wears: factors, durability, and nanomaterials. *Cellulose*, 1–31.

Kidile, S. (2019). Development of Ultraviolet Protective Fabric with Natural Herb. *International Journal of Engineering and Management Research*, 9.

Kocić, A., Bizjak, M., Popović, D., Poparić, G.B. and Stanković, S.B. (2019). UV protection afforded by textile fabrics made of natural and regenerated cellulose fibers. *Journal of Cleaner Production*, 228: 1229–1237.

Krysiak, Z.J. and Stachewicz, U. (2022). Electro-spun fibers as carriers for topical drug delivery and release in skin bandages and patches for atopic dermatitis treatment. *Wiley Interdisciplinary Reviews: Nanomedicine and Nanobiotechnology*, e1829.

Liu, J., Ma, X., Shi, W., Xing, J., Ma, C., Li, S. and Huang, Y. (2020). Anti-ultraviolet properties of β-cyclodextrin-grafted cotton fabrics dyed by broadleaf holly leaf extract. *Textile Research Journal*, 90(21–22): 2441–2453.

Louris, E., Sfiroera, E., Priniotakis, G., Makris, R., Siemos, H., Efthymiou, C. and Assimakopoulos, M.N. (2018, December). Evaluating the ultraviolet protection factor (UPF) of various knit fabric structures. *In: IOP Conference Series: Materials Science and Engineering*, 459(1): 012051.

Mondal, S. (2022). Nanomaterials for UV protective textiles. *Journal of Industrial Textiles*, 51(4_suppl.): 5592S–5621S.

Mothilal, B., Sampath Kumar, S.K., Prakash, C., Kumar, K.H., Venkatesh, B., Sivamani, S. and Karthikeyan, G. (2021). Musa Acuminata Leaves Extract Impedes Bacterial Growth and Ultraviolet Protection in Cotton Fabric. *Journal of Natural Fibers*, 1–10.

Nasreen, A., Umair, M., Shaker, K., Hamdani, S.T.A. and Nawab, Y. (2018). Development and characterization of three-dimensional woven fabric for ultraviolet protection. *International Journal of Clothing Science and Technology*, 30(4): 536–547.

Neale, R.E., Barnes, P.W., Robson, T.M., Neale, P.J., Williamson, C.E., Zepp, R.G. and Zhu, M. (2021). Environmental effects of stratospheric ozone depletion, UV radiation, and interactions with climate change: UNEP Environmental Effects Assessment Panel, Update 2020. *Photochemical & Photobiological Sciences*, 20(1): 1–67.

Newman, D.J. and Cragg, G.M. (2020). Natural products as sources of new drugs over the nearly four decades from 01/1981 to 09/2019. *Journal of Natural Products*, 83(3): 770–803.

Nisar, M.F., Uddin, A., Niaz, K., Hussain, S., Munawar, Q., Siddique, F. and Farooq, A. (2021). Ultraviolet Radiation A (UVA): Modulates the Production of Nitric Oxide (NO) to Combat COVID-19. Coronavirus Disease-19 (COVID-19): A Perspective of New Scenario, 192.

Premi, S., Han, L., Mehta, S., Knight, J., Zhao, D., Palmatier, M.A. and Brash, D.E. (2019). Genomic sites hypersensitive to ultraviolet radiation. *Proceedings of the National Academy of Sciences*, 116(48): 24196–24205.

Rachel, M.D.A. and Keerthana, M.R. (2018). A Progress of Artifact of UV Protective Bamboo Knit Garments of Men's T-Shirt by Green Tea and Cloves. *Journal Impact Factor*, 3: 135.

Rana, M.K., Barwal, T.S., Sharma, U., Bansal, R., Singh, K., Rana, A.P.S. and Khera, U. (2021). Current Trends of Carcinoma: Experience of a Tertiary Care Cancer Center in North India. *Cureus*, 13(6).

Rather, L.J., Zhou, Q., Ali, A., Haque, Q.M.R. and Li, Q. (2021). Valorization of agro-industrial waste from peanuts for sustainable natural dye production: Focus on adsorption mechanisms, ultraviolet protection, and antimicrobial properties of dyed wool fabric. *American Chemical Society Food Science & Technology*, 1(3): 427–442.

Roberts, W. (2021). Air pollution and skin disorders. *International Journal of Women's Dermatology*, 7(1): 91–97.

Sahar, A. and Ali, S. (2018). Treatment of Cotton Fiber with Newly Synthesized UV Absorbers: Optimization and Protection Efficiency. *Fibers and Polymers*, 19(11): 2290–2297.

Sankaran, A., Kamboj, A., Samant, L. and Jose, S. (2021). Synthetic and natural UV protective agents for textile finishing. Innovative and Emerging Technologies for Textile Dyeing and Finishing, 301–324.

Shah, J.N., Padhye, R. and Pachauri, R.D. (2021). Studies on UV protection and antimicrobial functionality of textiles. *Journal of Natural Fibers*, 1–12.

Singh, U., Mohan, S., Davis, F. and Mitchell, G. (2019). Modifying the thermomechanical properties of electro-spun fibers of poly-vinyl cinnamate by photo-cross-linking. *SN Applied Sciences*, 1(1): 1–7.

Souak, D., Barreau, M., Courtois, A., André, V., Duclairoir Poc, C., Feuilloley, M.G. and Gault, M. (2021). Challenging Cosmetic Innovation: The Skin Microbiota and Probiotics Protect the Skin from UV-Induced Damage. *Microorganisms*, 9(5), 936.

Soundharaj, S., Ramachandran, M. and Sivaji, C. (2022). The Role of Ultraviolet Radiation in Human Race. *Environmental Science and Engineering*, 1(2): 48–56.

Stankovic, S.B., Pavlović, S., Bizjak, M., Popović, D.M. and Poparić, G.B. (2022). Thermal Design Method for Optimization of Dry Heat Transfer through Hemp-Based Knitted Fabrics. *Journal of Natural Fibers*, 1–13.

Stanković, S., Popović, D., Kocić, A. and Poparić, G. (2017). Ultraviolet Protection Factor of Hemp/Filament Hybrid Yarn Knitted Fabrics. *Tekstilec*, 60(1).

Thompson, A.J., Hart-Cooper, W.M., Cunniffe, J., Johnson, K. and Orts, W.J. (2021). Safer sunscreens: Investigation of naturally derived UV absorbers for potential use in consumer products. *ACS Sustainable Chemistry & Engineering*, 9(27): 9085–9092.

Wang, S., Wang, L., Wu, M., Song, K. and Yu, Z. (2022). Dyeing of Silk Fabric with Natural Wall Nut Tree Wood Dye and Its Ultraviolet Protection Properties. *Journal of Natural Fibers*, 1–12.

Zayed, M., Ghazal, H., Othman, H. and Hassabo, A.G. (2022). Psidium Guajava leave extract for improving ultraviolet protection and antibacterial properties of cellulosic fabrics. *Biointerface Research in Applied Chemistry*, 12(3): 3811–3835.

Życzyński, N., Gazda, A. and Woźniak, J. (2019). IT support for the goods reallocation process in textiles-based fashion retail. *Fibres & Textiles in Eastern Europe*.

PART III

Case Studies with Challenges Associated with Super Smart Society

12

Super Smart Society

Proposal of An Innovative Digicircular Internet Platform Towards a More Sustainable, Resilient, and Human-centric Future

Fabio De Felice,[1] *Ilaria Baffo*[2] *and Antonella Petrillo*[1,*]

1. Introduction

From 1970 to 2017 the world population doubled (from 3.7 to 7.5 billion inhabitants), but the world consumption of materials increased fourfold, going from 26 to 104 Gt, with all the negative effects which follow in terms of, e.g., waste production (Khan et al., 2022). The international scientific community has long been addressing its interest towards possible ways to transform a potential waste into a resource creating a new value for waste (Mussatto et al., 2011; Saberian et al., 2021). In the light of circular economy revolution, digital transformation has become mainstream for

[1] Department of Engineering, University of Naples "Parthenope", Italy, Isola C4, Centro Direzionale Napoli, 80143 Napoli (NA), Italy.
Email: fabio.defelice@uniparthenope.it
[2] Department of Economics, Engineering, Society and Business Organization, University of Tuscia, Largo dell'Università, 01100 Viterbo (VT), Italy.
Email: ilaria.baffo@unitus.it
* Corresponding author: antonella.petrillo@uniparthenope.it

organizations, businesses, and industries, playing an essential role in the development of their business strategy (Nandi et al., 2021). In this scenario, it would be necessary to consider that the COVID-19 pandemic and the war conflict in Ukraine are imposing urgent reflection on new business models to diversify the sources of energy supply and maximize the use of resources (Camodeca et al., 2021). Therefore, the principles of the circular economy and bioeconomy seem to respond effectively to global challenges. Thus, it is necessary to critically rethink the culture of production going beyond culture of the waste (Blumberga et al., 2018; Batista et al., 2018). We are living in a digital transformation era which is radically changing all dimensions of global societies and economies including the sustainability paradigm (Parmentola et al., 2022). In particular, the emerging technologies (i.e., 5G, internet of things (IoT), Artificial Intelligence (AI), Machine Learning, Block Chain, Digital Twin, etc.) create innovations in information sharing, advanced data processing which drive a significant growth in industries according to Sustainable Development Goals (SDGs) (Preindl et al., 2020). Without digital, there is no green. In this scenario, digitalizing the supply chain is recognized as one of the successful factors of circular economy (Nayal et al., 2022). Recently, Physical Internet has emerged as an innovative paradigm toward a global efficient and sustainable logistics network (Li et al., 2022), also due to the current and upcoming reduction of the workforce in industrialized countries. The progressive aging of the population and the reduction of the nativity rate put scholars and industrialists in the right condition to find new technological solutions to maintain an adequate level of production (Sujitha et al., 2023).

Considering the above, the present study aims to focus on the coffee supply chain considering that coffee is the second-largest beverage consumed around the world and the second largest traded commodity after petroleum (Murthy et al., 2012). Consequently, Spent Coffee Grounds (SCGs) (a byproduct from the brewing process of coffee) are produced in huge quantities in many countries around the world and this makes the problem a global and urgent issue to face. The basic idea is a reconfiguration of the supply chain with a view to the physical internet of coffee according to the value of the derived byproducts to ensure the meeting between demand and offer of the byproduct through the analysis of quantities and real business opportunity. The research is therefore aimed at studying if and how new technologies can overcome the traditional obstacles to the reuse of spent Coffee Grounds, obstacles that until now have strongly limited the birth of new business opportunities. Through this research a 'digicircular' model is proposed like an integrated physical internet platform–PRISMA Platform (Physical internet RegeneratIve Sustainable MAterials)–to make the SCGs recovery activity with almost no waste. The platform's focus must be what and how all the 'actors' (from producer to consumer) involved in the coffee supply chain will be linked together (i.e., output, input, flows of energy, materials) to make the SCGs recovery activity with almost no waste. The use of the digital platform proposed by the authors, will greatly reduce the management costs of the byproduct and make the recovery process attractive from a business point of view.

The rest of the chapter is organized as follows: Section 2 summarizes the state of art and progress beyond it; Section 3 presents the methodology and physical internet vision proposed in the research; Section 4 develops the main features of the proposed platform, while Section 5 sums up the expected results and the main challenges of the proposal. Section 6 encapsulates the main potentialities and scientific, technological, and economic impacts of the proposed model. Finally, Section 7 outlines the main conclusions and future developments of the study.

2. Progresses Beyond the State of the Art

In recent years, circular economy models have gained attention from the food industry with the aim of recycling food waste and residues. In this scenario, byproducts and waste from coffee processing constitute a widely available, low-cost and good quality resource (Martinez-Saez et al., 2017). The reason of such interest in this waste must be found in its market volume. Just think that coffee is a growth market, just as they are constantly growing the discoveries of reusing SCGs. Growing demand for coffee, in the last 30 years, has resulted in the expansion of coffee production and exports. Global coffee production (in volume) has increased by more than 60% (ICO, 2019). The value of annual cross-border coffee exports (all forms, i.e., green, roasted, soluble) has more than quadrupled from USD 8.4 billion in 1991 to USD 35.6 billion in 2018. The COVID-19 pandemic is influencing coffee consumption trends and patterns. However, according to ICO (International Coffee Organization) it emerges that the Italian coffee market was valued at USD 3.16 billion in 2020, and it is projected to register a CAGR of 2.3% during the forecast period (2021–2026) (ICO 2020). According to Eurostat, Italy is the second-largest importer of green coffee beans in Europe, after Germany. The market is highly competitive, with key players, including Nestle SA, Kimbo SpA, and Luigi Lavazza SpA. The coffee industry generates huge amounts of waste during its processing: solid residues include coffee pulp, coffee husks, silver skin, and SCGs. It is estimated that one ton of green coffee produces 650 kg of SCG (Janissen et al., 2018). Nowadays, most of the SCGs produced worldwide is disposed of in landfills, with considerable disposal costs for the supply chain's actors and for the global environment (Lachman et al., 2022).

Disposing SCGs in landfills is unsafe, as for most organic wastes, because the risk of spontaneous combustion is quite high, and, moreover, an excessive production of harmful methane and carbon dioxide may occur contributing to the overall atmospheric pollution (Kookos et al., 2018; Vakalis et al., 2019).

However, SCGs have interesting characteristics that are related to the fixed carbon content and to the high content of valuable extractable compounds (Schmidt Rivera et al., 2020). By examining the state-of-the-art, several interesting and effective uses of coffee waste have been identified (i.e., biofuels, cosmetics, food, agriculture, eco-design, textiles, etc.) and the quality of these studies is very high (Dattatraya Saratale et al., 2020). In this regard, Starbucks investigated the use of SCGs as compost, Nespresso explored the use of SCGs as fertilizer in rice cultivation, Favini used SCGs to produce recycled paper or Kaffeeform produced cup of coffee using SCGs. Worldwide, there is also a growing attention among researchers and industries to utilize SCGs to produce chemicals and biofuels as an alternative to the petroleum-based products (Campos-Vega et al., 2015). In all the mentioned research, the supply chain linked to the collection of the SCG is indicated as the main critical issue for guaranteeing the success of the analyzed circular economy models (Battista et al., 2021; Mayson and Williams 2021). This is because in almost all the reuses tested, the reuse takes place by subjects who do not belong to the original coffee supply chain, and their inclusion in the supply chain is sometimes neither simple nor economic.

An efficient model has not yet been developed, capable of guaranteeing an adequate physical and digital connection between supply chain's actors; moreover, this issue does not only affect the coffee product but the whole agri-food sector. To overcome this important gap, the present research proposes a Physical Internet (PI or π) supply chain model (Shen et al., 2022). Fundamentally, PI is constructed to become a logistics networks, which aims to remove the high fragmentation of existing logistics networks based on interconnectivity and interoperation (Montreuil, 2018). The first PI proposal took place in 2011 thanks to Dr. Benoit Montreuil (Lin et al., 2022). Since the importance of problems related to the inefficiencies of logistics and the sustainable development of transport, several European projects have been developed in recent years (Treiblmaier et al., 2020). For example, the European Union has launched (1) the MODULUSHCA (Modular Logistics Units in Shared Co-Modal Networks) project; (2) the ALICE (Alliance for Logistics Innovation through Collaboration in Europe) platform; (3) the ICONET project a new ICT infrastructure and reference architecture to support Operations in future PI Logistics NETworks; (4) the SENSE project that aims to increase the level of understanding of the PI concept and the opportunities that it brings to transport and logistics. More recently, the PINPHA – PI for the post-covid PHARMA supply chain project is underway in Italy at LIUC University. Based on the existing applications, and the criticalities which concern the recovery chain of the SCGs, the present research aims at the integration of three elements, Physical Internet, Blockchain and Digital Technologies (i.e., IoT, AI, Big data, Digital Twin) in order to develop an integrated PI digital platform to enhance the entire coffee supply chain with particular attention to use of SCGs (Yadav et al., 2022). The aim is to make more economically, environmentally, and socially efficient and sustainable the way physical objects are transported, handled, stored, realized, supplied, and used throughout the world. It is remarkable to note that since the early 2000s several efforts have concentrated on the transition from the traditional linear production ("take–make–use–dispose") to a Circular Economy (CE), proposed as a promising strategy for dealing both with the current environmental issues and providing socio-economic benefits (Ilić et al., 2022; Govindan et al., 2018). The main progresses beyond the state of art that would carry the platform are summarized in Table 12.1.

3. Methodology: The Physical Internet Vision

Organizations and individuals are called to act for a great transition based on the implementation of innovation processes related to energy, digital transformation, the green economy, and the CE (Weckroth et al., 2022). These actions, if properly undertaken, all contribute to achieving the goals of sustainable development. To adequately address the task of supporting the transition of the CE and the development of innovative strategies capable of monitoring the physical, economic aspects, becomes fundamental and strategic to acquire useful and reliable information (Kumar et al., 2023). The collection of data and their re-elaboration today represents the starting point of every process, every production, logistics and organizational system today is looking for the most efficient, fast, and useful method of data collection and processing (Gligor et al., 2022).

However, there is a huge gap between the broad concept of CE and its practical implementation due to several types of barriers, which are lack of consistent and precise information about resources, products, and processes. Without a proper information flow, it is impossible to quantify circular initiatives, both in comparison with the actual linear situation or with circular alternative opportunities (Melnyk et al., 2022). A proper quantification of circular initiatives allows the assessment of economic, environmental, and social benefits and the preventative identification of potential barriers and relative solutions, monitoring the risk associated with circular investments and supporting the decision-making process (Howard et al., 2019). Since the transition to the CE is a dynamic process, the tools aimed at evaluating the path towards circularity must also be conceived as dynamic and updatable (Liu et al., 2021). Notably, the high fragmentation of different logistics networks represents a barrier to CE models. Figure 12.1 shows a scheme of flow within the coffee lifecycle.

Table 12.1 Progresses beyond the state of art.

Dimension	Gap of State of Art	Progress Enabled by PRISMA Project
Economic impacts	In general, there is any collaboration among consumers (individuals or organizations) to share products, reducing their underutilization and improving their value.	The PRISMA project intends to promote synergies between businesses and consumers along the entire value chain. What is waste for one industry or for citizens becomes a resource for other industries.
Environmental impacts	Lack of data about the characterization, localization, quantity of the coffee residue that could give indications on possible re-uses.	The project intends to favour methods and techniques to avoid the natural entropic reduction of the quality of the material following its reintroduction into the production cycle and therefore the increase in the number of lifecycles that it is able to sustain.
Social impacts	Lack of methods for analyzing impacts (positive or negative) on stakeholders throughout the Life Cycle of the product or service in question.	The project intends to contribute to the development of a management model to promote community well-being, quality of life, behaviors, practices, and activities of individuals and groups.
Logistics impacts	Lack of supply network to support stakeholder relationship and to promote new circular business model for good customer behaviors and business routines, and to support new actors that can be included in the original supply chain.	The project intends to develop a dynamic web-based system for the collection of coffee grounds (for example, from bars, consumers, etc.) to do so, it will be developing a real-time digital system for monitoring coffee grounds production, to collect relative data, produce an effective planning of the operating routes and filling a dashboard of performance indicators.
Technological impacts	Lack of a flexible and sustainable production systems due to some characteristics such as poor modularity, versatility, and adaptivity for the use of several different material/products.	The project intends to develop a digital platform based on the physical digital internet to promote models of industrial symbiosis and collaboration with citizens through new technologies.

To reverse this situation, PI follows the principle of the digital Internet (DI) to store, move, handle, and transport physical goods from one place to another in a sustainable and efficient way. Today transporters, logistics service providers, and owners of goods use closed logistics networks and in most cases with the sole purpose of shipping products (De Felice and Petrillo, 2021). Unfortunately, these closed networks are characterized by inefficiencies such as the no-load operation of the vehicles and non-optimal transit route. Other issues include lack of transparency, congestion,

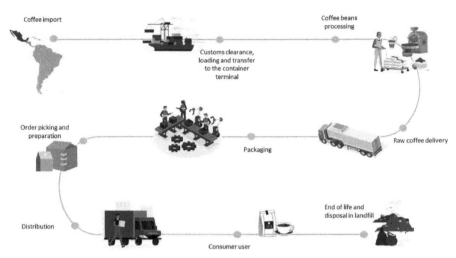

Fig. 12.1. Scheme of flow within the coffee lifecycle.

and negative environmental and social impacts. However, from a CE perspective, information is relevant and there can be no circularity of products/processes without the circularity of data (Mumtaz et al., 2018). To address these problems, it should be necessary to define environmental information and data to allow companies to structure and exchange them globally. Global standards are the tool that allows automation in the exchange of information, without which it is not possible to implement a CE (Niu et al., 2022). The ambition is that Europe is the first neutral continent from the point of view climate in the world by 2050. Figure 12.2 depicts the conceptual model, features, and principal operation of the PI network.

The research intends to investigate how PI, Blockchain, and Digital Technologies (i.e., IoT, AI, Big Data Analytics, and Digital Twin) can be

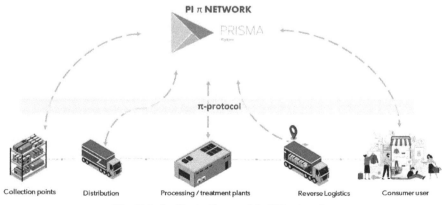

Fig. 12.2 An illustrative model of PI network.

integrated to manage coffee supply chains issues, starting from SCGs. The combined use of these technologies and approaches will allow to face the main challenges in the supply chain, distribution, reverse logistics of coffee and secondary products, as summarized in Table 12.2.

Table 12.2 Combined use of digital technologies: Opportunities and challenges.

Technologies	Opportunities and Challenges
BLOCKCHAIN (BC)	Securing supply chains against vulnerabilities require significant investment in human and resources. But these needs can increase the overall economic cost and lead times in supply chain operations. To mitigate such issues effectively the application of blockchain (BC) technology in supply chain data management can largely help.
INTERNET OF THINGS (IoT)	IoT is not a single technology but uses a collection of relevant technologies to realize interconnectivity and interoperation objectives between entities (e.g., objects, devices, organizations, and systems). In the IoT context, things are steadily transformed into smart entities by integrating ICT, with e.g., Radio Frequency Identification (RFID), Wireless Sensor Network (WSN), Global Positioning Systems (GPS) and Cyber-Physical Systems (CPS). In this context, IoT serves as a driver to digitize physical objects, enabling them to be represented on **digital platforms.** From the PI perspective, π-IoT is a framework for developing an IoT ecosystem that is dedicated to providing intelligent management services, especially in the π-nodes.
ARTIFICIAL INTELLIGENCE (AI)	In the field of logistics and supply chains, the volume of data is huge since the data are generated by various sources (e.g., embedded sensors, machines, devices, and customers) through many daily activities (e.g., loading, transportation, inventory control, and online shopping). In this context, AI and the availability of frequently updated data can support a more efficient optimization of transport routes in real time, taking dynamically into account, customer requests and traffic conditions, which influence the shipping operations.
BIG DATA ANALYTICS (BDA)	The coffee supply chain is characterized by a complex set of data, with enormous volumes and a high updating speed. In this context, the analysis of big data can support their structuring and comprehension also with the aim to developing decision support systems
DIGITAL TWINS (DT)	Digital twin is a virtual replica of a physical object that describes and stimulates the characteristics, states, and operations of its counterpart in a truthful and comprehensive manner. In this context, DT provides an environment for rendering, displaying, and predicting uncertainty in the logistics and supply chain environment.

To respond to the challenges indicated by the SCGs supply chain, we believe that a solution consisting of two supporting elements is necessary, in other words the PRISMA project will be based on:

1. A system to monitor, track and control the data generated by supply chain and logistics operations (BC + IoT);
2. A self-adaptive model (AI + BDA) which, on the basis of the variable characteristics of the waste, is able to meet and meet the real-time demand of the market (BDA), through suitable virtual models (DT), capable of reducing the consumption of resources.

Figure 12.3 illustrates the convergence and relationships of these technologies.

The PRISMA platform will give the opportunity to meet for all actors of the SC and to achieve their goals in all sustainability dimensions and at different levels. The final aim is to promote strong partnerships between public, private, and consumers and use renewable resources to create new products.

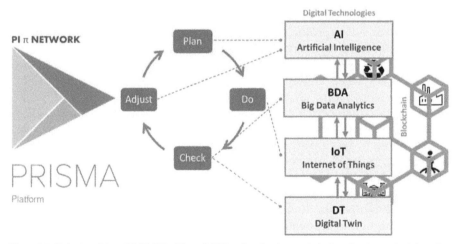

Fig. 12.3 Relationship of IoT, BD, AI and DT technologies and their roles in optimizing the performance of PRISMA platform.

4. Proposal of a PRISMA Platform: A Pilot Proposal

The aim of the research is to propose an integrated platform based on Physical Internet, Blockchain and Digital Technologies for a new sustainable economic 'circle' where the entire supply chain of coffee can recycle SCGs. The general architecture of the integrated physical internet digital platform for spent coffee grounds management is shown in Fig. 12.4.

In fact, SCGs contain large amounts of organic compounds (i.e., fatty acids, amino acids, polyphenols, minerals, and polysaccharides) that justify

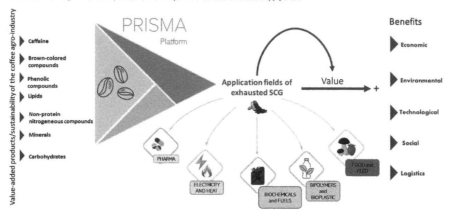

Fig. 12.4 An integrated physical internet digital platform for SCGs management.

its valorization. PI network is constructed as a network of logistics networks based on the interconnection and interoperation of π-nodes following the standardized π-protocols for handling, transporting, and storing π-containers. Key components of PI are: π-containers, π-nodes, π-transport, and π-ways (Tribe et al., 2022). The main contribution of this project is the proposal of a formal reference framework for the implementation of data governance systems for all actors involved in the supply chain of SCGs moving towards a digicircular economy model. The research will answer the following research questions: (1) What are the technical and methodological characteristics to develop an integrated framework of the PI? (2) How could the elements of the PI interact in a simulated environment? (3) How will the supply and distribution models for the coffee supply chain change? (4) How could the proposed circular business model and framework be generalized and applied to other products and fields? The platform, which is proposed with the ambition to innovate the CE sector, could be a tool for the private citizen, private companies, for the public body and for the environmental management body to pursue CE goals, to dispose of, to transport, to sell/ buy SCGs. Figure 12.5 shows a simplified scheme of the Platform.

The PRISMA platform intends to pursue the following goals:

1. GOAL#1 – Design a circular model for the supply chain of spent coffee grounds (SCGs) with a view to efficiency;
2. GOAL#2 – Identify new 'lives' for SCGs and promoe systems of "industrial symbiosis";
3. GOAL#3 – Define the main characteristics of the waste, in order to direct the process towards the most appropriate methods of recovery and recycling;

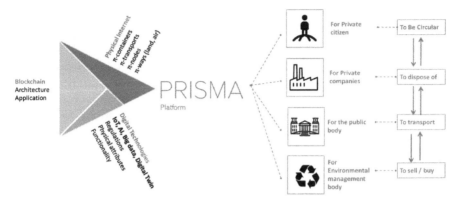

Fig. 12.5 Scheme of the PRISMA Platform.

4. GOAL#4 – Develop a sharing and matching platform capable of identifying the best use of SCGs moving from product to service logic by promoting procurement and transformation methods and systems;
5. GOAL#5 – Define the transport system for the collection/recovery of SCGs able to degrease physical flows and to increase data exchanges, to reduce CO_2 emissions and to maximize payloads, integrate public and private interests, and environmental and economic achievements.
6. GOAL#6 – Investigate the generalization of the proposed CE and its applicability to other food products and fields.

The research will be validated from an economic, technological, logistics, social, and environmental point of view:

- From the economic point of view, it is important to evaluate the costs along the entire life cycle of the product, from production to the disposal phase to increase the value of resources and secondary raw materials supporting the practices to ensure the long-term economic development;
- From the technological point of view, it is important contribute to the creation of a new business model based on physical digital internet to support all actors involved in the supply chain;
- From the logistics point of view, it is important to develop a dynamic web-based system for the round-trip spent coffee ground collection system (e.g., from bars, consumers, etc.) in order to implement a real-time digital system for data collection and data monitoring and routes planning;
- From the environmental point of view, it is important to assess a model based on a Life Cycle Thinking/Life Cycle Assessment perspective to contribute to the efficient and effective management of resources and reduce environmental impacts to ensure the health of citizens;

- From the social point of view, it is important to contribute to the development of a management model to promote community welfare, quality of life, behavior, practices and activities of people and groups.

The main novelty of the PRISMA is to propose a new business model to help all actors involved in the entire supply chain to focus on how their business should transform to increase its competitive advantage through digicircular initiatives. The model allows to allocate economic resources according to priority areas. More in details, the PRISMA project, based on the principles of the CE and the bioeconomy, aims to quantify the flows of resources and their value in different business scenarios and for different stakeholders (multi stakeholders) so that the exhausted coffee does not run out in landfill. Furthermore, it guides organizations/citizens by evaluating current practices and exploring future goals and strategies in the light of the principles of CE. Thus, the present research identifies challenges to achieve biocircular economy in a digital era defining a blueprint for a 'digicircular' transformation. The result is the proposal of an integrated PI digital platform to promote the biocircular economy implementation. An interdisciplinary tool to enable a model will be developed. In designing a physical Internet, we imagined making a verticalization on the coffee supply chain, since the potential is enormous, but our goal is to create a 'protocol' and then extend our platform to different domains.

5. Expected Results: A Real-time and Circular Economy Model

The expected results of the PRISMA can be summarized in the possibility of activating a process of disintermediation of SCGs produced by private citizens, public and private organizations. The implementation of an approach linked to the specificities and needs of the coffee sector for which that specific tool is created allow to correctly address the enhancement of the SCG (centrality of the unit of analysis). In fact, the adoption of a structured approach to production phases of goods and services (for example, including the main phases of the life cycle, from procurement to end-of-life) can allow the involvement of all key organizational figures during the evaluation process and the development of a circular awareness). Furthermore, the digitization of the tool itself on an online platform with effective functions able to correctly support the entire evaluation process can represent a further aspect to be considered (circular transformation increasingly enabled by a digital transformation). The PRISMA platform intends to involve both those who follow the processing of waste and those who produce them, with the possibility for these profits to take advantage of a dedicated delivery network. The spent coffee grounds will thus be destined to ensure the best possible enhancement. According to needs and perspectives, potential USERS OF THE PRISMA platform could be:

- B2B (companies that produce waste and companies that use scraps and waste in their production cycle);
- B2C (offer sales and/or assistance and private users);
- C2B (private users that produce waste and companies that use scraps and waste in their production cycle);
- Public body (that uses the platform for the planning of the collection of the SCGs from private citizens and companies).

6. Potentialities and Scientific, Technological, and Economic Impacts

The proposed research is oriented towards the targets of 17 Sustainable Development Goals (SDGs) adopted by the United Nations in 2015 as a universal call to action to end poverty, protect the planet, and ensure that by 2030 all people enjoy peace and prosperity (Govindan et al., 2020). In particular, the PRISMA project intends to achieve the following SDGs: (7) Affordable and Clean Energy, (9) Industry, Innovation, and Infrastructure (11) Sustainable Cities and Communities, and (12) Responsible Consumption and Production, (13) Climate Action (Fig. 12.6). In fact, SCGs are the moist solid residues of coffee brewing and in most cases, the disposal is done without any intermediate valorization actions for materials and energy recovery.

The coffee industry including production, commerce, consumption, and post-consumption creates significant quantities of waste and byproducts

Fig. 12.6 SDGs reachable from the PRISMA platform.

because coffee is one of the top traded products worldwide. The PRISMA platform will allow meeting the requirements of sustainability. The proposed model seeks to create not just industrial products, but complex industrial systems. It aims to implement sustainable productive systems in which material and energy flows are designed so that waste from one productive process becomes input to other processes, preventing them from being released into the environment, with a new economic model based on open industrial cycles. For Europe to leverage circular economy competitive advantages, it is essential to face the digital transformation process, addressing digitalization of agile and sustainable value networks, towards a data-driven circular-economy. However, several challenges are still standing for the next generation of B2B, C2B, and B2C platform for a fully digital industry and value chain, having direct impact on the implementation of circular principles at company and network level. Thus, the PRIMSA platform integrates sustainability assessment with digitalization and aims to address the following challenges:

- Challenge 1 – Lack of secure exchange of data and clear data ownership;
- Challenge 2 – Lack of metrics and measurement tools of the circular economy;
- Challenge 3 – Poor use and exploitation of IoT-enabled data streams.

The potential of the PRISMA platform applications are:

- NEW BUSINESS MODEL: Despite an abundant, regular production of the SCGs and the large quantities of material available, the lack of a structured business model capable of guaranteeing a competitive advantage for all the players in the supply chain emerges.
- MANAGEMENT AND OPTIMIZATION OF COSTS: One of the limits is certainly the costs associated with the management of the logistics chain. In fact, the value of the transported SCGs could be too low to be able to sustain important transfer and storage costs. The PRISMA project aims to reduce some of these costs through a computerized platform capable of sharing information and aggregating orders, transfers, and storage to reduce the impact of logistics operations on the by-product.
- TECHNOLOGICAL INNOVATION: The project aims to offer a technological solution capable of promoting the sustainable development of the supply chain by using a product that is widely usable and widely available both nationally and internationally. The PRISMA platform can reach such several players as to be able to guarantee significant and, above all, continuous quantities of availability over time, such as to allow the development of business and medium-term investment plans.

7. Conclusions

The aim of the present research was to design a digital platform based on Physical Internet, Blockchain, and Digital Technologies for a new sustainable economic 'circle' where the entire supply chain of coffee can recycle spent coffee grounds. Although the study is a pilot research project, the PRISMA platform has a very high potential scientific impact, thanks to the technological advancements promoted in the fields of physical internet. The platform is particularly suitable to promote the growing evolution of digital technologies within the production chains and represents substantial advantages in terms of competitiveness, productivity, and efficiency. This is even more true considering that the actions proposed within the strategies of the Sustainable Development Goals intend to promote a transition of Europe towards a competitive digital economy.

References

Batista, L., Bourlakis, M., Liu, Y., Smart, P. and Sohal, A. (2018). Supply chain operations for a circular economy. *Production Planning and Control*, 29: 419–424.

Battista, F., Zuliani, L., Rizzioli, F., Fusco, S. and Bolzonella, D. (2021). Biodiesel, biogas and fermentable sugars production from Spent Coffee Grounds: A cascade biorefinery approach. *Bioresource Technology*, 342: 125952.

Blumberga, A., Bazbauers, G. and Davidsen, P.I., (…), Gravelsins, A. and Prodanuks, T. (2018). System dynamics model of a biotechonomy. *Journal of Cleaner Production*, 172: 4018–4032.

Camodeca, R. and Almici, A. (2021). Digital transformation and convergence toward the 2030 agenda's sustainability development goals: Evidence from Italian listed firms. *Sustainability (Switzerland)*, 13(21): 11831.

Campos-Vega, R., Loarca-Piña, G., Vergara-Castañeda, H.A. and Dave Oomah, B. (2015). Spent coffee grounds: A review on current research and future prospects. *Trends in Food Science and Technology*, 45(1): 24–36.

Dattatraya Saratale, G., Bhosale, R., Shobana, S., Banu, J.R., Pugazhendhi, A., Mahmoud, E. et al. (2020). A review on valorization of spent coffee grounds (SCG) towards biopolymers and biocatalysts production. *Bioresource Technology*, 314: 123800.

De Felice, F. and Petrillo, A. (2021). Green transition: The frontier of the digicircular economy evidenced from a systematic literature review. Sustainability (*Switzerland*), 13(19): 11068.

Govindan, K. and Hasanagic, M. (2018). A systematic review on drivers, barriers, and practices towards circular economy: A supply chain perspective. *International Journal of Production Research*, 56(1–2): 278– 311.

Govindan, K., Shankar, K.M. and Kannan, D. (2020). Achieving sustainable development goals through identifying and analyzing barriers to industrial sharing economy: A framework development. *International Journal of Production Economics*, 227: 107575.

Gligor, D.M., Davis-Sramek, B., Tan, A., (…), Golgeci, I. and Wan, X. (2022). Utilizing blockchain technology for supply chain transparency: A resource orchestration perspective. *Journal of Business Logistics*, 43(1): 140– 159.

Howard, M., Hopkinson, P. and Miemczyk, J. (2019). The regenerative supply chain: A framework for developing circular economy indicators. *International Journal of Production Research*, 57(23): 7300– 7318.

ICO. (2019). Coffee Development Report 2019: Growing for Prosperity – Economic Viability as the Catalyst for a Sustainable Coffee Sector. ICO FR-01-19e, International Coffee Organization: London, UK.

ICO. (2020). Coffee Development Report. A Flagship Report of the International Coffee Organization. Available online: https://www.internationalcoffeecouncil.com/cdr2020.

Janissen, Brendan and Huynh, Tien. (2018). Resources, Conservation, & Recycling Chemical Composition and Value-Adding Applications of Coffee Industry By-Products: A Review. *Environmental Science*, 128: 110–117.

Ilić, M.P., Ranković, M., Dobrilović, M., (…), Gheța, M.I. and Simion, V.-E. (2022). Challenging Novelties within the Circular Economy Concept under the Digital Transformation of Society. Sustainability (*Switzerland*), 14(2): 702.

Khan, H., Weili, L., Khan, I. and Han, L. (2022). The effect of income inequality and energy consumption on environmental degradation: The role of institutions and financial development in 180 countries of the world. *Environmental Science and Pollution Research*, 29(14): 20632–20649.

Kookos, I.K. (2018). Technoeconomic and environmental assessment of a process for biodiesel production from spent coffee grounds (SCGs). *Resources, Conservation and Recycling*, 134 (July): 156–164.

Kumar, N., Sharma, B. and Narang, S. (2023). Emerging Communication Technologies for Industrial Internet of Things: Industry 5.0 Perspective. *Lecture Notes in Networks and Systems*, 421: 107–122.

Lachman, J., Lisý, M., Baláš, M., (…), Lisá, H. and Milčák, P. (2022). Spent coffee grounds and wood co-firing: Fuel preparation, properties, thermal decomposition, and emissions. *Renewable Energy*, 193: 464–474.

Li, M., Shao, S., Li, Y., (…), Zhang, N. and He, Y. (2022). A Physical Internet (PI) based inland container transportation problem with selective non-containerized shipping requests. *International Journal of Production Economics*, 245: 108403.

Lin, M., Lin, S., Ma, L. and Zhang, L. (2022). The value of the Physical Internet on the meals-on-wheels delivery system. *International Journal of Production Economics*, 248: 108459.

Liu, C., Zheng, P. and Xu, X. (2021). Digitalisation and servitisation of machine tools in the era of Industry 4.0: A review. *International Journal of Production Research*. doi: 10.1080/00207543.2021.1969462.

Martinez-Saez, N., García, A.T., Pérez, I.D., Rebollo-Hernanz, M., Mesías, M., Morales, F.J., Martín-Cabrejas, M.A. and del Castillo, M.D. (2017). Use of spent coffee grounds as food ingredient in bakery products. *Food Chem.*, 216: 114–122.

Mayson, S. and Williams, I.D. (2021). Applying a circular economy approach to valorize spent coffee grounds. *Resources, Conservation and Recycling*, 172: 105659.

Melnyk, S.A., Schoenherr, T., Speier-Pero, C., (...), Chang, J.F. and Friday, D. (2022). New challenges in supply chain management: Cybersecurity across the supply chain. *International Journal of Production Research*, 60(1): 162–183.

Montreuil, M. (2018). Toward a physical internet: Meeting the global logistics sustainability grand challenge. *Logistics Res.*, 3(2–3): 71–87..

Mumtaz, U., Ali, Y., Petrillo, A. and De Felice, F. (2018). Identifying the critical factors of green supply chain management: Environmental benefits in Pakistan. *Science of the Total Environment*, 640–641: 144–152.

Murthy, P.S. and Naidu, M.M. (2012). Sustainable management of coffee industry byproducts and value addition: A review. *Resources, Conservation and Recycling*, 66: 45–58.

Mussatto, S.I., Machado E.M.S., Martins, S. and Teixeira, J.A. (2011). Production, Composition and Application of Coffee and Its Industrial Residues. *Food Bioprocess Technol.*, 4(5): 661–672.

Nandi, S., Sarkis, J., Hervani, A.A. and Helms, M.M. (2021). Redesigning supply chains using blockchain-enabled circular economy and COVID-19 experiences. *Sustainable Production and Consumption*, 27: 10–22.

Nayal, K., Kumar, S., Raut, R.D., (...), Priyadarshinee, P. and Narkhede, B.E. (2022). Supply chain firm performance in circular economy and digital era to achieve sustainable development goals. *Business Strategy and the Environment*, 31(3): 1058–1073.

Niu, B., Dai, Z., Liu, Y. and Jin, Y. (2022). The role of Physical Internet in building trackable and sustainable logistics service supply chains: A game analysis. *International Journal of Production Economics*, 247: 108438. doi: 10.1016/j.ijpe.2022.108438.Parmentola, A., Petrillo, A., Tutore, I. and De Felice, F. (2022). Is blockchain able to enhance environmental sustainability? A systematic review and research agenda from the perspective of Sustainable Development Goals (SDGs). *Business Strategy and the Environment*, 31(1): 194–217.

Preindl, R., Nikolopoulos, K. and Litsiou, K. (2020). Transformation strategies for the supply chain: The impact of Industry 4.0 and digital transformation. Supply Chain Forum. *Int. J.*, 21(1): 26–34.

Saberian, M., Li, J., Donnoli, A., (...), Lockrey, S. and Siddique, R. (2021). Recycling of spent coffee grounds in construction materials: A review. *Journal of Cleaner Production*, 289: 125837.

Schmidt Rivera, X.C., Gallego-Schmid, A., Najdanovic-Visak, V. and Azapagic, A. (2020). Life cycle environmental sustainability of valorisation routes for spent coffee grounds: From waste to resources. *Resources, Conservation and Recycling*, 157: 104751.

Shen, X., Zhang, Y., Tang, Y., (...), Liu, N. and Yi, Z. (2022). A study on the impact of digital tobacco logistics on tobacco supply chain performance: Taking the tobacco industry in Guangxi as an example. *Industrial Management and Data Systems*, 122(6): 1416–1452.

Sujitha, R., Maheswari, B.U. and Raj, L.I.K. (2023). A Study on Impact of Industry 4.0 on Supply Chain Efficiency Among Manufacturing Firms. Lecture Notes in Mechanical Engineering, 385–396.

Treiblmaier, H., Mirkovski, K., Lowry, P.B. and Zacharia, Z.G. (2020). The physical internet as a new supply chain paradigm: A systematic literature review and a comprehensive framework. *International Journal of Logistics Management*, 31(2): 239–287.

Tribe, J., Hayward, S., van Lopik, K., Whittow, W.G. and West, A.A. (2022). Robust RFID tag design for reliable communication within the Internet of Things. *International Journal of Advanced Manufacturing Technology*, 121(5–6): 3903–3917.

Yadav, V.S., Singh, A.R., Raut, R.D., (...), Luthra, S. and Kumar, A. (2022). Exploring the application of Industry 4.0 technologies in the agricultural food supply chain: A systematic literature review. *Computers and Industrial Engineering*, 169: 108304. doi: 10.1016/j.cie.2022.108304.

13

5G Technology

Feasibility and Challenges from an International Point of View

Yousaf Ali,[1,]* *Amin Ullah Khan,*[3] *Muhammad Usama Hakeem*[2]
and *Ahmed Raza Qureshi*[2]

1. Introduction

This fifth-generation cellular network is 10–100 times faster than the fourth-generation network, i.e., what takes minutes over at 4G will be done in a matter of seconds if operated by 5G. This advancement in technology does not only show improvement in speed but in latency as well (Quliyev, 2022).

Globally speaking, several companies have started the rollout trails, and some have even commercially launched this technology for the masses in countries like Japan, Switzerland, and Norway. Ericsson, a leading networking company, states that by the year 2024 4.1 billion people will have access to 5G technology (Varga et al., 2022).

An evolutionary technology like 5G can do all the things that 4G can do but 5G will do it much faster. 5G will be 10–100 times faster than the

[1] School of Management Sciences, Ghulam Ishaq Khan Institute of Engineering Sciences & Technology, Topi, Swabi, KPK, Pakistan
[2] BS Engineering, Ghulam Ishaq Khan Institute of Engineering Sciences & Technology, Topi, Swabi, KPK, Pakistan.
[3] Department of Economics and Law, University of Macerata, Macerata, Marche, Italy
* Corresponding author: yousafkhan@giki.edu.pk

existing 4G. It will lower the latency and improve network densification. It will reduce energy usage by up to 90%, improving the capacity and increasing the number of devices connected from 10–100 times (Quliyev, 2022).

As compared to its predecessor, 4G, 5G operates at much higher frequencies. 4G operates between 700–2500 MHz whereas 5G operates above 25GHz. The much higher frequency means that the amount and rate of data being transferred are exponentially more. The more advanced wave spectrum means that more devices can be accommodated. One million per square kilometer for 5G compared to 4000 per square kilometer for 4G. This makes the 5G optimum mode of data transfer for high-density regions (Slimani et al., 2022). Table 13.1. below shows an overview of the overall characteristics' comparison of 3G, 4G, and 5G.

The guarantee of 5G has caught the consideration of business pioneers, strategy creators, and the media. In any case, how much of that promise is probably going to be acknowledged at any point soon? With the main genuine high band 5G organizations effectively live, this study sets out to take a reasonable view of how and where 5G availability could be conveyed and what it can achieve throughout the next 10 years.

In Pakistan, 5G technology will take a decent amount of time to become accessible to the public. Until now Zong is the only service provider that has conducted a speed test; that test showed a high speed (above 120mbps), but it also showed high latency (34 ms) (Khan et al., 2022). This means that this technology is nowhere near up to par compared to the technology being used or tested in foreign countries. There can be several reasons that could justify these results, reasons that are specific to Pakistan. However, when it comes to the complete rollout of this technology, the biggest issue will be the range of 5G signals. As discussed above it is stated that these signals have high frequency where higher frequency means lower range. A typical 4G tower emits signals up to a range of 10 miles whereas the maximum range of 5G signals is 0.2 miles or 1000 feet (Mishra and Varma, 2021). This means to cover the same area with 5G we need 50 antennas compared to only one 4G tower. This will be a big issue in a country like Pakistan as the most of the country consists of rural areas and providing services will require a lot of capital investment.

Table 13.1 Comparison of 3G, 4G, and 5G technology.

	3G	4G	5G
Bandwidth	2 mbps	200 mbps	> 1 gbps
Latency	100–500 ms	20–30 ms	< 10 ms
Average Speed	144 kbps	25 mbps	200–400 mbps

The period for this technology is dependent on various factors, firstly comes optimization of technology itself which is still in the development stages. As witnessed earlier in the development of 3G and 4G, 5G will come in phases and in urban areas first as the infrastructure will already be present followed by expansion towards rural areas.

In Pakistan, the basic infrastructure such as optical fiber is present, as was stated by PTA chairman Amir Azeem Bajwa recently. Bringing 5G into Pakistan could open several opportunities for businesses, establish a stable economy and resolve issues of high latency, reduce power consumption, and increase speed. 5G technology will allow Pakistan to keep up with the global technological developments such as self-driving cars and 8K video transmission (Hussain, 2019). Similarly, the Minister of Information Technology of Pakistan stated that it will take two years to improve the existing infrastructure such as optical fiber across the country. Moreover, he stated that once implemented it will place Pakistan in the top emerging economic markets and will possibly bring in $3.5 billion of investments (Ahmed, 2020).

2. Literature Review

5G is the 5th generation mobile broadband technology that is a new kind of network technology that aims to virtually connect different machines and devices (Daly et al., 2020). Companies like Qualcomm, Ericsson, and Huawei played an important role in laying its foundation and are now the center of applications and research for many organizations.

Initially, Stephens et al. (2019) saw technological advancements as a basic driver affecting the context of how future governments would operate. They additionally see that future governments will investigate around the globe to execute their plans. Furthermore, another study highlights the overall image of the benefits of bringing the 5G innovation into the communication framework and presents potential plans for operation by utilizing it. As per the researchers, the capacity reached by such a network is adequate to give wireless and fast Internet access to many users (Ahmed et al., 2018). Similarly, another report portrays the current market serious trends and revolutionary networks that are a must for the change from 4G to 5G. As depicted, the development in potential benefits that the carriers of remote communications may get through the dispatch of the new framework shows the struggle for the rapid deployment of this framework (Deloitte, 2018).

2.1 5G rollout procedure

The technical aspects of 5G will incorporate conveying a peak speed of 20 Gigabits per second (Gbps) per user in a low mobility situation latency of <1 ms, essentially higher region traffic capacity (1000x LTE), and a higher number of gadgets and devices connected (Shafi et al., 2017).

According to the International Telecommunication Union (ITU), 5G will enable the following:

1. Ehanced Mobile Broadband (eMMB)—an improved user experience that will address the increasing demand for bandwidth.
2. Massive Machine Type Communications (mMTC)—this will allow technologies such as self-driving cars or others that use advanced artificial intelligence (AI) to flourish.
3. Ultra-reliable and low-latency communications (URLLC)—applications that require remote access but minimum delay, remote operations, and unmanned aerial vehicles (UAVs).

To successfully uptake 5G, the deployment of an ultra-dense network will take place because 5G signals have very limited range and penetration capacity (Malta Communication Authority, 2019). Ultra-dense network system means placing antennas in close proximity. According to the suggested report, the deployment will be a spiral, initiating from a small region and extending to the full coverage of 5G.

By cost modeling, the difference between data traffic demand and network cost can be compared. The costs of 5G infrastructure heavily depend on base station price, interest rates, and cost of wavelength licenses. Recent cost modeling of 5G concluded that in the earlier stages of rollout impactions of Radio Access Network (RAN), it can reduce costs by 63% compared to traditional methods. Furthermore, British research showed that between 2020 and 2030, 90% of Britain's population will have access to 5G technology with a drop in quality and speed in rural areas due to capital and support issues. The focus was to study the effect of the spiral rollout which means increasing the radius of this technology radially from a localized area (Cheng, et al., 2022). Similarly, in the case of Pakistan, it can be assumed that the services can have a better impact by getting initiated in populated cities like Islamabad and then moving on to the lesser populated areas. Although it will be a slow process it will prove to be efficient in providing consistent services for the consumer bodies (Oughton and Frias, 2018).

2.2 Prerequisite investments

For paving the path toward the 5th generation MBB, mobile network operators must invest in the following areas.

2.2.1 Spectrum

All signals emitted by 3G, 4G, and 5G lie in a specific frequency range. To broadcast such signals, operators need licenses from the government to carry out efficient operations as this is important to prevent signal distortion and overlap in the future (Khalil et al., 2017). As per the agreements, specific ranges can be used as per license as the cost of these licenses varies and depends on government policies and taxes.

2.2.2 Radio Access Network (RAN)

It is the network system that connects existing or new broadcast sites with each other. As the introduction of 5G will result in an increasing number of sites, the network must be updated accordingly to support the influx of new sites. This is a hardware and manpower investment that will come in handy for operations in the future (Dahlman et al., 2018).

2.2.3 Transmission and cell sites

Existing cell sites must be upgraded and due to the lower range on 5G signals, new sites must be deployed. This will involve a bigger investment by any service provider as compared to the previous technologies (Reja and Varghese, 2019).

2.2.4 Backhaul support

This means backup sites that normally remain inactive and come into action once an active site is down due to maintenance or other reasons (Ge et al., 2014).

Rather than directly upgrading, operators should take an evolutionary approach and upgrade their existing technology a For instance, in Pakistan service providers like Ufone do not have access to 4G LTE technology so it would be better to first upgrade to 4G, followed by LTE, and eventually to 5G. This will allow them to slowly develop technology and will provide incremental revenue. Operators have two options, i.e., firstly, to directly invest in 5G with hopes of high commercial prospects, and secondly, to upgrade the existing technology and invest in infrastructure for 5G and not expect high revenue from 5G technology (Grijpink et al., 2018). In Pakistan, as the market for 5G will be initially limited (this statement is based on past experiences with 3G and 4G technology), the operator would be able to benefit from using the second option.

2.3 Cost and power optimization of sites

Most of the provincial populace involves low-pay clients and accordingly the provisioning of the internet in these zones with the best QoS assurances would be an overambitious practice. The central issue is that the provincial zones are denied internet, so the main issue is that of accessibility rather than execution. Subsequently, this issue must be tended to in the initiation stages of the project (Okundamiya et al., 2022).

As discussed earlier, 5G antennas consume more power on average as compared to 4G and 3G antennas. In Pakistan, power shortages are common, especially in rural areas and to address this issue, the use of solar panels was proposed. As small cell sites can be placed on lampposts within cities, it was suggested that those signal boosters or antennas should be powered by solar cells. This sort of setup will be easier to execute as, in cities like Islamabad and Lahore, solar-powered streetlamps are common. Therefore, installing solar cells for 5G antennas can be done easily (Ihemelandu et al., 2022). Furthermore, the on/off mechanism in rural areas can also be used such that at times of high bandwidth usage, 5G signals can be turned on and can remain closed for the rest of the days. Data needed for high consumption hours will be available with each service provider. This could be a rather radical power-saving mechanism (Khalil et al., 2017).

2.4 Cloud computing

Cloud computing is developing at an extremely fast rate. However, it cannot be implicated due to slow internet speeds. Companies worldwide have moved their data storage away from physical hard drives to cloud networks. Accessing this infinite amount of data needs extremely fast internet and since the information technology (IT) sector in Pakistan is lacking cloud computing technology, uploading, and downloading data from cloud networks takes a lot of time. With 5G and speeds up to 1 Gb/s, even the largest files will be accessible in seconds which means that one can use high-profile software without installing them. This will allow one to save money in terms of investment in hardware and power usage (Yang, 2022).

3. Methodology

The following methods and tools are used to analyze and check the feasibility of bringing 5G to Pakistan.

3.1 Fuzzy set theory

The fuzzy set theory was a concept developed and introduced by (Zadeh, 1965). It is used for multicriteria decision-making analysis and complex applications (Khan and Ali, 2021). A term called linguistic variable is used in fuzzy logic. The linguistic variable is an estimation or a generalization of numbers that characterize the state of approval and relates them to numerical values. The linguistic variables are unclear and tend to possess uncertainty in the collected data. For this purpose, a Likert scale is used to depict the level of agreement or disagreement based on the linguistic variable (Albaum, 1997). The Likert scale of the linguistic variables used in this project ranges from very low, low, high, and very high and utilizes the fuzzy set theory to convert the collected data into a numerical representation. The significant addition of fuzzy set theory to research analysis is that it eliminates the vagueness from the linguistic variables which helps in effective data communication. The distribution pattern for the fuzzy numbers, chosen for this study is the triangular fuzzy numbers (TFN), such as, if an individual in a survey ranks a criterion as "very high", it would refer to (0.7, 0.9, 1.0) values and so on (Afful-Dadzie et al., 2014). The fuzzy set theory is utilized for the analysis of FUCOM and the TOPSIS technique in the current research study.

3.1 Fuzzy FUCOM

Finding out the weights of the criteria is a big problem in multi-criteria decision-making. For determining the weights of criteria for a complicated decision-making process, an appropriate model is chosen (Pamučar et al., 2018). It is evident to pay close attention to determining the weights of criteria as they influence the outcome of decision-making. The list of criteria factors to be analyzed is given in Fig. 13.1.

The steps involved in the analysis of FUCOM are given as followed (Khan and Ali, 2022; Pamučar et al., 2018):

Step 1. The starting point of the methodology involves the pairwise comparison of the criteria factors, carried out by the experts.

Step 2. The evaluation received is averaged and the criteria factors are ranked accordingly with the factor having the highest significance weight kept at the top-most ranked spot and the other factors are arranged in descending order, as shown in Equation (13.1).

$$Ci_1^m > Ci_2^m > \ldots > Ci_n^m \tag{13.1}$$

Where m represents the number of experts and n represents the rank of the criteria factor.

Step 3. The weights obtained must meet the two conditions of mathematical transitivity after prioritization. The conditions are as follows:

i. The weight coefficient ratio must be equal to the criteria comparative significance as depicted in Equation (13.2).

$$\frac{w_i^m}{w_{i+1}^m} = \varphi \frac{i}{i+1} \tag{13.2}$$

ii. Furthermore, the second condition states that the mathematical sensitivity must be satisfied by the weight coefficients, as shown in Equation (13.3).

$$\frac{\varphi_i^m}{(i+1)} \times \frac{\varphi_{i+1}^m}{(i+2)} = \frac{\varphi_i^m}{(i+2)} \tag{13.3}$$

Step 4. The final step involves the calculation of the criteria factor weights as shown in Equations (13.4 to 13.7). These equations are subjected to the LINDO software for final calculations.

$$\left| \frac{W_i^m}{W_{i+1}^m} - \frac{\varphi_i^m}{(i+1)} \right| \leq \chi, \forall_k \tag{13.4}$$

$$\left| \frac{W_i^m}{W_{i+1}^m} - \frac{\varphi_i^m}{(i+1)} \otimes \frac{\varphi_{i+1}^m}{(i+2)} \right| \leq \chi, \forall_k \tag{13.5}$$

$$\sum_{k=1}^m W_k^m = 1, \forall_k \tag{13.6}$$

$$W_k^m \geq 0, \forall_k \tag{13.7}$$

Step 5. Evaluation of the final values weight coefficients of criteria are concluded as $(W_1^m, W_2^m, ..., W_n^m)^T$.

Fuzzy TOPSISTo conduct this study, the questionnaire was designed in two parts: (1) Criteria assessment, i.e., which criteria is the best or second best to judge the Quality of Services (QoS) of the internet in Pakistan; (2) Alternative assessment is a relative ranking of each criterion to all the alternatives. The criteria were the major problems identified to assess the QoS of internet in Pakistan while the alternative was the probable causes of those problems. The criteria and alternatives are listed below in Figs. 13.1 and 13.2, respectively.

The TOPSIS technique can be described as a technique that evaluates an alternative that should have the shortest distance from the positive ideal solution and must have the longest distance from the negative ideal solution. That type of alternative is considered perfect for the assessment. The steps involved in the fuzzy TOPSIS are as followed (Tariq et al., 2021):

Step 1. The data is collected from the experts in form of the weights given to the criteria factors along with the alternative ratings based on the Likert scale. The decision-makers are represented as D_m (m=1, ..., k).

Fig. 13.1 List of criteria factors chosen for the analysis.

Step 2. The aggregation is carried out for the criteria weights and alternative ratings by utilizing Equations (13.8) and (13.9).

$$[o_l = \tfrac{1}{k}[o_l^1 + o_l^2 + \ldots + o_l^k]] \tag{13.8}$$

$$[c_{nl} = \tfrac{1}{k}[c_{nl}^1 + c_{nl}^2 + \ldots + c_{nl}^k]] \tag{13.9}$$

Here o depicts the weights of the l criteria, where E_1 (l = 1, ..., g). Furthermore, depicts the nth alternative's rating, i.e., F_n (n= 1, ..., h) with respect to the lth criterion as given by the decision-makers.

Step 3. Linear scale transformation is used to evaluate the normalized decision matrix A. Equations (13.10), (13.11), and (13.12) depict the process.

$$A = [a_{nl}]_{g \times h} \tag{13.10}$$

$$a_{nl} = \left(\tfrac{c_{nl}}{d_l^+}, \tfrac{g_{nl}}{d_l^+}, \tfrac{d_{nl}}{d_l^+}\right) \text{ and= } d^{nl} = max_n(\text{benefit criteria}) \tag{13.11}$$

$$a_{nl} = \left(\tfrac{c_l^-}{d_{nl}}, \tfrac{c_l^-}{g_{nl}}, \tfrac{c_l^-}{c_{nl}}\right) \text{ and } = max_n c_{nl} \text{ (cost criteria)} \tag{13.12}$$

Step 4. This step involves the multiplication of the elements of the normalized fuzzy decision matrix A with the weights of the evaluation criteria to obtain the weighted normalized matrix B. The final weighted normalized matrix is given in Equation (13.13), and the calculation is depicted in Equation (13.14).

$$B = [b_{nl}]_{g \times h} \tag{13.13}$$

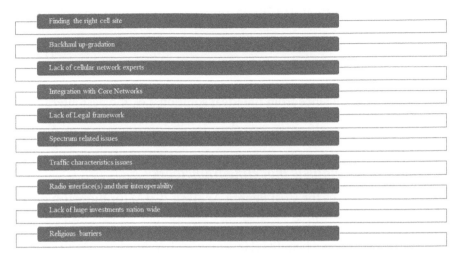

Fig. 13.2 List of alternatives to be analyzed by Fuzzy TOPSIS.

$$b_{nl} = a_{nl} \times w_l \tag{13.14}$$

Step 5. The negative and positive ideal solutions are represented by I^- and I^+, respectively and are depicted in Equations (13.15) and (13.16), respectively.

$$I^- = \{b_1^-, b_l^-, \dots, b_g^-\} \tag{13.15}$$

$$I^+ = \{b_1^+, b_l^+, \dots, b_g^+\} \tag{13.16}$$

Here $b_n^+ = (1,1,1)$ and $b_l^- = (0,0,0)$

Step 6. To calculate the distances J_n^+ and J_n^- of the alternatives, the values b_n^+ and b_n^- are considered. The distances are depicted in Equations (13.17) and (13.18).

$$J_n^+ = \sum_{n=1}^h J_b(b_{nl}, b_l^+) \tag{13.17}$$

$$J_n^- = \sum_{l=1}^h J_b(b_{nl}, b_l^-) \tag{13.18}$$

It must be kept in mind that in the above equations, J represents the relative distance between the fuzzy numbers and is evaluated using the vertex method. Post-evaluation values depict the distance from the positive ideal solution and negative ideal solution J_n^-. It is further expressed as in Equation (13.19).

$$J(q, p) = \sqrt{\frac{1}{3}\left[(c_q - c_p)^2 + (g_q - g_p)^2 + (d_q - d_p)^2\right]} \tag{13.19}$$

Step 7. Lastly, Equation (13.20) is utilized to calculate the closeness coefficient,

$$R_n = \frac{J_n^-}{J_n^+ + J_n^-}$$ (13.20)

The value of the closeness coefficient is utilized to rank the alternatives as it evaluates the distance from the ideal solution, i.e., the best alternative is at a minimum distance from the positive ideal solution and is at the maximum distance from the negative ideal solution.

3.4 Cost-benefit analysis modeling framework

The next step involves the Cost Benefit Analysis (CBA) for the setup of 5G technology in a developing country like Pakistan. CBA is a method that analyzes the efficiency of a proposed project through systematic anticipation of social costs and social benefits (Haveman and Weimer, 2001). It predicts the usefulness of a project and provides a brief analysis of the cost and benefits gained from a project. It works on the principle that states that if *"Benefits/Costs > 1"* then the project is feasible as the benefits outweigh the costs of the project.

The cost model will only analyze the capital investment required to rollout 5G in Pakistan. The cost model will exclude the maintenance cost as they fall under the operational expenditure. The model is depicted in Equation (13.21).

$$Capex_{5G} = C_{macro\ cell} + C_{small\ cell} + C_{backhaul} + C_{core}$$ (13.21)

Here, *Capex* is the sum of costs of all assets that consist of Brown macro cell upgrades, green field small cell deployments, fiber backhaul, and core upgrade costs. The 5G technology relies on wireless technology and hence there will be no need to access it by any tower antenna or implementation of special cell sites. Thus, the cost of paying rent for a cell site, and the cost of expensive and tall tower antennas are significantly reduced in comparison to the 4G technology. The reason for that is due to the previously installed sites of Macro cells in the popular cities in Pakistan and therefore, these macro cells only need upgradation. Lastly, it's important to mention that the cost mostly comprises small cells and backhaul upgradation.

4. Results and Discussion

4.1 Key results from fuzzy FUCOM and fuzzy TOPSIS

The experts responded to the survey that was carried out for the analysis. In the first stage of the fuzzy FUCOM analysis, the experts ranked reliability as the most significant issue (criteria factor) followed by hands-off support

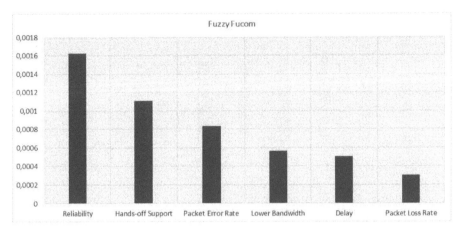

Fig. 13.3 Ranking of criteria factors via fuzzy FUCOM analysis.

and pack error rate. The results of fuzzy FUCOM are depicted in Fig. 13.3. which shows the ranking of criteria factors. From these results, it can be assumed that the installation of the 5G technology would require a reliable setup to function efficiently. Previously, the old setup of 3G and 4G setup in Pakistan have proved to be very unreliable with power outages and meager services and that's why this could be the main issue that needs to be handled by the government along with the telecom companies in a developing country like Pakistan. The rest of the criteria factors ranked hands-off support, Packet error rate, lower bandwidth, delay, and packet loss rate in descending order.

Furthermore, in the second stage, alternatives were analyzed via the fuzzy TOPSIS technique. The technique analyzed 10 alternatives based on six distinct criteria. The result concluded "proper cell site position" as the significant cause, followed closely by Backhaul upgradation, traffic characteristics issues, and integration with core networks as the top-ranked alternatives. The alternative ranking can be justified by the fact that proper location identification is extremely important when it comes to signals/network broadcast and especially if the area is densely occupied or less occupied. This kind of situation becomes a challenge for the installation of the setup and requires a huge amount of surveying, cost assessments, coverage testing, etc. Therefore, is the most probable cause of the problems that arise during the installation of a new 5G setup. Furthermore, the reasons to conduct this sort of analysis were to identify the major issues in the IT and telecommunication infrastructure. What were the possible causes behind them? This research survey paves the path to 5G rollout, by first overcoming

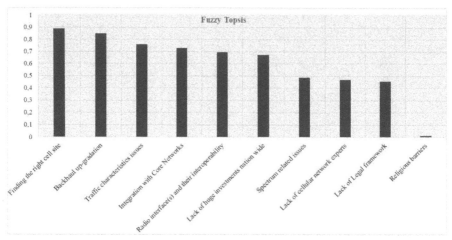

Fig. 13.4 Ranking of alternatives via Fuzzy TOPSIS technique.

the mentioned challenges in our IT and telecommunication infrastructure. Similarly, investments and improvizations to be made in proper cell site location, backhaul upgradation, minimizing traffic characteristic issues, and proper integration of core networks. Lastly, the ranking of "religious barriers" alternative, however, does not show any significant importance. The results are mentioned in Fig. 13.4.

4.2 Key results from cost benefit analysis

The cost model has four entities. However, the capital expenditure comprises mainly several small cells. Popular cities in Pakistan have already 4G macro sites installed and for upgradation to 5G, they only need to be upgraded without getting fully replaced. The upgradation process is discussed briefly in the section of the 5G deployment model to cover the maximum area with minimized expenditure. One small cell cost 10,000 USD while the macro cell costs 50,000–100,000 USD and the new tower erection will cost approximately 200,000 USD.

Furthermore, the same model in the UK is depicted to deliver 15 billion GBP in benefits at a cost of 3 billion GBP. The model suggests that smart production and smart urban areas equal the largest benefit. So, it would be wise to invest and start the spiral 5G implantation in these respective areas (Daly et al., 2020). Due to the lack of huge investments in MBB in Pakistan, this project plans to implement 5G in smart production and smart urban areas which hugely comprise heavily populated, developed, and infrastructurally stable cities like Karachi, Islamabad, Faisalabad, and Lahore. However, there are certain areas in smart rural, smart public services, and smart urban clusters which will be funded by the public. Such areas include municipal buildings, hospitals, healthcare, smart automotive, and education.

4.3 Benefits from 5th generation technology in Pakistan

The gains are numerous from bringing in the 5th generation technology but the only problem is the lack of huge investments. 5G would bring an electronic revolution in automotive, transportation, online gaming, supply chain of goods, and manufacturing of products.

4.3.1 Manufacturing industry

5G will bring a lot of advancements in the manufacturing industry as new applications will be promoted. It would bring advancements in the following areas.

- 4D printing
- Supply chain and blockchain
- Mechatronics and robotics
- Wearables and implantable
- Graphics Designing
- Aviation and heavy mechanical tools

Not only these but maintenance processes will be improvized when a part is to be replaced. If a part is showing wear or tear and needs maintenance over a certain period, robots could replace it, thus negating the requirement for human help.

4.2.3 Healthcare

5G will bring opportunities for doctors to perform surgery remotely rather than being present in the OR. Robot-assisted surgery will become a daily routine. Certain patients could benefit from this technology by wearing devices that will monitor their condition without the need to visit a doctor or medical expert unless issues are detected. The diagnosis will be heavily improvized and become more accurate.

4.3.3 Gaming and IT industry

With Pakistan having enormous talented gamers and streamers, they are still facing a lot of challenges and issues with lower bandwidth, high latency, and packet losses. A lot of talented gamers are forced to leave the country to pursue their goals. 5G technology will provide them with 1ms latency and significantly low packet losses that will help them to play competitive games and stream smoothly. According to a research study, the mobile gaming industry revenue is growing at 42% per year and reports suggest that it could generate 10 billion USD per year if quality internet services like 5G are provided (Tomić, 2018). Moreover, another report suggests that in Pakistan software houses generate a 2 billion USD revenue which is growing at a 30% increase per year (Daily Pakistan, 2022). More opportunities in AI and data sciences will become possible in the future.

4.3.4 Transportation and automotive

Driverless vehicles have been around for some time but the only thing stopping them from being implemented in Pakistan is the lack of quality internet connectivity. The driverless vehicle will be a significant addition to the automotive and transportation industry. It will provide better, and faster navigation tools and transport authorities will be improved.

4.3.5 E-banking and other businesses

Based on the previous year's progress, the MBB subscribers in Pakistan have grown from 1 to 30% (Malik, 2018). The same study highlights a research survey carried out by Arthur D. Little in the countries that extensively used 4G in the past decade. The survey depicts that organizations have seen a 67% increase in productivity, costs have been reduced by 47% and 75% of people agree that 4G has helped their organization both directly and indirectly. Furthermore, the State Bank of Pakistan states that the transactions made by mobile phone banking have grown by 195% over the last few years. Mobile banking mostly includes apps like Uber, Careem, Easypaisa, JazzCash, and Omni. This report also suggests that online transactions could generate more than 4 billion USD if 5G is introduced in Pakistan. Moreover, it could help provide employment to more than 300,000 people. Only the IT and gaming industry would generate a revenue of 10 billion USD in the future excluding other areas (Daily Pakistan, 2022). Hence, it can be justified that the revenue generated by the 5G technology after its implementation would exceed the capital expenditure of the 5G, thus becoming a bringer of good news for the country's economy.

4.4 Benefits for subcategories

There are multiple subcategories that could benefit extensively from 5th generation technology in Pakistan. From research conducted by (Daly et al., 2020), these categories include:

- Factories
  With access to 5G technology, factories would be able to provide better surveillance which would improve the site's security. Moreover, the technical workforce would become more efficient as it would be able to monitor the supply chain in real-time. This would also improve equipment's lifetime as the factory's maintenance predictions would become more accurate and well-timed. These features will lead to much

more productive and efficient factories, thus resulting in an increase in the GDP.

• Energy
• Energy-based machinery would be able to enhance their work performance with the introduction of 5G technology. This is because it will be able to provide better grid management with improved sensors and efficient surge detection, thus allowing plants to utilize energy equipment more efficiently and eventually reduce pollution due to the incorporation of electric cars and other environment-friendly products.

• Airports
5G technology would allow airports to use its AI-based features more effectively. These features may include: AI-guided repairs, AI passenger and luggage detection systems, Autonomous airside vehicle systems, etc. These features would reduce the waiting time and make air travel more secure. Additionally, autonomous systems would reduce congestion and minimize air pollution. Such measures would result in a GDP increase due to better productivity and efficiency.

• Ports
Like factories, 5G can make ports have better inventory management, surveillance, and remote-control machinery. It will help in reducing the waiting time and improving site security. Such better operational efficiency will result in lower carbon emissions and growth in the GDP as ports would become more productive.

• Mining
Mining has been one of the most environmentally harmful industries. However, with 5G technology mining firms can partake in improving air quality control and reducing carbon emissions. This can be done with the introduction of 5G-based features like drone monitoring which does not produce pollution when compared to conventional mining machinery.

• Logistics and Transport
The ogistics and transport sector would be able to not only improve productivity but can introduce better business models with 5G technology. 5G will let them bring autonomous airside vehicle systems into the market and improve their fleet management. As a result, there will be an improvement in security and safety and better supply-chain efficiency.

4.5 *Recommendation and deployment model of 5G in Pakistan*

The model focuses on the feasible rollout of 5G in Pakistan. It is not possible to launch it nationwide all at once. However, the approach used in this project is a spiral deployment of the 5G technology. 5G coverage will be restricted to a small region to a certain mile radius which will grow after a certain period depending on its local region performance. The same approach was used in Malta by a group of researchers focused on upgrading their 4G infrastructure to use their macro cells for 5G services. It resulted in saving the cost of macro cells and the capital cost comprised only of small cells and the cost of backhaul (Malta Communication Authority, 2019). Furthermore, as discussed earlier a minimum of 50 antennas are needed to cover an area of 10 square miles, this number does not consider the fact of penetration losses (loss of signal strength when going into buildings). These small stations will have to be deployed on lamp posts, tall buildings, etc. to provide good quality signals. The deployment process will initially provide coverage to metropolitan regions and not rural areas.

4.5.1 Deployment of 5G in Islamabad as a test case scenario

Furthermore, to see the feasibility of implementing the 5G technology in a densely populated city, a test case of the capital city Islamabad is considered. A few features of the city include:

- The total area of Islamabad is 90 miles squared (Source: Google Maps).
- A single 5G antenna covers an area of 0.125 miles squared (Marco, 2020).
- The total antennas needed will be 720 for the implementation of 5G.

Islamabad is divided into 60 sectors, each sector with an area of 1.5 square miles. To cover an entire sector with 5G, a total of 12 antennas will be needed. Similarly, a small-cell site will cost $10,000 on an average for installation. But a macro site will cost $50,000–$100,000 in terms of equipment and a new tower erection will cost roughly $200,000 dollars (Marco, 2020). According to the cell mapper, currently, there are 10–12 macro sites in Islamabad for each service provider. Currently, the service providers are using 10–12 macro sites to provide 4G signals in Islamabad. Due to the difference in signal range, a total of 720 small cell sites will be needed to cover the Islamabad region. Cost of upgrading existing sites will then be = 75k * 12 = $900k and the cost of erecting 708 new macro sites will become $141.6 million.

It can be stated that a total of 720 small sites are required in Islamabad which would cost = $7.2 million but for backhaul support, macro sites will be needed at least for proper functioning. It would be more feasible to upgrade current 4G sites to support 5G. The cost of upgrading existing 25 5G macro sites is 25 * 75k = $1.9 million. Small cell sites need a connection to optical fiber cable (OFC) which has already been installed in the entire city, so the cost of installation is zero. However, the existing OFC belongs to internet service provider Nayatel and companies would have to negotiate a deal to use their fiber.

ConclusionA comprehensive feasibility study was deemed necessary to ensure the effective and efficient installation of the 5G technology across a developing country such as Pakistan. The current study analyzed the cost, coverage, benefits, and barriers in the rollout of 5G MBB in Pakistan. The evaluation was carried out by evaluating the problems in the IT and telecommunication infrastructure and how to overcome those issues for paving the path for 5th generation technology. The study utilized multicriteria decision-making (MCDM) techniques such as fuzzy FUCOM and fuzzy TOPSIS to analyze the criteria factors and the alternatives based on distinct criteria, respectively.

The introduction of 5G in Pakistan will become significantly important in terms of benefits such as productivity of industries, construction, supply chain, environment, healthcare, education, security, and dealing with a lot of digital issues. This study briefly analyzed the feasibility of 5G deployment in the country and concluded that the 5G rollout should be initially restricted to smart production, smart urban, and logistics sectors and then gradually extended to other sectors.

In the current study, a slow and gradual rollout is proposed, which may, at first sight, look like it is ignoring rural areas. The reason is that providing 5G access to rural areas instantly will not be feasible as such regions have lower users and will provide minimal revenue. The 5G technology will therefore benefit the IT industry the most because urban regions have a much more developed IT setup as compared to rural areas. Similarly, the current devices that support 5G are very expensive and most people cannot afford them, thus providing a service that cannot be used properly could prove to be a financial disaster. The study concludes that the spiral rollout will deliver the best results as 5G coverage will expand slowly over time and by the time it will reach rural areas, the devices will become more affordable. The setting up of 5G technology in the urban areas is recommended because such a trend was observed for both 3G and 4G technologies in the past. Lastly, the spiral rollout will mean that service providers will get a return on their investments quicker and the number of initial investments will also drop.

This study proves to be the first of its kind to have been conducted in a developing country such as Pakistan. The study also takes pride in the novelty of the hybrid MCDM techniques application in such an area, i.e., Fuzzy FUCOM and fuzzy TOPSIS. The study will prove a pathway for the policymakers and the government institutions to take the necessary steps to set up 5G technology across the country, thus forming a smooth foundation for the rollout of 5G in the country. The study also faced limitations in terms of data collection and cost analysis due to the presence of a lesser number of professionals in the area. The study can be further expanded to other countries along with the consideration of other MCDM techniques.

References

Afful-Dadzie, E., Nabareseh, S. and Oplatková, Z.K. (2014). Fuzzy VIKOR Approach: Evaluating Quality of Internet Health Information. Warsaw, IEEE.

Ahmed, N., Radchenko, A., Pommerenke, D. and Zheng, Y.R. (2018). Design and evaluation of low-cost and energy-efficient magneto-inductive sensor nodes for wireless sensor networks. *IEEE Systems Journal*, 13(2): 1135–1144.

Ahmed, K. (2020). Pakistan set to launch 5G connectivity by 2022 after loud clear test call. Arab News, 16 November.

Albaum, G. (1997). The Likert scale revisited. Market Research Society. *Journal*, 39(2): 1–21.

Cheng, X., Hu, Y. and Varga, L. (2022). 5G network deployment and the associated energy consumption in the UK: A complex systems' exploration. *Technological Forecasting and Social Change*, 180(1): 1–24.

Dahlman, E., Parkvall, S. and Skold, J. (2018). 5G NR –the Next Generation Wireless Access Technology: The Next Generation Wireless Access Technology. *San Diego: Elsevier Science & Technology*.

Daily Pakistan. (2022). Daily Pakistan. [Online] Available at: https://en.dailypakistan. com.pk/23-Apr-2022/ict-exports-surge-to-near-dollar-2-billion-in-9m-fy22.[Accessed 5 August 2022].

Daly, A., Nickerson, C. and Stewart, J. (2020). Cost Benefit Analysis on Full 5G deployment UK Results, London: Analysis Mason.

Deloitte. (2018). 5G: The Chance to Lead for a Decade. London: Deloitte Development LLC.

Ge, X., Cheng, H., Guizani, M. and Han, T. (2014). 5G wireless backhaul networks: Challenges and research advances. *IEEE Network*, 28(6): 6–11. doi:10.1109/mnet.2014.6963798.

Grijpink, F., Menard, A., Sigurdsson, H. and Vucevic, N. (2018). The Road to 5G: The Inevitable Growth of Infrastructure Cost, New York: Mckinsey & Company.

Haveman, R. and Weimer, D. (2001). Cost–Benefit Analysis, Pergamon: Oxford.

Hussain, J. (2019). Pakistan on short list of 5G-ready countries with Zong's successful trial. Dawn, 23August.

Ihemelandu, J.C. et al. (2022). A Survey of RF Energy Harvesting Techniques for Powering 5G Mobile Devices. International Journal of Information Processing and Communication (IJIPC) , 7(1): 1–25.

Khalil, M., Qadir, J., Onireti, O., Imran, M. and Younis, S. (2017). Feasibility, architecture, and cost considerations of using TVWS for rural Internet access in 5G. 2017 20th Conference on Innovations in Clouds, Internet and Networks (ICI).

Khan, A.U. and Ali, Y. (2021). Sustainable supplier selection for the cold supply chain (CSC) in the context of a developing country. Environment, Development and Sustainability, 23(9): 13135–13164.

Khan, A.U. and Ali, Y. (2022). Enhancement of resilience and quality of cold supply chain under the disruptions caused by COVID-19: A case of a developing country. Australian Journal of Management, (Special issue): 1–25.

Khan, F., Memon, A.L. and Shah, R. (2022). A qualitative analysis on customer churn of zong and jazz cellular MNOS: A case study of Hyderabad, Pakistan. International Research *Journal of Modernization in Engineering Technology and Science*, 4(2): 903–912.

Malta Communication Authority, 2019. 5G Demand and Future Business Models Towards a Feasible 5G Deployment, Floriana: Valletta Waterfront, Pinto Wharf.

Malik, Y. (2018). Get ready for 5G. Dawn, November.

Marco, F. (2020, June). Cost, Performance and Energy Consumption of 5G Fixed Wireless Access versus Pure Fiber-based Broadband in Sweden. Retrieved from ECONSTOR.

Mishra, L. and Varma, S. (2021). Seamless health monitoring using 5G NR for Internet of Medical Things. *Wireless Personal Communications*, 120(3): 2259–2289.

Okundamiya, M.S., Wara, S.T. and Obakhena, H.I. (2022). Optimization and techno-economic analysis of a mixed power system for sustainable operation of cellular sites in 5G era.

Optimization and Techno-Economic Analysis of a Mixed Power System for Sustainable Operation of Cellular Sites in 5G Era, 47(39): 17351–17366.

Oughton, E.J. and Frias, Z. (2018). The cost, coverage and rollout implications of 5G infrastructure in Britain. *Telecommunications Policy*, 42(8): 636–652. doi: 10.1016/j.telpol.2017.07.009.

Pamučar, D., Stević, Ž. and Sremac, S. (2018). A new model for determining weight coefficients of criteria in mcdm models: Full consistency method (fucom). *Symmetry*, 10(9): 393.

Quliyev, N. (2022). 5G Technologies are creating a new world order. *Norwegian Journal of Development of the International Science*, 82(1): 62–68.

Reja, V.K. and Varghese, K. (2019). Impact of 5G Technology on IoT Applications in Construction Project Management. Banff Alberta, ISARC.

Shafi, M. et al. (2017). 5G: A tutorial overview of standards, trials, challenges, deployment, and practice. *IEEE Journal on Selected Areas in Communications*, 35(6): 1201–1221.

Slimani, A. et al. (2022). A miniaturized high isolation printed diplexer model based on band pass filter technique for 4G-LTE/5G-Sub 6 GHz applications. *Journal of Instrumentation*, 17(7): 1–10.

Stephens, M., El-Sholkamy, M.M., Moonesar, I.A. and Awamleh, R. (2019). Future Governments: Actions and Insights – Middle East North Africa. Vol. 7, Emerald.

Tariq, H. et al. (2021). Sustainable production of diapers and their potential outputs for the Pakistani market in the circular economy perspective. *Science of The Total Environment*, 769(1) 1–14.

Tomić, N.Z. (2018). Economic model of microtransactions in video games. *Journal of Economic Science Research*, 1(1): 17–23.

Varga, P. et al. (2022). Converging Telco-Grade Solutions 5G and beyond to Support Production in Industry 4.0. *Applied Sciences*, 12(15): 1–44.

Yang, X. (2022). On the New Mode of Education and Teaching in the Age of 5G. Cham, Springer.

14

Metal-air Batteries for Wearable Electronics:

A Case Study for Modern Society

Arpana Agarwal[1],* and *Chaudhery Mustansar Hussain*[2]

1. Introduction

Smart clothing, smart watches, bandages, and various flexible wearable healthcare equipment, sensors, etc. are few of the innovative concepts that have been imagined and have become an important part of our modern society (Zhang et al., 2016; Nishide and Oyaizu, 2008; Liu et al., 2017). Widespread interest in science and technology has been generated by the development of such downsized, more adaptable, stretchy, movable apparel with electronic devices. For energizing such electronic devices, efficient power systems are required that are compatible to the human skin. Looking at the demands of modern society, traditional metal-air batteries (MABs), because of being bulky and non-flexible remain unsuitable for fabricating today's wearable electronic devices. Consequently, flexible gear with MABs have emerged as innovative and promising candidates for fabricating such new generation devices owing to their ability of long-lived power supply possessing increased specific capacities, high energy density, economic

[1] Department of Physics, Shri Neelkantheshwar Government Post-Graduate College, Khandwa-450001, India.
[2] Department of Chemistry and Environmental Science, New Jersey Institute of Technology, USA.
* Corresponding author: agrawal.arpana01@gmail.com

viability, long life, accessibility of raw materials and environmental responsibility, flexibility, and outstanding adaptability to irregular geometric surfaces of human skin (Peng et al., 2017; Ji et al., 2019). Such flexible clothing with MABs should possess encouraging electrochemical performances even under various deformable circumstances including rolling, bending, twisting, folding, stretching, etc., which significantly affects their wearability.

Flexible wearable MABs mainly consist of supple electrolytes either gel or polymer-based electrolyte and flexible electrode materials mainly metal anode, and air cathode (Vassal et al., 2000; Yang and Lin, 2003; Wu et al., 2006; Balaish et al., 2015; Zhang et al., 2014; Kuboki et al., 2015). It should be emphasized that the air electrode is essential to the battery's operation and affects both the mechanical flexibility required for ductile batteries as well as the electrocatalytic capabilities. Oxygen reduction/evolution reaction (ORR/OER) on air cathode, and cation dissolution/deposition on anode, are the fundamental operating principles of MABs. Among the various available MABs, zinc-air battery (ZAB), lithium-air battery (LAB), and aluminium-air battery (AAB) are considerably more popular.

Apart from the flexible battery components, significant work has also been carried out to achieve MAB's flexibility by modifying their structural design either in planar sandwich-type structure formed by sandwiching the electrolyte in between the electrodes or fiber-type structures where the battery components are coaxially arranged. Figure 14.1 schematically illustrates wearable flexible MABs (ZABs, LABs, and AABs) along with few advantages offered by them. It is worth noting that while sandwich structures allow uniaxial bending, they cannot be employed to examine other

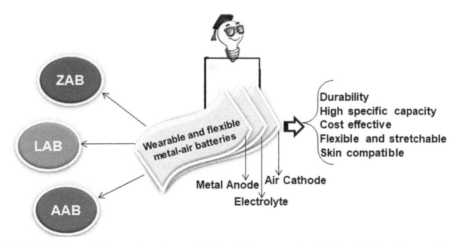

Fig. 14.1 Schematic illustration of wearable flexible MABs (ZABs, LABs and AABs) along with their advantages.

deformation circumstances such as twisting (Weng et al., 2016). In contrast to this, fiber architectures are used to address this disadvantage. However, both the structures have their own advantages and have been immensely employed for fabricating wearable devices of modern society.

Accordingly, the present chapter demonstrates several case studies for apparel with MABs that are critically relevant to satisfy the need of our modern society, with a focus on the investigation of various flexible components of ZABs, LABs, and AABs required for the manufacture of workable and wearable MABs.

2. Flexible Components of Wearable Metal-air Batteries

Upcoming clothing of the modern world is often carried directly on curvy and sensitive human skins and can withstand a variety of deformable circumstances. As a result, breakthroughs in flexible and stretchy materials with exceptional mechanical and electronic qualities are crucial. Malleable components of MAB include either an electrode/electrolyte or both. The electrolyte facilitates passage of metal ions/oxygen species while charge/discharge operations and unfortunately, repetitive mechanical deformations of the flexible battery during actual applications result in leaking of the typical liquid electrolyte. Furthermore, because air oxygen acts as one of the reactants and energy sources, the MAB has an open structure. In these instances, leakage difficulties include battery failure and contamination of environment. To circumvent such stability and safety issues, designing flexible electrolytes serving as separator/ion conductor is highly essential and helps in minimizing leakage difficulties.

The utilization of a gel or polymer strategy in the building of flexible electrolytes for MABs is a useful mechanism owing to their outstanding flexibility and mechanical properties under numerous deformations. Pliable anode electrodes mainly the metallic foils/sheets, in addition to the flexible electrolyte, are critical for the construction of a wearable MAB. However, because MABs rely on the ORR and OER, the search for a supple air cathode has become significantly more vital. Unfortunately, many flexible substrates have limited porosity with weak electrochemical catalytic activities, decreasing specific capacity. Accordingly, scientific community is looking for stretchable and highly conductive materials with exceptional catalytic activity that suit the expectations enforced by the fabrication of a flexible MAB. As a result, conductive polymers, and carbonaceous materials (carbon nanofibers (CNF), carbon cloths/textiles, carbon nanotubes (CNT), graphene-based materials, etc.), have been immensely utilized as flexible electrode substrates for a wearable MAB. Polymers are naturally ductile and carbon-based materials are desirable because of their inexpensiveness, good electrical conductivity, and mechanical durability under a variety of deforming situations. In this section, various case studies of wearable flexible MABs will be discussed that are highly important for our modern society.

2.1 Zinc-air batteries

ZABs are among the most fascinating and potential power sources, particularly MABs. So far, various researchers have designed and fabricated these batteries and utilized them for practical applications. Fu et al. (Fu et al., 2015) were the first to propose a planar sandwich structured rechargeable flexible ZAB where a poly(vinyl alcohol) (PVA)-based electrolyte is inserted in between the Zn anode and carbon cloth serving as cathode filled with a bifunctional catalyst (LaNiO$_3$/NCNT). The constructed flexible ZAB can tolerate a variety of deformable situations and exhibits unwavering behavior for 120 cycles and outstanding energy density (581 Wh kg^{-1} at 25 A kg^{-1}).

Three-dimensional hierarchical CoNi alloy/NCNSAs/CC-800 flexible ZAB with an improved mechanical cycle capability, an elevated energy density of 98.8 mW cm^{-2}, and a capacity of 879 mAhg^{-1} was also reported which when series assembled can power 51-red LED to illuminate the 'HBU' indication, indicating ZABs as a feasible alternative for storage of electrochemical energy (Zhang et al., 2020). Ji et al. (Ji et al., 2019) have fabricated a flexible wearable ZAB using a binder free air-cathode made up of atomically transition metals and porous carbon flake array.

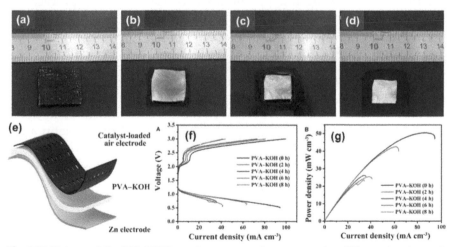

Fig. 14.2 Pictures of the PVA-KOH membrane upon air exposure after different time intervals ranging from 0–24 h ((a–d): 0, 6, 12, and 24 h. (e) Schematic representation of flexible ZAB. Polarization results (f) and power density results (g) of PVA-KOH-based flexible ZABs. (Reprinted from Fan et al. (Fan et al., 2019) under Creative Commons Attribution License (CCBY).)

Fan et al. (Fan et al., 2019), have fabricated a flexible ZAB by sandwiching the PVA-KOH membrane serving as a polymer electrolyte between Co_3O_4- loaded air-cathode and Zn anode. They have also examined the stability of the electrolyte in environmental conditions including ionic conductivity and retention capability. Figure 14.2(a–d) illustrate the pictures of the PVA-KOH membrane upon air exposure after different time intervals ranging from 0–24 h. Figure 14.2(e) schematically illustrates the ZAB which clearly indicates the excellent battery bendability. Polarization and the power density results of the fabricated flexible ZAB after air exposure of polymer membrane are illustrated in Figs. 14.2(f) and 14.2(g), respectively.

The same group have also fabricated durable flexible ZAB using porous nanocomposite gel-polymer electrolyte (Fan et al., 2019). Guan et al. (Guan et al., 2017) utilized carbon arrays containing hollow Co_3O_4 nanosphere for fabricating flexible solid state ZABs. Fu et al. (Fu et al., 2019) presented a wearable band-aid composed of tiny ZAB that serves as a flexible substrate and is air-permeable. The device is exceedingly flexible and rechargeable, with a cellulose electrolyte based membrane layered between Co_3O_4-loaded carbon cloth cathode and Zn metal anode made from Zn powder, carbon nanofiber, carbon black, and a polymer binder. When twisted all around the finger, this band aid can brighten up the LED under bending and has a high specific capacity. Jing et al. (Jing et al., 2016) showed the efficacy of a rechargeable air cathode in a flexible ZAB with a three-dimension nanoarchitecture morphologically resembling the human hair. To construct such an air electrode, nanoscale hair-like catalysts were precisely grown laterally on a flexible stainless-steel mesh. Chemical vapor deposition was used to create the nitrogen-doped multiwalled CNT (NCNT), proceeded by iron catalyst electrodeposition and calcination. Using electrodeposition and annealing, the NCNT was then constructed with Co_3O_4 nanopetals. The prepared air cathode displayed exceptional flexibility even under 360° twisting condition. The manufactured flexible ZAB has an encouraging energy density (847.6 Wh kg^{-1} at 25 mAcm^{-2}) and a steady cycle performance, signifying its applicability for smart wearable devices.

Park et al. (Park et al., 2015) developed the first fiber-structured flexible ZAB using a Zn foil based spiral anode and placing it on a gelatin and KOH electrolyte loaded template and chilling it to bridge the electrolyte. The resultant Zn anode was then coiled with an air cathode containing a Fe/N/C catalyst before being enclosed in a rubber cable. At 0.1 mAcm^{-2} and 0.92 V discharge voltage, the battery displayed a voltage plateau. Using cross-stacked porous CNT sheets simultaneously serving as a gas diffusion/ catalyst layer, and current collector, Xu et al. (Xu et al., 2015) showed the fabrication of a flexible, stretchy, and wearable fiber-shaped ZAB. The ZAB is made up of a PVA/PEO/KOH electrolyte enfolded around the Zn metal spring electrode before being covered with RuO_2 and multiwall CNT sheets to create an air electrode. To minimize additional electrolyte water evaporation, the produced fiber-shaped ZAB might be enclosed in a punched tube. At 1A g^{-1}, it revealed a discharging and charging voltages of 1.0 V and 1.9 V for more than 30 cycles. The discharge potential remained consistent upon bending the battery at varied angles. Alongside flexibility, the MAB was also allowed to be stretched by 10% without any visible structural failure or performance degradation.

Shi et al. (Shi et al., 2019) have manufactured ZAB utilizing the CNT-based framework (CNTF) obtained from metal-organic crystals, namely, CoFe and CoNi-MOFs. Two routes were followed: (1) to obtain CoFe-based CNTF, CoNi-based CNTF, and Co-based CNTF and (2) to produce CoFe-MOFs on three- dimension graphene foam designated as CoFe-CNTF@3DG. For the first route, initially metallic salt mixtures were utilized to produce a bimetallic MOF crystal (serving as a precursor) followed by annealing to pyrolyze MOFs and initiate the growth of CNTs. The obtained product was then made to undergo acid treatment to obtain the desired CNTFs. To grow the CoFe-CNTF@3DG electrode, the same annealing process is employed

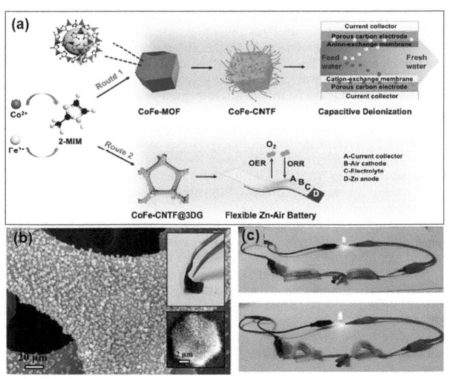

Fig. 14.3 (a) Pictorial representation of overall growth routes of CoFe-CNTF and CoFe-MOFs on 3D graphene foam with their applicability in ZAB. (b) SEM scans of CoFe-CNTF@3DG. Insets shows magnified SEM scans and picture of bent CoFe-CNTF@3DG. (D) LED powered by ZABs assembled in series under two bending conditions. (Reprinted from Shi et al. (Shi et al., 2019) under Creative Commons Attribution License (CC BY)).

for the *in-situ* production of CoFe-MOFs on a 3D graphene foam. The overall growth process of CoFe-based CNTF and CoFe-CNTF grown on a 3D graphene with the usages for ZAB is illustrated in Fig. 14.3(a). Scanning electron microscopy (SEM) scans of CoFe-CNTF@3DG are depicted in Fig. 14.3(b), which clearly reveals its excellent flexibility. Utilizing this electrode, ZABs were fabricated which when connected in series even under various bending conditions can glow a LED bulb as shown in Fig. 14.3(c).

Zhiqian et al. (Zhiqian et al., 2017) have developed a flexible ZAB by adding multiwall CNTs into a Zn anode to improve its electrochemical performance. The ZAB also consists of a paper separator coated with poly(acrylic acid) and PVA to enhance the electrolyte storage capacity. The fabricated battery is extremely flexible and can endure several bending circumstances.

Meng et al. (Meng et al., 2016) showed the construction of a ZAB with strong electrochemical activities and excellent capacity (774 mAhg^{-1} at 10 mAcm^{-2}), utilizing a novel stretchable electrode comprising of Co4N, carbon network, and carbon cloth (Co4N/CNW/CC) made via the pyrolysis of ZIF-67 on a polypyrrole nanofiber network developed on carbon cloths (ZIF-67/PNW/CC), with nitrogenous gases generated during polypyrrole pyrolysis boosting the ZIF-67 conversion to Co4N. The resultant air electrode was very flexible and had a large specific surface area. With a low overpotential (0.31V), a positive midway potential (0.8V), and consistent current density retention, both OER and ORR displayed good and long-term catalytic activity (20 h).

Niu et al. (Niu et al., 2019) have fabricated an efficient flexible long-lived ZAB utilizing metal-heteroatom-doped carbon nanofibers, particularly, Zn/Co-N@PCNFs. Figure 14.4(a) shows the overall growth process for synthesizing Zn/Co-N@PCNFs with their micromorphologies prior to and after the carbonization process The stimulated cross-sections as well as molecular structure diagrams are depicted in Figs. 14.4(b) and 14.4(c), respectively. Figure 14.4(d) gives digital pictures of the grown nanofibers before and after carbonization. Design of a knittable ZAB utilizing the atomic layer Co$_3$O$_4$ nanosheets [27] has also been reported.

Qu et al. (Qu et al., 2017) have demonstrated an electrochemical approach for synthesizing an air cathode for a flexible ZAB array possessing controllable output voltage and current. They built a layer-by-layer Co$_3$O$_4$ nanosheet-based ZAB array that exhibits a sustained electrochemical response at 100% strain. After reorganizing the electrode array, this manufactured ZAB might get discharged under bending circumstances, producing voltages ranging from 1 to 4 V. Employing nonporous carbon fiber sheets (bifunctional catalytic electrodes), Liu et al. (Liu et al., 2016) showed the production of a flexible ZAB. Shinde et al. (Shinde et al., 2018) presented a flexible ZAB made up of three-dimensional C2N aerogels being a bifunctional oxygen electrode.

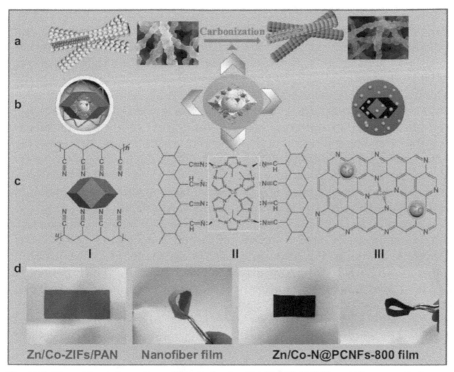

Fig. 14.4 Pictorial representation of the synthesis of Zn/Co–N@PCNFs. (a) Micromorphologies of nanofibers before and after carbonization. Diagrams of simulated cross-section (b) with their molecular structures (c). (d) Photographs of the nanofiber film before and after carbonization. (Reprinted from Niu et al. [26] under Creative Commons Attribution 4.0 International License. Simulated)

2.2 Lithium-air batteries

Apart from ZABs, LABs are also very popular for fabricating wearable electronic devices for the modern society. Lei et al. [31] have demonstrated the fabrication of a flexible LAB operating under an ambient air condition with an in situ prepared gel electrolyte which effectively lessens the Li metal anode rusting and hence improves cycle performance. Jaradat et al. (Jaradat et al., 2021) have reported the construction of an air-breathing flexible LAB. Liu et al. (Liu et al., 2019) reported the method of shielding the Li anode for a safe and flexible LAB operating in ambient air. A flexible zeolite electrolyte was also reported to be employed to fabricate LAB (Chi et al., 2021). Zou et al. (Zou et al., 2018) have utilized a gel polymer electrolyte for fabricating an LAB that can work under severe conditions. The gel polymer employed was highly flexible, flame-resistant, and dendrite impermeable. An extremely safe multifunctional flexible LAB was also fabricated by Shu et al. (Shu et al., 2019) utilizing a quasi-solid electrolyte.

A sandwich architecture (Liu et al., 2015) Li-O$_2$ battery was also constructed by layering flexible and conductive paper-based ink cathode, glass fiber membrane, and Li foil anode. Even after 1000-time folds, no discernible change was found in the charging-discharging patterns. Relying on an integrated framework, Li et al. (Li et al., 2021) showed that a flexible LAB works under extreme circumstances. By employing a Co$_3$O$_4$@ MnO$_2$ cathode, a composite Li metal electrode sheathed in a gel electrolyte, they created an exceptional battery design. A composite Li anode manufactured using a simple rolling technique significantly decreases the aging fracture of the Li electrode. After this, the gel electrolyte encompasses the composite Li electrode, to lower the interface resistance of electrode/electrolyte and hence serves as a protective layer against the corrosion of the Li anode. This, in turn, leads to improved cycle stability.

Liu et al. (Liu et al., 2015) designed a flexible LAB using a Li anode and TiO$_2$ nanowire-based arrays seeded on carbon textiles (TiO$_2$ NAs/CT) as cathode as shown in Fig. 14.5(a) depicts the schematic of LAB with SEM image of discharged cathode (Current density: 100 mAg^{-1}) in Fig. 14.5(b). The fabricated LAB shows excellent performance even under a twisting angle of 360° as shown in Fig. 14.5(c). They have also examined the battery behavior with recovered cathodes and the SEM image of 10th recovered cathode is depicted in Fig. 14.5(d) with its terminal discharge voltage

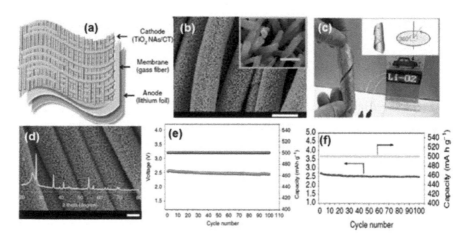

Fig. 14.5 (a) Pictorial representation of LAB made up of a Li metal, TiO$_2$ NAs/CT as anode and cathode electrodes and membrane. (b) SEM images of discharged TiO$_2$ NAs/CT cathode (Current density: 100mAg^{-1}; scale bar, 10 mm). Inset depicts high magnification SEM image (Scale bars: 500 nm). (c) Twisting behavior of the fabricated LAB. (d) SEM scan of the 10th-recovered cathode (Scale bar, 5 mm). Inset shows the respective X-ray diffraction profile. (e) Dependency of discharge voltage of 10th recovered TiO$_2$ NAs/CT cathode at terminals on cycle number. (f) Dependency of fabricated LAB's terminal discharge voltage on cycle number corresponding to twisting angle of 360°. (Reprinted from Liu et al. [39] under Creative Common License.)

dependency on the cycle number in Fig. 14.5(e). Under a 360° twisting angle, the dependency of terminal discharge voltage of th fabricated LAB on cycle number is represented in Fig. 14.5(f). Dai et al. [40] conducted a thorough evaluation of graphene-based electrodes for different flexible electrochemical devices. Remarkably, a wearable LAB based on bamboo slips has also been described [41], where the electrodes are knitted with no air diffusion layer, allowing oxygen to easily access the cathodes from both sides, giving the battery a record energy density.

It has also been established that a highly flexible wearable, foldable, and stretchy LAB with better electrochemical properties can be constructed and stay stable under repeated deforming circumstances (Wang et al., 2016). This LAB is prepared using a polymer electrolyte, Li-array anode, and a rippling air cathode made of aligned multiwalled carbon nanotube sheets. Liu et al. [43] presented an incredibly thin, lightweight, and wearable LAB with an exceptional battery performance even after numerous repeated deformations.

2.3 Aluminium-air batteries

AABs have also reported to be very useful for powering electronic devices and for industrial purposes. Chen et al. (Chen et al., 2021) reported a flexible AAB consisting of a PVA/LiCl/PEO interpenetrating composite electrolyte, whereas Shui et al. (Shui et al., 2022) have employed an ionic liquid gel polymer electrolyte for constructing a pliable LAB. Design of a cellulose paper inspired flexible wearable AAB was reported by Wang et al. (Wang et al., 2019), employing an Al foil enveloped within the paper substrate at some stage of producing the paper which serves as the anode and the oxygen reduction ink was used to accumulate the cathode on the paper substrate. Herein, manufacturing an Al foil embedded cellulose paper-based substrate, highly pure Al foil was sandwiched between layers of the paper pulp that were uniformly dispersed across a stainless-steel mesh. The constructed three-layered anode structure consists of paper pulp/Al foil/paper pulp where the paper pulp is then removed from the stainless-steel mesh and evenly compressed to eliminate any remaining moisture content. Finally, drying the structure at 60°C for 30 minutes ensures structural stability. The developed anode structure had a very thin overall thickness of 0.57 ± 0.02 mm, which allowed for less ionic resistance and thus increased its power efficiency. The MnO_2 and multiwalled CNTs used in the ORR-based air-breathing cathode, on the other side, served as the ORR catalyst and catalyst support, respectively, along with a binder, all of which were distributed and sonicated in an ethanol water solvent. Following that, this ready-made ink was sprayed onto an Ag grid (area: 5×5 cm^2) put on paper substrate. The limitation of the cathode area is made possible by such a design. The same group has also shown how to make an inexpensive paper-based screen-printed AAB with high energy density that can be used to power electrical devices. On cellulose paper, they have printed ink that contains Al microspheres, CNTs, a cellulose binder, and oxygen reduction inks (Wang et al., 2020). They also noted that a paper-based AAB for the miniwatt market was influenced by green energy technology in another work (Wang et al., 2019). For this, PMMA shells were surrounded by an Al foil, a filter or cellulose paper, and carbon paper, which served as the anode, electrolyte substrate, and cathode, respectively. Valisevskis et al. (Valisevskis et al., 2020) demonstrated a textile inspired flexible and eco-friendly AAB comprising of fabric coated with Ag serving as the cathode.

Paper inspired Al/Polyaniline (PANI) air batteries were designed by Cao et al. [50] as shown in Fig. 14.6(a), where the Al anode, PANI/Fe/AQDS cathode, preservation film, separator of anode and cathode and gel electrolyte of NH_4Cl, TEA and $NaNO_3$, all are produced on paper. The cathode is made from filter paper and exhibits a discharge capacity of 50 mAhcm^{-2}. Paper-based gel electrolyte ensures the eco-friendliness of the fabricated AAB. Figure 14.6(b) shows the polarization curves of the Al anode in various electrolyte solutions comprising of TEA, NH_4Cl, and $NaNO_3$; and a mixture of $NaNO_3$ and TEA. Figure 14.6 clearly shows that the electrolyte containing a mixture of TEA and $NaNO_3$ significantly affects the polarization curve of the Al electrode while for other, it shows a similar kind of behavior. Avoundjian et al. (Avoundjian et al., 2017) fabricated more economic compact flexible AAB where Kim wipes were used as the paper substrate, and folded aluminium served as the anode with 1.5 M KOH serving as the electrolyte. This 9 cm^2 AAB may deliver promising current (17.4 mA) and power (3 mW).

Despite such advances in the fabrication of wearable flexible MABs, a few challenges remain, including (i) creation of an interphase film of electrolyte because of the reaction between an anode material and an

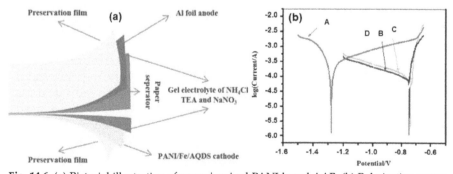

Fig. 14.6 (a) Pictorial illustration of paper inspired PANI-based AAB. (b) Polarization curves of Al electrodes in several electrolyte solutions (The scanning rate: 1 mV/s); (NaNO$_3$ + TEA) (A); TEA (B); NH$_4$Cl (C); NaNO$_3$ (D). (Reprinted from Cao et al. (Cao et al., 2019)under Creative Common Licence.)

electrolyte, resulting in irreversible loss in battery efficiency (Mengeritsky et al., 1996); (ii) dendrite growth on the anode electrode, causing severe short circuit in the MAB and thus deteriorating battery efficiency (Zhang et al., 2016); and (iii) desired electrolyte, and (iv) stability of materials where cathodic reaction occurs in the MAB. All these factors weaken the cyclability of MABs which hampers battery performance (Thotiyl et al., 2013).

Conclusion

In conclusion, the present chapter provides few case studies of wearable MABs particularly ZABs, LABs, and AABs, which are extremely important for our modern society. All the flexible battery components (flexible electrolytes, either gel or polymer; and flexible electrodes) and structural designs (planar sandwich and fiber-type structure) have been critically discussed. Flexible wearable MABs that have been discussed exhibit outstanding electrochemical performances even under twisting, folding, rolling, bending, and stretching conditions. Moreover, collective efforts are further required to come out of the hurdles that limit their utility for fabricating electronic devices of the modern society and hence undertake the commercialization of flexible wearable MABs.

References

Avoundjian, A., Galvan, V. and Gomez, F.A. (2017). An inexpensive paper-based aluminum–air battery. Micromachines, 8: 222.

Cao, H., Si, S., Xu, X., Li, J. and Lan, C. (2019). A novel flexible aluminum//polyaniline air battery. Int. J. Electrochem. Sci., 14: 9796–9804.

Chen, L., Li, B., Zhu, L., Deng, X., Sun, X., Liu, Y., Zhang, C., Zhao, W. and Chen, X. (2021). RSC Adv., 13: 39476–39483.

Chen, X., Zhong, C., Liu, B., Liu, Z., Bi, X., Zhao, N., Han, X. Deng, Y., Lu, J. and Hu, W. (2018). Atomic layer Co_3O_4 nanosheets: The key to knittable Zn–air batteries. Small, 14: 1702987.

Chi, X., Li, M., Di, J., Bai, P., Song, L., Wang, X., Li, F., Liang, S., Xu, J. and Yu, J. (2021). A highly stable and flexible zeolite electrolyte solid-state Li–air battery. Nature, 592: 551–557.

Dai, C., Sun, G., Hu, L., Xiao, Y., Zhang, Z. and Qu, L. (2020). Recent progress in graphene-based electrodes for flexible batteries. InfoMat,, 2: 509–526.

Fan, X., Liu, J., Ding, J., Deng, Y., Han, X., Hu, W. and Zhong, C. (2019). Investigation of the Environmental Stability of Poly(vinyl alcohol)–KOH Polymer Electrolytes for Flexible Zinc–Air Batteries. Front. Chem., 7: 678.

Fan, X., Liu, J., Song, Z., Han, X., Deng, Y., Zhong, C. and Hu, W. (2019). Porous nanocomposite gel polymer electrolyte with high ionic conductivity and superior electrolyte retention capability for long-cycle-life flexible zinc–air batteries. Nano Energy, 56: 454–462.

Fu, J., Lee, D.U., Hassan, F.M., Yang, L., Bai, Z., Park, M.G. and Chen, Z. (2015). Flexible high-energy polymer-electrolyte-based rechargeable zinc–air batteries. Adv. Mater., 27: 5617–5622.

Fu, J., Zhang, J., Song, X., Zarrin, H., Tian, X., Qiao, J., Rasen, L., Li, K., Chen, Z. (2016). A flexible solid-state electrolyte for wide-scale integration of rechargeable zinc–air batteries. Energy Environ Sci., 9: 663–670.

Guan, C., Sumboja, A., Wu, H.J., Ren, W.N., Liu, X.M., Zhang, H., et al. (2017). Hollow Co3O4 nanosphere embedded in carbon arrays for stable and flexible solid-state zinc–air batteries. Adv. Mater., 29:1704117.

Jaradat, A., Zhang, C., Singh, S.K., Ahmed, J., Ahmadiparidari, A., Majidi, L., Rastegar, S., et al. (2021). A High Performance Air Breathing Flexible Lithium Jaradat,Air Battery. Small, 17: 2102072.

Ji, D., Fan, L., Li, L., Peng, S., Yu, D., Song, J., et al. (2019). Atomically transition metals on self-supported porous carbon flake arrays as binder-free air-cathode for wearable zinc–air batteries. Adv. Mater.31: 1808267.

Ji, D., Fan, L., Li, L., Peng, S., Yu, D., Song, J., Ramakrishna, S. and Guo, S. (2019). Atomically transition metals on self-supported porous carbon flake arrays as binder-free air cathode for wearable zinc– air batteries. Adv. Mater., 31: 1808267.Vassal, N., Salmon, E., Fauvarque, J.F. (2000). Electrochemical properties of an alkaline solid polymer electrolyte based on P (ECH-co-EO). Electrochim. Acta., 45: 1527–1532.

Jing, F., Hassan, F.M., Li, J., Lee, D.U., Ghannoum, A.R., Liu, G., Hoque, M.A. and Chen, Z. (2016). Flexible rechargeable zinc-air batteries through morphological emulation of human hair array. Adv. Mater., 28: 6421–6428.

Kuboki, T., Okuyama, T., Ohsaki, T. and Takami, N. (2015). Lithium-air batteries using hydrophobic room temperature ionic liquid electrolyte. J. Power Sources, 146: 766–769.

Lei, X., Liu, X., Ma, W., Cao, Z., Wang, Y. and Ding, Y. (2018). Flexible lithium–air battery in ambient air with an in situ formed gel electrolyte. Angew. Chem. Int. Ed., 57: 16131–16135.

Li, J., Wang, Z., Yang, L., Liu, Y., Xing, Y., Zhang, S. and Xu, H. (2021). A flexible li–air battery workable under harsh conditions based on an integrated structure: A composite lithium anode encased in a gel electrolyte. ACS Appl. Mater. Interfaces, 13: 18627–18637.

Liu, Q., Wang, Y., Dai, L. and Yao, J. (2016).Scalable fabrication of nanoporous carbon fiber films as bifunctional catalytic electrodes for flexible Zn–air batteries. Adv. Mater., 28: 3000–3006.

Liu, Q.C., Li, L., Xu, J.J., Chang, Z.W., Xu, D., Yin, Y.B., Yang, X.Y., Liu, T., Jiang, Y.S., Yan, J.M. and Zhang, X.B. (2015). Flexible and foldable Li–O$_2$ battery based on paper-ink cathode. Adv. Mater., 27 8095–8101.

Liu, Q.C., Liu, T., Liu, D.P., Li, Z.J. Zhang, X.B. and Zhang, Y. (2016). A flexible and wearable lithium–oxygen battery with record energy density achieved by the interlaced architecture inspired by bamboo slips. Adv. Mater., 28: 8413–8418.

Liu, Q.C., Xu, J.J., Xu, D. and Zhang, X.B. (2015). Flexible lithium–oxygen battery based on a recoverable cathode. Nat. Commun., 6: 1–8.

Liu, T., Feng, X.L., Jin, X., Shao, M.Z., Su, Y.T., Zhang, Y. and Zhang, X.B. (2019). Protecting the Lithium Metal Anode for a Safe Flexible Lithium–Air Battery in Ambient Air. Angew. Chem. Int. Ed. Engl., 9: 18240–18245.

Liu, T., Xu, J.J., Liu, Q.C., Chang, Z.W., Yin, Y.B., Yang, X.Y. and Zhang, X.B. (2017). Ultrathin, Lightweight, and Wearable Li–O$_2$ Battery with High Robustness and Gravimetric/Volumetric Energy Density. Small, 13: 1602952.

Liu, W., Feng, K., Zhang, Y., Yu, T., Han, L., Liu, G., Li, M., Chiu, G. Fung, P., Yu, A. (2017). Hair-based flexible knittable supercapacitor with wide operating voltage and ultra-high-rate capability. Nano Energy, 34: 491–499.

Meng, F., Zhong, H., Bao, D., Yan, J. and Zhang, X. (2016). In situ coupling of strung Co$_4$N and intertwined N–C fibers toward free-standing bifunctional cathode for robust, efficient, and flexible Zn–air batteries. J. Am. Chem. Soc., 138: 10226–10231.

Mengeritsky, E., Dan, P., Weissman, I., Zaban, A. and Aurbach, D. (1996). Safety and Performance of Tadiran TLR-7103 Rechargeable Batteries J. Electrochem. Soc., 143: 2110.

Nishide and Oyaizu, K. (2008). Toward flexible batteries., Science, 319: 737–738.

Niu, Q., Chen, B., Guo, J., Nie, J., Guo, X., Ma, G. (2019). Flexible, Porous, and Metal–Heteroatom-Doped Carbon Nanofibers as Efficient ORR Electrocatalysts for Zn–Air Battery. Nano-Micro Lett., 11: 8.

Park, J., Park, M., Nam, G., Lee, J.S., Cho, J. (2015). All-solid-state cable-type flexible zinc–air battery. Adv. Mater., 27: 1396–1401.

Peng, T., Chen, B., Xu, H., Zhang, H., Cai, W., Ni, M., Liu, M. and Shao, Z. (2017). Flexible Zn- and Li-air batteries: Recent advances, challenges, and future perspectives., Energy Environ. Sci., 10: 2056–2080.

Qu, S., Song, Z., Liu, J., Li, Y., Kou, Y., Ma, C., Han, X., Deng, Y., Zhao, N., Hu, W. and Zhong, C. (2017). Electrochemical approach to prepare integrated air electrodes for highly stretchable zinc–air battery array with tunable output voltage and current for wearable electronics. Nano Energy, 39: 101–110.

Serial nos. of refs to be inserted within [ ], e.g., (Mengeritsky et al., 1996).

Shi, W., Xu, X., Ye, C., Sha, D., Yin, R., Shen, X., Liu, X., Liu, W., Shen, J., Cao, X. and Gao, C. (2019). Bimetallic Metal-Organic Framework-derived Carbon Nanotube-based Frameworks for Enhanced Capacitive Deionization and Zn–Air Battery. Front. Chem., 7: 449.

Shinde, S.S., Lee, C.H., Yu, J.-Y., Kim, D.-H., Lee, S.U. and Lee, J.-H. (2018). Hierarchically designed 3D holey C$_2$N aerogels as bifunctional oxygen electrodes for flexible and rechargeable Zn–air batteries., ACS Nano., 12: 596–608.

Shu, C., Long, J., Dou, S.X. and Wang, J. (2019). Component-Interaction Reinforced Quasi-Solid Electrolyte with Multifunctionality for Flexible Li–O$_2$ Battery with Superior Safety under Extreme Conditions. Small, 15: 1804701.

Shui, Z., Chen, Y., Zhao, W. and Chen, X., (2022). Flexible Aluminum–Air Battery Based on Ionic Liquid-Gel Polymer Electrolyte., Langmuir, 6: 10791–10798.

Thotiyl, , Freunberger, S.A., Peng, Z. and Bruce, P.G. (2013). The Carbon Electrode in Nonaqueous Li–O$_2$ Cells. J. Am. Chem. Soc., 135: 494.

Vališevskis, A., Briedis, U., Juchnevičienė, Z., Jucienė, M. and Carvalho, M. (2020). Design improvement of flexible textile aluminium–air battery. J. Text. Inst., 111: 985–990.

Wang, L., Zhang, Y., Pan, J. and Peng, H. (2016). Stretchable lithium–air batteries for wearable electronics. J. Mater. Chem. A, 4: 13419–13424.

Wang, Y., Kwok, H.Y., Pan, W., Zhang, H. and Leung, D.Y. (2019). Innovative paper-based Al–air batteries as a low-cost and green energy technology for the miniwatt market. J. Power Sources, 414: 278–282.

Wang, Y., Kwok, H.Y., Pan, W., Zhang, Y., Zhang, H., Lu, X. and Leung, D.Y. (2019). Combining Al–air battery with paper-making industry, a novel type of flexible primary battery technology, Electrochimica Acta., 319: 947–957.

Wang, Y., Kwok, H.Y., Pan, W., Zhang, Y., Zhang, H., Lu, X. and Leung, D.Y. (2020), Printing Al-air batteries on paper for powering disposable printed electronics. J. Power Sources, 450: 227685.

Weng, W., Chen, P., He, S., Sun, X. and Peng, H. (2016). Smart electronic textiles. Angew. Chem. Int. Ed., 55: 6140–6169.

Wu, G.M., Lin, S.J. and Yang, C.C. (2006). Preparation and characterization of PVA/PAA membranes for solid polymer electrolytes. J. Membr. Sci., 275: 127–133.Balaish, M., Peled, E., Golodnitsky, D. and Ein-Eli, Y. (2015). Liquid-free lithium–oxygen batteries. Angew. Chem., 127: 446–450.

Xu, Y., Zhang, Y., Guo, Z., Ren, J., Wang,Y. and Peng, H. (2015). Flexible, stretchable, and rechargeable fiber-shaped zinc–air battery based on cross-stacked carbon nanotube sheets. Angew. Chem., 127: 15610–15614.

Yang, C.C. and Lin, S.J. (2003). Preparation of alkaline PVA-based polymer electrolytes for Ni–MH and Zn–air batteries. J. Appl. Electrochem., 33: 777–784.

Zhang, H., Ma, X., Lin, C. and Zhu, B. (2014). Gel polymer electrolyte-based on PVDF/fluorinated amphiphilic copolymer blends for high performance lithium-ion batteries. RSC Adv., 4: 33713–33719.

Zhang, J., Fu, , X. J., Jiang, G., Zarrin, H., Xu, P., Li, K., Yu, A. and Chen, Z. (2016). Laminated cross-linked nanocellulose/graphene oxide electrolyte for flexible rechargeable zinc–air batteries. Adv. Energy Mater., 6: 1600476.

Zhang, W., Li, Z., Chen, J., Wang, X., Li, X., Yang, K. and Li, L. (2020). Three-dimensional CoNi alloy nanoparticle and carbon nanotube decorated N doped carbon nanosheet arrays for use as bifunctional electrocatalysts in wearable and flexible Zn–air batteries. Nanotechnology, 31: 185703.

Zhang, X., Wang, X.-G., Xie, Z. and Zhou, Z. (2016). Recent Progress in Rechargeable Alkali Metal–Air Batteries. Green Energy and Environment, 1: 4.

Zhiqian W.,, Xianyang, M., Zheqiong, W. and Mitra, S. (2017). Development of flexible zinc–air battery with nanocomposite electrodes and a novel separator. J. Energy Chem., 26: 129–138.

Zou, X., Lu, Q., Zhong, Y., Liao, K., Zhou, W. and Shao, Z. (2018). Flexible, Flame-Resistant, and Dendrite-Impermeable Gel-Polymer Electrolyte for Li–O$_2$/Air Batteries Workable Under Hurdle Conditions. Small, 14: 1801798.

15

Lifecycle Assessment of Alternative Building Materials

Marco Ruggiero,* Cinzia Salzano, Marta Travaglioni, Francesco Colangelo and *Ilenia Farina*

1. Introduction

In the last few decades, increasing awareness of more sustainable design has led many researchers to investigate possible recycling and recovery operations in the field of building materials and technologies. These studies have long demonstrated the feasibility of producing sustainable materials using recycled aggregates. Particularly, they have extensively focused their attention on how to use waste materials in place of commonly used virgin materials. This replacement leads to important benefits regarding the reduction of the use of raw materials, which unfortunately are increasingly scarce, the reduction of energy-intensive processes, and the reduction of emissions harmful to the environment, resulting in improvement of human health and well-being of the ecosystem. Among these studies, several cases can be investigated including the use of industrial aggregates

Department of Engineering, University of Naples "Parthenope", Centro Direzionale, Naples, Italy.
Emails: cinzia.salzano001@studenti.uniparthenope.it; marta.travaglioni@collaboratore. uniparthenope.it; francesco.colangelo@uniparthenope.it; ilenia.farina@uniparthenope.it
* Corresponding author: marco.ruggiero1@studenti.uniparthnope.it

(Massoudinejad et al., 2019), agricultural (Raut and Gomez, 2017; Chippagiri et al., 2021), construction and demolition waste (Colangelo and Cioffi, 2017), recycled aggregates (Xiao et al., 2007; Liu et al., 2019; Saribas et al., 2019; Meng et al., 2020) and light-weight aggregates (LWAs) (Akan et al., 2021; Carrajola et al., 2021; Ibrahim et al., 2020; Ren et al., 2020; Ren et al., 2020; Li et al., 2017) consisting of waste materials as good candidates for the partial replacement of natural aggregates (Ferraro et al., 2020). There is an increasing need to carry out a broad-spectrum analysis of the use of these aggregates from a qualitative, environmental, and performance point of view. This work focuses on the production of LWAs through the cold-bonding pelletization process. The waste materials used in this regard are fly ash resulting from the incineration of industrial solid waste and blast furnace slag. The process of cold-bonding pelletization, in addition to not involving any type of gaseous emission, takes place at room temperature and this allows a significant saving in energy intake.

Particular attention was paid to Life Cycle Assessment (LCA) which allowed the assessment of the environmental impacts of these light-weight aggregates produced. LCA is a methodology that in recent years is increasingly gaining ground, not only in the material engineering industry but also in the new generation industries with focus on, for example, electric vehicle batteries (Tang et al., 2020), eco-design processes (Leonardi et al., 2022), bio-ethanol production (Soleymani et al., 2021), nanotechnologies in modern farming (Younis et al., 2021), and the medical industry in relation to the disposal of medicines in the Covid-19 era (Nabavi-Pelesaraei et al., 2022).

This methodology is then applied by all cutting-edge companies who want to identify the processes involved in the lifecycle of a product, from each component to packaging, distribution and the end of the life cycle. This allows these companies to adapt to European legislation and regulations for the elimination of emissions, helping to make their product more sustainable and therefore, given the needs of the market, more competitive. That's why having a sustainable product is an opportunity for every company.

2. Materials and Methods

2.1 *Artificial lightweight aggregates*

The experimental activity consisted in the application of the cold pelletizing process to produce LWAs using three different mixtures. It was carried out in the laboratories of the University of Naples 'Parthenope'. The binding system is made of three main components:

- Portland cement (EMC II/A-L 42.5R);
- Fly ash from solid municipal waste incinerated (MSWI): these materials are classified, according to European regulations, as

hazardous materials (https://eur-lex.europa.eu/legal-content/IT/TXT/?uri=celex%3A32000D0532)and it is mandatory to carry out a preliminary treatment for their use or disposal. In the three mixtures listed above it is possible to find the same percentage by weight of fly ash, equal to 80%;

- Granulated ground blast furnace slag (GGBFS).

As can be seen from Table 15.1, the accurate composition of these three components has been determined. In particular, the analysis conducted, that is the X-ray fluorescence (XRF) implemented at room temperature thanks to the spectrophotometer BRUKER Explorer S4, has shown that calcium oxide (CaO) is present about 24% in fly ash, 67% in cement because of limestone decomposition, and about 17% in blast furnace slag. Another component that can be found in significant quantities is silica (SiO_2), present for 3% in fly ash and for 17% in cement because of the decomposition of clay. The latter also provides the presence in the cement of alumina (Al_2O_3) for 4% and iron oxide (Fe_2O_3) for 3%. However, alumina is present in fly ash in a quantity of 1.5%, while iron oxide is the most common component in GGBFS with a quantity of about 25%.

The treatment before the use of fly ash consists in a first washing phase, necessary to remove soluble salts, which obviously follows a second drying phase to remove water. The washing phase is not as simple as it may seem: it is necessary to guarantee a process economy and an optimization of the quantity of water through a reduced water consumption in a process that requires a counter-current operation with two washing phases.

To choose the most effective washing mode possible, several studies have been taken into consideration that provide optimization tests of the pre-treatment process of fly ash. In particular, (Colangelo et al., 2018) demonstrates, with the same retention time in ionized water at room temperature, that the use of two-step washing allows the removal of soluble salts by optimizing water consumption, by means of a liquid/solid (L/S) ratio lower than a single-phase washing condition.

So, it was chosen in this study to apply a two-step wash. In particular, the L/S ratio used is 2.5:1 while the chosen water retention time is the same for both phases and is equal to 90 minutes.

After the washing process, as mentioned above, it is necessary to carry out a second pretreatment process that is a drying process necessary to remove the quantities of water introduced into the fly ash. This process is carried out by using an active furnace for a time of 24 h at a temperature of 45°C.

After the processes of pre-treatment of fly ash, the three mixtures were prepared, which are characterized by a constant amount of fly ash of 80% by weight, while cement and ground-granulated blast furnace slag, are present in a percentage by weight of, respectively, 5% and 15% in mixture

Table 15.1 Chemical composition of main components (wt%).

Compounds (wt%)	Fe_2O_3	CaO	CO	SiO_2	Al_2O_3	SO_3	ClO	MgO	ZnO	TiO_2	Na_2O	K_2O	Mn_2O_3	Cr_2O_3	NOx
FA	0.86	24.31	16.35	2.62	1.53	8.57	21.20	1.09	2.85	0.36	13.87	6.41			
GGBFS	25.53	17.48	11.29		8.93			7.94					3.44	1.84	10.07
CEM II	3.41	67.16		16.65	4.21	5.34		1.71				1.54			

A, 10% for both components in mixture B and 15% and 5% in mixture C. These values are shown in Table 15.2.

Cold bonding pelletization was carried out thanks to a pilot-scale disc granulator formed by a plate of 80 cm diameter that has the characteristic of being rotating and tipping. The granulator allows a great variability of operation thanks to the possibility to vary the speed of rotation and the angle of inclination between large limits. In the present case the rotation was set at 45 rpm while the inclination was assigned an angle of 45°. In particular, the three components (FA, GGBFS, and CEM II) were inserted into the machine and, by progressive addition of quantities of water (0.99 L in LWA A; 0.11 L in LWA B; 1.04 L in LWA C), it has been possible to carry out the mixing until reaching particles of considerable diameters (Fig. 15.1) formed thanks to the formation of inter-particles bonds between the fine particles, the LWAs. As a result were gained about 4 kg of aggregates for each one of the three different mixtures. As usual, to accomplish the most suitable mechanical conditions, the light aggregates obtained were cured for 28 days at a relative humidity of 95% and at room temperature.

Table 15.2 The different mixtures of lightweight aggregates (wt%).

Mixture (wt%)	LWA A	LWA B	LWA C
FA	80	80	80
GGBFS	15	10	5
CEM II	5	10	15

Fig. 15.1 Result of artificial lightweight aggregates after cold bonding pelletization (LWA).

2.2 Lifecycle assessment

The LCA is an internationally recognized scientific procedure and standardized by UNI EN ISO 14040. This methodology aims to assess the impact, and therefore positive and negative effects, of processes, resources, and performance on the environment. This methodology is already widely used by major multinationals to know the environmental impacts that processes and resources can have during their lifecycle with the aim of generating huge benefits regarding the sustainability of the planet. Through this methodology it is possible to make precise decisions to determine at which stage of the life process of a good or service intervene to reduce environmental pressure. In this case, the LCA is the most useful and effective tool in the implementation of a sustainable natural resources' strategy and one on waste prevention and recycling of waste. LCA is structured into 4 phases:

1. Definition of the goal and scope;
2. Life Cycle Inventory (LCI) analysis;
3. Life Cycle Impact Assessment (LCIA);
4. The interpretation of LCIA.

In this case, it was decided to carry out a medium-term impact assessment. For this purpose, the hierarchical perspective has been assessed as the default method and indicators at the midpoint level are established at the intermediate level along the mechanism.

3. Analysis for Optimizing Sustainability of LWA

3.1 Goal and scope definition

Defined in the previous chapter the methodology of the LCA and its phases, it has been applied to this experimental work and an environmental impact analysis related to the production processes of the light aggregates concerned has been carried out. Having produced three different mixtures, it will therefore be necessary to carry out a study aimed at the comparability of the environmental impacts related to each of them. Figure 15.2 shows the production process described above and for each step of the process it's possible to detect all the inputs and outputs.

To carry out an LCA study, it is essential to define the functional unit and the boundaries to attribute to the system. Regarding the first point, as is also suggested by many studies on the application of LCA analysis on artificial aggregates, a functional unit of 1 kg has been chosen. This quantity is chosen in such a way as to avoid complications regarding the implementation of the data. As regards the second point, it was decided to exclude from the analysis the use and end-of-life phases and to consider

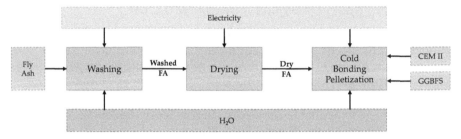

Fig. 15.2 Production process flow chart of lightweight aggregates.

the manufacture, transport, and disposal of raw materials (but excluding the disposal of the contaminated water used for the washing phase) including emissions and energy related to the production processes of the light aggregates concerned. A "cradle-to-gate" approach has therefore been chosen to consider these boundaries, as illustrated in Fig. 15.3.

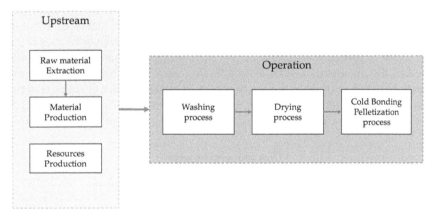

Fig. 15.3 System boundaries.

3.2 Lifecycle inventory (LCI)

The second phase of the LCA analysis is defined as LCI analysis and its purpose is to acquire all the data about the inputs and outputs previously illustrated through literature studies, experimental analyses and databases. In this regard, the literature has provided data on GGBFS, with an allocation coefficient of 24% while from databases have been obtained all the data on cement, deionized water, and electricity. It's slightly different when it comes to obtaining data on fly ash, as these were found in the Environmental Declaration 2020 (https://a2a-be.s3.eu-west-1.amazonaws.com/a2a/2020-06/DA%202020_TMV%20Acerra.pdf?null) and have an allocation coefficient of 1.5%.

3.3 *Results of life-cycle impact assessment (LCIA)*

In this section, the objective is to assess the environmental effects of the production process of artificial aggregates described above. When we talk about environmental effects, we are referring to precise parameters that concern possible damage to the ecosystem, such as climate change, possible damage to human health, such as photochemical ozone formation, and any damage to available resources, such as fossil depletion. Modeling was performed using the Recipe 2016 Midpoint (H) method. However, the results obtained through this modeling cannot be correlated with each other as each has its own unit of measurement. It was therefore necessary to carry out a normalization of these values. This normalization is necessary to obtain a reference unit by which it is possible to make a comparison between the various impacts. Therefore, once the modeling and normalization of the values has been performed, it has been noted that the production of LWA determines a greater impact on climate change as can be seen from Fig. 15.4. The main origin of climate change is the greenhouse effect. This phenomenon acts by imprisoning the heat from the sun's rays blocking the possibility of re-entry into space. This effect is the basis of what is commonly called global warming. CO2 produced by human activities is the main factor of global warming. In 2020, the concentration in the atmosphere exceeded the pre-industrial level by 48% (before 1750). The global warming potential (GWP) is used to compare different gases. In fact, it measures the heating effect of each gas per unit mass during a time interval of 20 or 100 years compared to carbon dioxide. More precisely, it provides, in the same

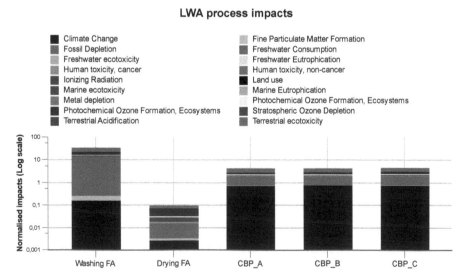

Fig. 15.4 Normalized results by ReCiPe 2016 v1.1 (H).

time frame, the additional radiative forcing caused by the emission of 1 kg of greenhouse gases compared to the forced radiative caused by the release of 1 kg of CO_2. Then the unit of measurement provided by the GWP is defined as CO_2 equivalent and expresses the impact on global warming of a certain amount of greenhouse gas compared to the same amount of carbon dioxide (CO_2).

From Fig. 15.5 we can analyze climate change in reference to the processes used to produce LWA. It can be seen that among all the processes, the one that determines by far the greatest impacts is the one related to washing with a production of 76 kg of CO_2 eq. compared to 104 kg of CO_2 eq. produced by the entire production process. The same applies to inorganic emissions, mainly caused by CO_2, where washing is responsible for emissions of 18 kg of CO_2 eq. on the 46 kg of CO_2 eq. total. This is caused by the fly ash allocation factor which we remember to be of' 1.5%.

As far as long-term emissions are concerned, there is a premise: when we talk about long-term emissions, we are talking also about certain issues that can take place over a time horizon of more than 100 years referred to in the Recipe Midpoint (H). For this reason, we must refer not to when the emission causes its impact but to when the emission is released into the environment. Here again, the washing process is the most impactful and for this reason the emissions of FA, water, and electricity in Fig. 15.6 have been analyzed in detail.

Figure 15.6 shows that the values for water and electricity emissions are practically negligible compared to those for fly ash which account for 99% of the kg of CO_2 eq. the contribution made by the washing process.

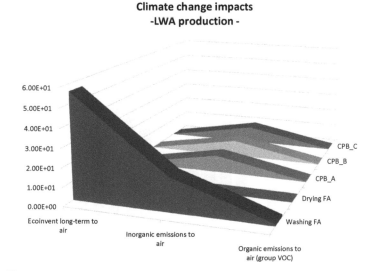

Fig. 15.5 Impacts of LWA process on the climate change through air emissions.

Climate Change contribution of washing process

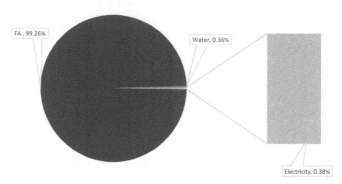

Fig. 15.6 Contribution of the main components to the washing process.

In Fig. 15.7, however, it is possible to note the contributions related to the individual components of the three mixtures produced in this study.

The graph in Fig. 15.7 shows that the trends in the impacts of the components are almost similar for the three mixtures. The most impactful contribution, in fact, is provided in all 3 cases by fly ash with values equal to 9 kg of CO_2 eq. for mixture A and mixture C and 8,7 kg CO_2 eq. for mixture B. The second most impacting component is GGBFS with values of 0,6 kg CO_2 eq. for mixture A, 0,4 kg CO_2 eq. for mixture B and 0,2 kg of CO_2 eq. for mixture C. Therefore, it is clear from these analyses that the least impacting mixture of the three is mixture B (Fig. 15.8). It should be borne in mind that electricity was not considered in this calculation as the electricity supply is the same for the three processes and therefore, contributing in a similar way in all threecases, has no relevance in this comparison.

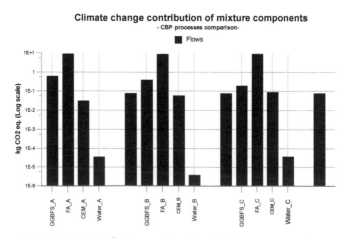

Fig. 15.7 Climate change contribution of mixture components.

Climate change impacts
- CPB process -

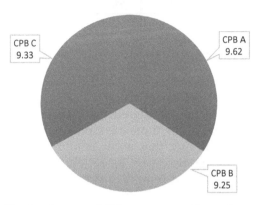

CPB C
9.33

CPB A
9.62

CPB B
9.25

Fig. 15.8 Impacts of CBP processes on climate change.

From what has been said so far and from the graphs shown, it can be concluded that the most important process for the environmental impacts is the cold-bonding pelletization because the washing and drying processes are the same for all types of lightweight aggregates products. Therefore, because of the process of cold-bonding pelletization relative to the mixture B is the least impacting one, it follows that the mixture that contributes to the least environmental impacts is precisely the mixture B.

3.4 Discussion of interpretation of LCIA

Having analyzed in detail all the components of the production processes of the three mixtures covered by this study and having identified and described how they affect environmental impacts, in this paragraph are approached discussion and interpretation of the results examined.

If we start from the consideration that 1 kg of GGBFS affects with an impact of 4,27 kg of CO_2 eq. /kg and 1 kg of cement has an impact of 0,68 kg of CO_2 eq./kg we could soon deduce that the process of cold bonding pelletization concerning the mixture A is the most impacting because it is the mixture that, with the same FA content, contains the largest amount of GGBFS (15%) and the least amount of cement (5%) among the three. On the other hand, for the same statement just performed, the mixture C, always with the same FA content, having the lowest amount of GGBFA (5%) and the greatest amount of cement (15%) should be the less impacting of the three.

The reality of the situation, however, is slightly different. In fact, as shown in Table 15.3, the three components do not impact equally and

Table 15.3 Percentage contribution of main constituents of lightweight aggregates mixtures.

Mixture	LWA A	LWA B	LWA C
FA	92.73%	94.36%	96.03%
GGBFS	6.13%	4.16%	2.12%
CEM II	0.32%	0.66%	1.00%

particularly, the cement is more impacting in the mixture C than the mixture A and, conversely, GGBFS is less impacting.

In addition, it should be considered the contribution of water that, although not equal to that of the three components, is still not negligible. It should be considered, in fact, that water determines a variation of the impacts due to the quantities necessary for cold bonding pelletization.

More specifically, mixture A requires a water quantity of 0.99 L for the formation of lightweight aggregates with an impact of 3.52×10^{-5} kg CO_2 eq., mixture B requires a quantity of water equal to 0.11 L with an impact of 3.83×10^{-6} kg CO_2 eq., while the mixture C needs a quantity of water equal to 1.04 L with an impact of 3.72×10^{-5} kg CO_2 eq.

Therefore, mixture B has a lower impact of an order of magnitude than the other two.

Figure 15.9 summarizes the total impacts of the three mixtures.

It can therefore be concluded, as noted, that considering the contribution of all the key processes involved in the production of the light aggregates covered by this study, the least impacting mixture is mixture B with 34.6 kg CO2 eq. compared to mixture A with 35 kg CO2 eq. and mixture C with 34.7 kg CO2 eq.

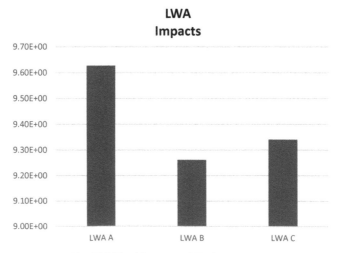

Fig. 15.9 Total impacts of LWA mixtures.

4. Conclusions

This study focuses on the objective of making a comparison in relation to the environmental impact caused by three mixtures of lightweight aggregates made with equal amounts of fly ash and varying the quantities of cement and GGBFS. This analysis was carried out using the methodology of the Life Cycle Assessment (LCA) and using a set of environmental indicators, through which the environmental impact related to the different types of mixture was assessed.

Particularly, all the processes necessary for producing LWAs, from the pretreatment of fly ash to cold-bonding pelletization, were analyzed. It is recalled that the pretreatment of fly ash is necessary because of its classification as hazardous materials due to the different substances present in them such as heavy metals, chlorides, and sulfates. Moreover, this pretreatment has been shown to be the most impactful process, assuming values equal to 76 kg of CO_2 eq. /kg. Of this impact, 99% is due to the use of fly ash and only 1% to that of water.

Regarding electricity, it should be noted that this has not been taken into account because the value used, in addition to being low is also the same for all mixtures, so it makes no sense to take it into account for a comparison.

As for the other processes analyzed, namely drying and cold bonding pelletization, they affect in a completely opposite way. While the drying process only affects values of 0.13 kg of CO_2 eq/kg, with therefore little significant impact, the cold bonding pelletization processes affect for values equal to 9.63 kg CO_2 eq./kg, 9,26 kg CO_2 eq./kg and 9,34 kg CO_2 eq./kg for the three mixtures A, B, and C, respectively.

A final comparison of all factors shows that the least impacted mixture is B with a value of 34,6 kg of CO_2 eq./kg compared to 35 kg of CO_2 eq./kg for mixture A and 34,7 kg of CO_2 eq./kg for mixture C.

It can be concluded that, since the granulation process is at room temperature and free of gaseous emissions, this process has undisputed advantages in terms of environmental impact. It is therefore possible to say that the process of cold bonding pelletization to produce LWAs is advantageous in terms of environmental sustainability.

References

Akan, A.E., Ünal, F. and Koçyiğit, F. (2021). Investigation of Energy Saving Potential in Buildings Using Novel Developed Lightweight Concrete. *Int. J. Thermophys.*, 42: 4. doi:10.1007/s10765-020-02761-1.

Carrajola, R., Hawreen, A., Flores-Colen, I. and de Brito, J. (2021). Fresh properties of cement-based thermal renders with fly ash, air lime and lightweight aggregates. *J. Build. Eng.*, 34: 101868. doi:10.1016/j.jobe.2020.101868.

Chippagiri, R., Gavali, H.R., Ralegaonkar, R.V., Riley, M., Shaw, A. and Bras, A. (2021). Application of sustainable prefabricated wall technology for energy efficient social housing. *Sustainability*, 13: 1–12. doi:10.3390/su13031195.

Colangelo, F. and Cioffi, R. (2017). Mechanical properties and durability of mortar containing fine fraction of demolition wastes produced by selective demolition in South Italy. *Compos. Part B Eng.*, 115: 43–50. doi:10.1016/j.compositesb.2016.10.045.

Colangelo, F., Petrillo, A., Cioffi, R., Borrelli, C. and Forcina, A. (2018). Lifecycle assessment of recycled concretes: A case study in southern Italy. *Sci. Tot. Environ.*, 615: 1506–1517, doi: 10.1016/j.scitotenv.2017.09.107.

Dichiarazione Ambientale by RINA S.p.A. for A2A. Ambiente Termovalorizzatore di Acerra. (2020). Available online: https://a2a-be.s3.eu-west-1.amazonaws.com/a2a/2020-06/DA%202020_TMV%20Acerra.pdf?null= (accessed on 9 November 2020)

EUR-lex. European waste catalog. Available online: https://eur-lex.europa.eu/legal-content/IT/TXT/?uri=celex%3A32000D0532 (accessed on 13 December 2020).

Ferraro, A., Colangelo, F., Farina, I., Race, M., Cioffi, R., Cheeseman, C. and Fabbricino, M. (2020). Cold-bonding process for treatment and reuse of waste materials: Technical designs and applications of pelletized products. *Crit. Rev. Environ. Sci. Technol.*, 1–35. doi:10.1080/10643389.2020.1776052.

Ibrahim, M., Ahmad, A., Barry, M.S., Alhems, L.M. and Mohamed Suhoothi, A.C. (2020). Durability of Structural Lightweight Concrete Containing Expanded Perlite Aggregate. *Int. J. Concr. Struct. Mater.*, 14: 50. doi:10.1186/s40069-020-00425-w.

Leonardi, S., Perpignan, C., Eynard, B., Baouch, Y. and Robin, V. (2022). Life Cycle Assessment in an Ecodesign Process: A Pedagogical Case Study. *Proceedings of the Design Society*, 2: 2323–2332.

Li, S., Liu, K. and Sun, L. (2017). Effect of different kinds of lightweight aggregates on performance of composite insulation materials. *J. Build. Mater.* 20: 245–250. doi:10.3969/j.issn.1007-9629.2017.02.015.

Liu, C., Fan, J., Bai, G., Quan, Z., Fu, G., Zhu, C. and Fan, Z. (2019)., Cyclic load tests and seismic performance of recycled aggregate concrete (RAC) columns. *Constr. Build. Mater.*, 195: 682–694. doi:10.1016/j.conbuildmat.2018.10.078.

Massoudinejad, M., Amanidaz, N., Santos, R.M. and Bakhshoodeh, R. (2019). Use of municipal, agricultural, industrial, construction, and demolition waste in thermal and sound building insulation materials: A review article. *J. Environ. Health Sci. Eng.*, 17: 1227–1242. doi:10.1007/s40201-019-00380-z.

Meng, E., Yu, Y., Zhang, X., and Su, Y. () Seismic experiment and performance index studies on recycled aggregate concrete filled steel tube frames infilled with recycled hollow block filler walls. Struct. Concr. doi:10.1002/suco.202000254.

Nabavi-Pelesaraei, A., Mohammadkashi, N., Naderloo, L., Abbasi, M. and Chau, K. W. (2022). Principal of environmental life cycle assessment for medical waste during COVID-19 outbreak to support sustainable development goals. *Science of the Total Environment*, 827: 154416.

Raut, A.N. and Gomez, C.P. (2017). Development of thermally efficient fibre-based eco-friendly brick reusing locally available waste materials. *Constr. Build. Mater.* 133: 275–284. doi:10.1016/j.conbuildmat.2016.12.055.

Ren, M., Liu, Y. and Gao, X. (2020). Incorporation of phase change material and carbon nanofibers into lightweight aggregate concrete for thermal energy regulation in buildings. *Energy*, 197: 117262. doi:10.1016/j.energy.2020.117262.

Saribas, I., Goksu, C., Binbir, E. and Ilki, A. (2019). Seismic performance of full-scale RC columns containing high proportion recycled aggregate. *Bull. Earthq. Eng.*, 17: 6009–6037. doi:10.1007/s10518-019-00687-0.

Soleymani Angili, T., Grzesik, K., Rödl, A. and Kaltschmitt, M. (2021). Life cycle assessment of bioethanol production: A review of feedstock, technology, and methodology. *Energies*, 14(10): 2939.

Tang, B., Xu, Y. and Wang, M. (2022). Life Cycle Assessment of Battery Electric and Internal Combustion Engine Vehicles Considering the Impact of Electricity Generation Mix: A Case Study in China. *Atmosphere*, 13(2): 252.

The serial nos in the references to be enclosed within [ ], e.g., [22] to maintain consistency with other chapters.

Xiao, J., Sun, Y. and Falkner, H. (2006). Seismic performance of frame structures with recycled aggregate concrete. *Engineering Structures*, 28: 1–8. doi:10.1016/j.engstruct.2005.06.019.

Younis, S.A., Kim, K.H., Shaheen, S.M., Antoniadis, V., Tsang, Y.F., Rinklebe, J. and Brown, R.J. (2021). Advancements of nanotechnologies in crop promotion and soil fertility: Benefits, lifecycle assessment, and legislation policies. *Renewable and Sustainable Energy Reviews*, 152: 111686.

16

Towards (An Aggregated) Territorial Digital Twin

From Smart-village to Smart-territory via the Territorial System of Digital Twin

Cédrick Béler,[1] *Gregory Zacharewicz,*[2,*]
Paul-Antoine Bisgambiglia,[3] *Bastien Poggi,*[4]
Florent Poux[5] and *Antoine-Santoni Thierry*[6]

1. Introduction

The concept of a Smart City has gained significant attention in recent years as urban populations continue to grow and the need for efficient and sustainable urban management becomes increasingly pressing. A Smart City is defined as a city that utilizes technology to improve the quality of life for its citizens, enhance sustainability, and optimize the use of resources (Boubekri, 2017). An intelligent city or smart city, therefore, not

---

[1] <cedrick.beler@enit.fr>; https://orcid.org/0000-0002-0439-724X
[2] https://orcid.org/0000-0001-7726-1725
[3] <bisgambiglia_pa@univ-corse.fr>; https://orcid.org/0000-0003-3026-2258
[4] <poggi_b@univ-corse.fr>; https://orcid.org/0000-0002-3253-6868
    <florent@learngeodata.eu>; https://orcid.org/0000-0001-6368-4399
[5] <antoine-santoni_t@univ-corse.fr>; https://orcid.org/0000-0002-6645-9311.
[*] Corresponding author: Gregory.Zacharewicz@mines-ales.fr

only places the citizen at the center of its development strategy, thanks to its information technology (IT) infrastructure for data-driven management but also for the implementation of digital services, often interactive or participative (information systems, communications, storage, security, responsiveness, sensor networks, etc.). Its constant objective would be supervising its natural, energy, human, and economic resources. This new approach has been tested for several years by major European cities such as Amsterdam, London, Stockholm, Barcelona, Paris, Manchester, Padua, and Santander (Zubizarreta et al., 2015). However, the heterogeneity of trajectories, political visions, and situations raises questions about the implementation of the smart city. Is it reality, marketing, or utopia? At a time of unprecedented climatic upheaval on a human scale and in a digital divide between urban and rural areas, is it not essential to ask the question of the territories on the outskirts of large cities and rural areas?

The 'smartification' of cities can create a digital divide between geographical areas, especially rural ones. In recent years, research on smart cities has attracted the attention of academics and practitioners. Smart, in this case, refers to initiatives by government officials, citizens, and other stakeholders, primarily using technology, to make cities more livable and places more 'intelligent', which refers to the continuum within the local government (Gil-Garcia et al., 2015). One key aspect of a Smart City is the creation and use of a digital twin.

On the industrial domain, emerged the concept of Digital Twin which is a virtual representation of a physical asset or system, such as a building, infrastructure, or city (Gao, 2018). It is created by collecting and analyzing data from various sources, such as sensors and internet of things (IoT) devices, and using this data to create a detailed, 3D models of the asset or system. The Digital Twin can then be used for a variety of purposes, such as simulation, prediction, and optimization (Jiang et al., 2019).

Coupling those two concepts, a digital twin for a smart city is a virtual replica of a city that can be used to simulate and analyze the city's physical and functional characteristics. It can be used for a wide range of purposes, such as urban planning, resource management, and emergency response. The digital twin is created by collecting and integrating data from a variety of sources, such as sensor networks, aerial imagery, and demographic data. The data is then used to create a detailed model of the city, which can be used to simulate different scenarios and analyze the impact of different decisions.

The objective of this chapter is to focus on Smart villages and smart territories (Komorowski and Stanny, 2020). The contribution presents the current development in relation to smart villages to provide direction to get a larger scale territory that goes beyond the frontier of a given city of village. The first part is dealing with recalls on experiences and feedback from the proof of concept in Cozzano. Then territorial services in large rural areas that consider medical and air quality cases are presented. Thereafter, the major required concepts of a territorial System of digital twins including human, system of systems, and 3D environments are dealt with. Finally, there are some conclusions.

2. Experience Feedback

2.1 *Feedback from the proof of concept in Cozzano (Smart Paesi – Corsica)*

The program Smart Village–Smart Paesi

The program Smart Village–Smart Paesi: Emergence of Smart Territories is a scientific action of the Università di Corsica Pasquale Paoli/CNRS, launched 2017 in partnership with the IT company SITEC and EDF island energy systems within a small rural village of Cozzano (300 inhabitants) in central Corsica. Until July 2021, financially supported by the European Fund FEDER and the Collectivité de Corse, the project was to propose a proof of concept of a Smart Village at the scale of a commune by relying on its assets and its strategy.

The village of Cozzano is in the center of Corsica, surrounded by mountains at an altitude of 650 meters and forests of pine, chestnut, and oak. The community has worked for many years to maintain services and activities but has also embarked on a renewable energy strategy. The village of Cozzano aims to become a positive energy zone that produces more energy than it consumes through biomass boilers, river hydroelectric plants, and solar panels (with a self-consumption system) on municipal buildings.

During the construction of the project in 2015, without a validated definition of the smart village concept, four basic elements were proposed to design it: sustainable development (in all its components), IT/digital infrastructure, education/citizenship, and agricultural activities. The smart village system is based on an information system with five actions: collect, store, analyze, predict, and visualize to ensure data-driven management.

The Smart Paesi program builds a mindset for information systems (IS) data management along four axes:

1. Environmental and infrastructure monitoring via wireless sensor networks;
2. Information system integrating real-time monitoring and prediction tools for decision support;
3. System optimization (energy, water, public buildings, lighting, services) and connectivity/data visualization;
4. Development of services and places for actors/experimenters (farms, livestock, fire brigades, etc., follow-up) and public awareness.

The Smart Paese is based on the collection of data on the environment of a village thanks to a network of sensors or connected objects using LoRA/LoRaWAN technology: weather, air, water and soil quality, energy consumption, and geolocation. LoRa technology was chosen for the telemetry activity because of its range of wavelengths allowing communications over several kilometers, its ability to overcome obstacles, and the low energy consumption of the connected objects using this technology. All this information is stored in a flexible and scalable database for analysis and prediction. Indeed, the data is used by a prediction-optimization algorithm called Smart Entity (Antoine-Santoni et al., 2019). This algorithmic structure must predict the evolution of the systems observed in the village and propose different levels of restitution by visualization (open data, dashboard for the municipality, information for the population) to optimize the sustainable development strategy.

Among these initiatives, a saffron grower's organic farm has been connected to a network of LoRaWAN wireless sensors to collect various data (soil quality, irrigation, weather). A field visualization dashboard allows him to monitor and manage his crop and identify the optimal environmental conditions for the operation. Similarly, collaborative work has been initiated between the saffron farmer and researchers to determine the saffron's flowering time, using the Smart Entity prediction tool. Pig and goat farmers use geolocation systems, integrating their farms into the Smart Village process (Antoine-Santoni and Poggi, 2022).

2.1.1 Information system deployed and related digital services

The Smart Paesi program builds its thinking around data-based management in an information system (IS) along 4 main lines:

- An infrastructure for monitoring the environment and the buildings of the municipality through a network of wireless sensors; the long-range communication technology LoRa/LoRaWAN has been chosen;
- An IS integrating real-time monitoring with predictive tools for decision support within the framework of ambient intelligence;
- System optimization (energy, water, public buildings, lighting, and services) and data interfacing/visualization;
- Development of services for experimental actors (monitoring of farms, livestock, firemen, etc.) and for public awareness.

Implementing these primary objectives made it possible to validate a proof of concept (POC) within a Living Lab integrating an effective and efficient IS. In line with the philosophy of the Smart Village definition, many parallel projects have been carried out in the municipality in a logic of third places, awareness-raising actions, and integration of citizens in the issue of sustainable development and reduction of the digital bill.

The Centre d'Immersion et d'Innovation Numérique par la Technologie Ubiquitaire (CIINTU) is a third-party space that was designed in parallel with the implementation of the Smart Village. The need to create a place to store equipment and accommodate researchers appeared early in the scientific program. In this context, a building was renovated, and a new place was made available to researchers and inhabitants. This building has become the third place materializing the Smart Village but also a co-working space for people passing through the village who do not have an internet connection. It is equipped with a broadband connection, workspace, printer, various computer equipment and a small FabLab (3D printer, digital embroidery machine, robots, etc.). It currently hosts many activities: training sessions for various organizations, computer training for the population, workshops with the school, co-working, and researchers from the University of Corsica.

The Town Hall has a comprehensive website with a wealth of regularly updated information; it has a permanent communication system for social networks (Facebook, Instagram). It has a notice board on the front of the town hall providing information to residents and visitors: practical information, promotional films, events, and data on the Smart Village. The digital advisor, recruited in 2021 following a national strategy to deploy mediation agents in the most isolated areas, is located at the CIINTU. He assists the population of Haut-Taravo who may be suffering from illiteracy, whatever their profile, in getting to grips with digital tools and understanding their uses. To do this, it organizes group workshops as well as individualized support in various areas: teaching users to use the various computer equipment, making them independent in surfing the Internet, training them in writing and sending emails, word processing, and managing digital content (creation, storage, sharing), enabling them to fully master the use of a smartphone, in particular, to install and use applications, and giving them the keys to understanding digital vocabulary. In addition, it also has a role to play in facilitating specific uses such as social networks and digital communication tools, searching for a job on the Internet and applying online, carrying out administrative procedures online, knowing how to make purchases from the Internet, and understanding how digital school tools work, for users who must follow their children's schooling.

His role in digital mediation is central in the municipality. Moreover, the person recruited is a former member of the Smart Paesi program, which makes it much easier to explain the program.

The school is very much part of the Smart village philosophy. The two classrooms in the commune are well equipped with digital equipment, including multi-touch boards, digital tablets, and headsets, which allow many activities to be carried out with the children. This has been achieved by responding to calls for innovative projects from the CARDIE unit. The CARDIE unit, in collaboration with the INSPE of Corsica, proposes to accompany the schools and establishments of our academy in a development approach based on innovation, supported by teachers and teacher-researchers of the Università di Corsica.

In Cozzano, the school created a school newspaper. The children worked as little journalists for a year, making it possible to create three newspapers in the village, broadcasting scientific information on sustainable development (e.g., sorting or water management). The following year, the newspaper was transformed into a web radio. This year, robotics workshops are being offered to help children discover algorithms and programming.

Citizens are actors in the feedback process, thanks to the activity of 'mediators'. Indeed, students from the University of Corsica, holders of a master's degree in engineering, come regularly for several weeks of immersion to exchange with the population and collect the population's feelings to imagine the evolution of the Smart Paesi. Indeed, many subjects are shared with the population around the scientific project and digital, health, and energy savings.

2.1.2 Conclusions and learnings on the smart-Paese project

The development of smart cities must be community-driven, even before investments in technological solutions are made. Citizens are an invaluable source of information for understanding problems and needs regarding the quality and adequacy of urban services provided (Antoine-Santoni et al., 2022). The success of a smart city is determined mainly by the degree of involvement of citizens in the creation of the smart city itself. Therefore, dialogues between citizens and policymakers, urban planners, and managers are fundamental for planning and action strategies, especially for collecting feedback to adjust and improve current and future policies. Urban services refer to services to/on the territory and governance services.

Today, local authorities focus on designing, implementing, and managing solutions for cities characterized by shared information and a networked approach. However, it is necessary to question these models of intelligent territories in the face of climate disruption and more local crises, such as the digital or health divide. To enable this dialogue between users and decision-makers for efficient decision-making, IT and digital technology represent a significant opportunity.

3. Territorial Services in Large Rural Areas: Medical and Air Quality Cases

Whether city or village is concerned, there are boundaries to the developed smart-systems that do not easily allow sharing information (or assets) between these systems. In this section, two research works are presented to demonstrate the need to overcome the frontier of a commune. The first one is about the necessary coordination to improve the medical desert situation while the second research is about collecting data from sensors on a large area so that municipalities can take decision regarding air quality.

3.1 Improving rural medical desert through ICT by better healthcare pathways

3.1.1 Medical desert statement

Rural areas are often more sparsely populated than urban areas, and are generally characterized by a lower population density, a smaller number of services including healthcare, and a more agricultural or natural resource-based economy. Some rural areas are relatively isolated and may be difficult to access, while others are more connected to urban centers and may have more developed infrastructure. In consequence, a rural area is a geographic region that is not located in or near a city or suburb frequently experiences healthcare resources shortage. So, as part of smart village features, making access to healthcare resources easier thanks to digital support is an important objective.

In more detail, in rural areas, it exists medical desert that is an area that lacks access to healthcare services. This can occur for a variety of reasons, such as a shortage of healthcare providers or geographical isolation. Medical deserts can have serious consequences for the people who live in them, as they may not have access to necessary medical care or preventive services. In some cases, medical deserts can lead to higher rates of illness and mortality, as well as increased healthcare costs due to the need for more expensive forms of care (such as emergency department visits or hospitalizations) for conditions that could have been treated more efficiently and effectively in a primary care setting.

3.1.2 Solutions

To address medical deserts and increase access to healthcare services in underserved areas, there are several solutions. Some of these strategies include:

1. Increasing the number of healthcare providers in the area: This can be done by developing more attractivity for recruiting and retaining more doctors, nurses, and other healthcare professionals to work in the area, or by using telemedicine and other technology to connect patients with healthcare providers remotely.
2. Improving transportation: Many people living in medical deserts may have difficulty getting to healthcare providers due to a lack of transportation options. Providing transportation assistance (such as

buses or vans) or improving public transportation can help to make it easier for people to access healthcare services.
3. Building or renovating healthcare facilities: Building or renovating healthcare facilities in underserved areas can help to improve access to care. Creating telemedicine spots can facilitate access to care as well. This can be especially important in rural areas, where the nearest healthcare facility may be a significant distance away.
4. Increasing community outreach and education: Providing education and outreach to the community can help to raise awareness about the importance of preventive care and encourage people to seek medical treatment when needed.

To deal with points 1, 2, and 3, it can be required to evaluate the costs and benefits for the population. Modeling and simulation can help with that. Authors in (Starkiene, 2013) described the maldistribution of human resources for health as a worldwide phenomenon that may appear in different dimensions. Shortages and imbalances in the distribution of the health workforce are social and political problems that, along with the socio-economic inequality and reduce the access of the population to the health services (Oliveira et al., 2017). This issue is more significant in areas with higher proportions of low-income and minority residents, such as rural areas, tend to suffer most from lower physician supply. The study (Sbayou et al., 2019) has chosen regions where there is a physicians' shortage and then they exposed and approach that combines Business Process Modeling and Notation (BPMN) with DEVS (Developers) for managing healthcare environment in French regions that suffer from physicians' imbalances issue. It is conceivable that smart villages can provide data to build such models and simulations.

Based on the data collected on the ground through the smart village digital twin, a model of the territory with services and population needs can be set. Then some healthcare processes or pathways can be described informally by text description or by semi-formal languages (Eshuis et al., 2010) according to the needs of the patients. A patient pathway process is generally defined as a sequence of events that uses patient status as inputs to produce outputs. Besides, a business process is an activity or set of activities that will accomplish a specific goal. It is also considered as a sequence of performed steps that drive information to produce and/or provides services. The business process needs to be managed and controlled to drive the calls to healthcare resources if available; it attests the need to use a business process management (BPM) methodology.

In this context, the use of BPM in healthcare sector can be a solution to anticipate population needs by simulation and then ensure coordination at run time between healthcare stakeholders. Among different languages to model BPM processes, one of the most useful one is BPMN. However, one issue that can be faced by healthcare organizations while modeling with BPMN is that BPMN gives a static representation of the system, and as healthcare environment is becoming more dynamic and volatile, and follows more complex processes, there is a need to combine BPMN with an agent based model in terms of ensuring resource allocation, in the other hand, managing the availability of resources in healthcare sector is a very challenging problem with very little research attention.

3.1.3 Dordogne case study

In the works of (Sbayou et al., 2019), the Dordogne County (fr. Department) located in southwest of France was chosen and considered as a representative case of "medical desert" zone among the counties in Nouvelle Aquitaine region where the low population, as with most difficulties in accessing to health services. Figure 16.1 presents the map of this rural department. This map has been generated with QGIS thanks to INSEE French demographical and healthcare resources data from the yellow pages. It was coupled with a simulation platform to instantiate simulation agents (Zeigler, 2019). It can be observed in Fig. 16.1 that a patient located in the far north of the department may have access to a medical doctor (Fig. 16.1, arrow 1) in its neighborhood. Nevertheless, in the frame of a medical pathway that includes appointments with specialists' doctors (Fig. 16.1, arrow 3) and radio (Fig. 16.1, arrow 2) the distance is more important, and the waiting time can be impacted due to the limited number of resources. When a great part of this territory has a low level of population and almost no specialist healthcare resources, it leads to patient pathways that are not performant.

Fig. 16.1 Patient Pathway steps in Dordogne Case study.

The study of (Sbayou et al., 2019) has reported a modeling and simulation approach that can be applied in the healthcare sector of population needs in a rural region, it can help decision support of decision- makers to install new healthcare services on one side and citizens on the other side as self-decision- maker on the building of their medical pathways. The idea presented in the study was to demonstrate the interest of coupling a healthcare patient pathway workflow modelled with BPMN with different healthcare resources organized in social networks described with DEVS based models as described in Fig. 16.2. Such an approach can be proposed to study a territory in terms of sufficient or insufficient resources/services available in a specific area.

Figure 16.2 depicts a modeling and simulation architecture and the social networks that can be used to get the information about the healthcare service required. The easiness/difficulty can depend on the physical location of the patient (first dimension), on the patient connection to healthcare services through the general practitioner (second dimension), and on the personal connection to people around the patient including family and friends (third dimension), both can provide support in searching for and reaching healthcare services. Then the architecture sets the patient in the different dimensions (layers) to attempt to identify and reach the appropriate resource. At the end, after having identified the resources, a BPMN process is used to follow the steps of the required medical pathway.

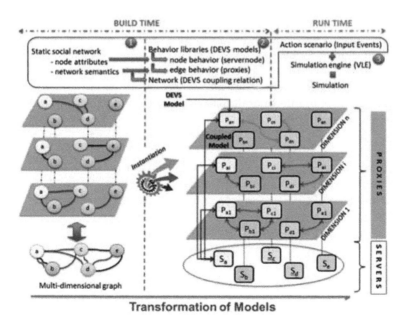

Fig. 16.2 Modeling and Simulation Platform.

The idea in this study of the Dordogne territory was to face the problems regarding barriers to healthcare access and to anticipate solutions thanks to simulation. Results were interesting but mostly generated from estimated data. It can be assumed that thanks to the data provided directly from population inputs and updates, the identification of a problem and the anticipation of new healthcare solutions by providing new resources can be improved. Nevertheless, to reach the goals some barriers will have to be overcome. Real and fresh data, confidential data management, public services description, and integration in the database.

Finally, it can be integrated as one feature of a future Smart village digital twin. It can be interesting as well to apply this architecture in other sectors (transportation, education networks, employment...) to make it functional for any service-oriented needs.

3..1.4 Foresight conclusions and learnings

As lessons learned, there is a clear need to have a global and privacy respectful architecture to collect data and orchestrate processes for medical services. There is initiative in the domain including the French National Health Data System (SNDS) that represents a considerable advance in analyzing and improving the health of the population and provide database for analyses. Nevertheless, this database is more dedicated for a posteriori analysis rather than real-time decision support.

In addition, even if data are securely anonymized and managed at the local level, the information cannot stay only inside one city, it must go beyond the frontier of a commune, especially in rural areas to gather information and propose better support to build a patient pathway optimized according to the profile and location of the patient.

It leads at the end to the need of an Interterritorial continuum where the local data will have to interoperate with other data from other digital twins in the neighborhood.

3.1.5 Air quality monitoring in the Pyrenees

Air pollution is a major societal issue with significant health, social, environmental, and economic consequences, causing millions of deaths per year and loss of healthy life years. Numerous studies have been conducted and today regulations have determined exposure thresholds for different air pollutants (PM, VOCs, NOx, etc.) (World Health Organization, 2021). However, it is still quite difficult to know the exact composition of the atmosphere, especially due to lack of measuring points, particularly in rural areas. In France, ATMO, the association in charge of monitoring air quality, has only 720 measurement stations (fixed and mobile) to cover more than 550,000 square kilometers. The fixed stations mainly cover large cities, which are more prone to air quality issues. For example, in the Hautes-Pyrénées (4,464 km2), there are only two permanent ATMO stations located in Tarbes and Lourdes, but no mountain area has an official measuring point. The principle of air quality prediction, a bit like weather forecast, relies on the injection of the measure of these stations into predictive models, but it is difficult for them to be precise in singular topologies (mountainous area in particular). In addition, these models are not sophisticated enough to detect or find the origin of local pollution phenomena (e.g., controlled burning done at more than 50 km of the stations). At best, some induced effects can be perceived. To address this problem of insufficient measurement resolution, there are public and inexpensive capture solutions. It is even possible to build low-cost stations using microsensors (Poupry et al., 2022).

3.1.6 BOLDAIR project

The BOLDAIR project started in 2019 and was financed by the French region Occitanie and the CCPVG inter-municipality (Communauté de Communes Pyrénées Vallées des Gaves). CCPVG is a medium to high mountain area, very touristy, including several ski resorts and a large part of which falls within the protected area of the Pyrenees National Park. Argelès-Gazost, its administrative seat, is also an important transit point since it is at the crossroads of three valleys. The CCPVG, aware of the importance of good air quality, has voluntarily committed to a Territorial Climate-Air-Energy Plan (PCAET) to give the inhabitants the opportunity to reflect on the future of their territory and to question the sustainability of well-being in the valleys. Several challenges have been established and the place of air quality has been defined as a priority. A second challenge identified in the context of this PCAET concerns governance, including the issue of monitoring and observation of pollutant emissions. In this rural context, the main emissions considered are those from the agricultural sector, the transport sector (especially during periods of high tourist affluence), biomass combustion, distinguishing outdoor air quality (controlled burning, methanization, etc.) and indoor air quality (inefficient wood heating equipment).

The goal of the BOLDAIR project is to enable municipalities to gain a better understanding of the air quality in their areas, which are often not well-instrumented, and to communicate this information to the public. Additionally, it provides decision-makers with tools to implement and evaluate measures to improve air quality. These tools will enable us to detect episodes of pollution, identify their causes, and potentially find and store solutions based on the given context. The Laboratoire Génie de Production (LGP) of Tarbes was responsible for the development of the BOLDAIR project. The objective was to create a decision-making system that would allow elected officials and citizens to monitor the levels of different pollutants and take appropriate action. The resulting architecture, illustrated in Fig. 16.3, consists primarily of an observation and decision-making system.

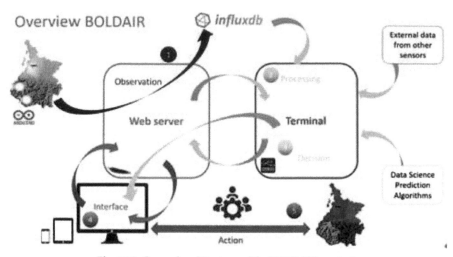

Fig. 16.3 General architecture of the BOLDAIR project.

The observation system involves the monitoring of various physical pollutants (PM2.5, PM10, etc.) using a network of Low Cost Measurement Stations (LCMS) (Poupry et al., 2022). To achieve multiple measuring points at a reasonable cost, the solution is based on low-cost sensors (LCS). To address the limitations of LCS (precision, reliability, etc.), the architecture employs triple redundancy. The measurement station aggregates measurements from three sub-stations called Smart-Sensors (SmS that are based on a microcontroller with WiFi connection. These three SmS are connected to a micro-server based on a Raspberry Pi Zero (node A on Fig. 16.4), forming an autonomous measuring station that can communicate with the decisional system. The station also has local memory to store measuring points and perform some data processing. One of these processing capabilities is to provide a reliable measure from the three

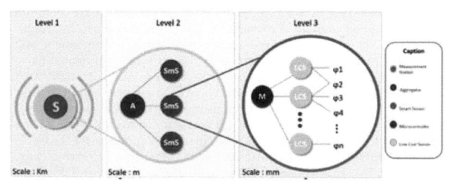

Fig. 16.4 Low-Cost Measurement Stations (LCMS) and Smart-Sensors (SmS).

captures. The station is also capable of self-analysis to detect any internal component failures.

The decision system provides a web-based tool to elected officials, giving them access to measurements from the LCMS, alerts if thresholds are exceeded, and instructions to follow in case of detected pollution episodes. The web application integrates open-source cartography capabilities (Open-Street-Map and IGN). The underlying architecture that aggregates spatio-temporal information from the LCMS is a network of nodes with a recursive aggregation strategy, enabling spatial scaling on a territory. Each LCMS corresponds to a specific place (latitude, longitude) from a municipality. An inter-communality may have interest in the data from several villages, connecting to the aggregating nodes of each of these villages. Recursively, a region or country may want to process the information from that village. The proposed architecture does not force a bottom-up inclusion, and every intermediary node can share its data to any other nodes if desired.

This aggregation or ascendant characteristics of the proposed architecture is synthetized in the following picture. So 'Observation' and 'Decision' are materialized with three types of nodes (Fig. 16.5). LCMS nodes are shown in gray. Green nodes enable recursive aggregation, receiving and sending spatio-temporal measurements to and from other nodes, and storing the information. Blue nodes, or HMI (Human Machine Interface) nodes are like green nodes but include user applications and interfaces. These nodes provide the necessary applications, algorithms, and processing capabilities for users, particularly for decision-makers (e.g., elected officials responsible for monitoring air quality).

This logic of aggregation has been the foundation of the concept of scaling spatially to cover large areas. If the data stream is properly managed and controlled (by respective owners in the physical world of the digital counterpart), it is believed that this type of architecture could be highly resilient, particularly if the network connection is possible without the Internet via direct connection between nearby locations. This is not an alternative to the current Internet, but rather a complementary approach. In

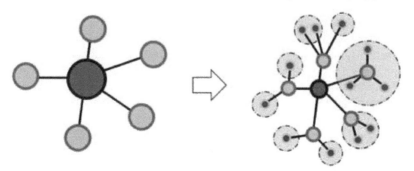

Fig. 16.5 'Observation' and 'Decision' information nodes.

the same way that people tend to prefer short food circuits, it is particularly relevant to consider short data circuits. In other words, data should be stored close to where it is produced, and if a resilient smart-territory exists in the future, it is believed that it will be based on a distributed cloud/edge architecture distributed in each home, village, town, city... For now, the architecture uses the Internet to reach other villages, but ideally each village should be physically connected with a direct connection (Village-to-Village) to avoid using the global network when not necessary. Existing peer-to-peer and meshed networks are very interesting alternatives from the perspective of resilience and trust.

3.1.7. Perspectives of the BOLDAIR project

Most of the project involved designing, constructing, and deploying an LCMS as an efficient and scalable data sensor architecture. However, it is not enough to simply provide information; it is essential to involve the population to create greater awareness, effect behavior change, and alter practices. This project seeks to develop a decision system for local authorities that integrates technological data from sensors, human reports, and expert analyses into collective decision-making and action tools. This system will take the form of a digital twin, which is "a dynamic representation of a physical system using data, models and interconnected processes that provides access to knowledge from past, present and future states for system management" (Camara et al., 2021) This digital twin will support the incremental development of public and private knowledge bases to help decision-makers and citizens understand the state of the territory and make informed decisions in real time to improve air quality (Endsley et al., 2003).

3.2 Synthesis

Experiments conducted in villages and larger territories have shown that developing a smart territory is more than just collecting and gathering information about the area. For the tool to be accepted and utilized to its full potential, it is essential to involve human participation and inclusion in the process (Boy, 2020). Inhabitants, elected officials, and developers should be brought together to discuss and propose the most useful services. Moreover, user-friendly interfaces between humans and machines must be developed. Maps and cartography have long been a universal representation that everyone knows how to use. With the advancement of ICT, 3D representation, and 3D virtual worlds are even more capable of accurately reproducing reality. It is important to use these to provide engaging interfaces that will encourage participation, and enhance ergonomics and usability. 3D worlds promise to be an excellent interface that could be easier and more natural to use than classic computer interfaces. Integrating all these information systems, through a distributed architecture that is private by design and resilient, into a 3D virtual world interface, could be seen as a genuine digital twin of the territory, which could be the backbone of smart villages, cities, and territories.

One important aspect of a smart territory is about being able to model the current situation but also to be able to foresight possible future to anticipate and take advised decisions. These aspects are fundamental in the definition of a digital twin and therefore simulation capabilities must be added to the whole proposition. Indeed, the notion of digital twins offers an exciting opportunity to test new scenarios and the resilience of a territory. Based on multi-scale modeling, this digitalized approach to the territory makes it possible to imagine different scenarios for anticipating, resolving crises or even to optimize the territorial and societal operations.

4. Towards a Territorial System of Digital Twins

A Digital Twin is essentially composed of 5 components as defined by and extended by (Xia et al., 2022). A Physical Part (the Scene), A Virtual Model, a Connection, which is the gathering device used for obtaining data for integrating virtual and physical spaces (Deng et al., 2021), Data from the physical part, and a Service (target application, e.g., smart cities). The use of a Digital Twin in a Smart City allows for improved decision-making and resource management. By having a detailed representation of the city and its systems, city officials and managers can analyze data and simulate potential scenarios to determine the most efficient and sustainable course of action (Jiang et al., 2019). Furthermore, the Digital Twin can be used to monitor the performance of the city and its systems, allowing for early detection of issues and proactive maintenance (Gao, 2018).

While the concept of a Smart City and a Smart Village share similarities, there are distinct differences in the application of Digital Twins in these contexts. A Smart City typically has a larger population, more advanced infrastructure, and more resources at its disposal compared to a Smart Village. As such, the implementation and use of a Digital Twin in a Smart City may be more complex and require more advanced technology. However, the principles of a Digital Twin can still be applied to a Smart Village. The process of creating a Digital Twin for a Smart Village would involve collecting and analyzing data from various sources, such as sensors and IoT devices, to create detailed, 3D models of the village and its systems. This Digital Twin can then be used for simulation, prediction, and optimization of the village's resources and infrastructure.

One key difference between a Smart City and a Smart Village in terms of Digital Twins is the scale and complexity of the systems being modeled. A Smart City has a larger and more diverse set of systems and infrastructure, such as transportation networks and public utilities, whereas a Smart Village typically has a more limited and simpler set of systems. As a result, the Digital Twin of a Smart Village may be less complex than that of a Smart City, and mostly focuses on trying to get an efficient virtual model. This part needs a direct gathering device, where recent work has shown a high level of accuracy in capturing data of the physical part in point clouds formats.

In summary, while the principles of a Digital Twin can be applied to both Smart Cities and Smart Villages, the implementation and use of a Digital Twin in a Smart Village may be less complex and require less advanced technology due to the smaller scale and simpler set of systems present in a village. On another consideration this digital twin will exist only by and humans. It will be made as a system of systems, based on an ecosystem, distributed, at the edge. Section II showed that to achieve the smartification of the territory, one must consider not only cities and villages but the whole territory. We detail in the following the major aspects to be implemented in the digital twin of villages.

4.1 Human centered and participatory dimensions of the digital twin

To effectively implement a digital twin in rural areas, it is necessary to consider the human centered and participatory dimensions of the project. Unlike the digital twins of smart cities ("cyborg city") where the human is considered as an object and not a subject (Cathelat B. et al., 2019), it seems preferable to develop a 'community' approach, placing the human at the center, involving every user from the beginning as soon as the design starts. Uses and practices of ICT should result in the definition of services co-constructed with the population and supported by the smart-village and its digital twins. Then it will be able to improve the social links between inhabitants but also e-governance with officials. To conduct such a project of smart society in rural areas, several dimensions need to be developed. Implementation of digital twins must be integrated with a service-oriented architecture that will support local services (professionals, neighbors, individuals...) via a local exchange system mechanism, as well as institutional services (e-governance, e-administration, security, and crisis management).

The dynamic modeling of territories must be carried out collectively to limit the cost but also to involve users (crowdsourcing). The development of (distributed) functional internal engines is a challenge, especially to ensure resilience and ensure respect of the privacy of actors (3D engine and service orchestrator). To include everyone, universal interfaces with the architecture must be developed, whether in the form of applications on computers, mobile or tablet devices, or via immersive approaches enabled by the 3D dimension of the project provided by the digital twin.

A major lock here, beyond capture, is proposed to be updated by limiting the number of sensors and citizen participation. The acceptability and appropriation of digital technologies is an important aspect to be considered in the digital twin, and this will be addressed through a strong hypothesis which is since 3D technologies bring a more appropriable interface, being closer to reality, and thus reducing the digital divide (the acceptability of immersive interfaces is also to be evaluated). Delegation functionalities should also be included to address the fact that some people may not be able to use the system (such as those in strong illectronisme situations). Training aspects must also be considered, regarding the use of the architecture through the interfaces and aspects related to cyber security, so that bringing digital technology to those who need it does not open the door to maliciousness.

4.2 Territorial digital twins, a system of digital twins

A system of systems (SoS) is a collection of multiple, independent systems that work together to achieve a common goal or set of goals (Boardman, 2006). These systems may be from different organizations or different domains and may have different levels of complexity and functionality. The SoS is characterized by its ability to integrate the capabilities of its constituent systems in a way that enables new or enhanced capabilities that are not possible with the individual systems alone. Examples of SoS include transportation networks, power grids, and military systems. The smart village is clearly made of sub systems. An SoS differs from a single monolithic system in that it consists of multiple autonomous systems linked by interfaces and protocols. Each system within the SoS is responsible for its own internal behavior and operations but can also interact and share information with other systems within the SoS to achieve common goals. This makes SoS more adaptable and resilient, but it can also make it more complex and harder to manage.

A system of ecosystems is a collection of multiple, interrelated ecosystems that work together to maintain a balance within an environment or region. Ecosystems are defined as a community of living and non-living things and their interactions with each other and the environment. In the context of smart villages and territories, a system of ecosystems refers to the interconnections and interdependencies among different ecosystems including human, physical and digital systems within a region. It is the source of data for the smart village and or territory. The territorial digital twin is a system of digital twins that will possess the following features.

4.2.1 *Autonomous and composable*

It is believed that a smart territory, which is a system of systems and ecosystems, must have a highly distributed global architecture that is federated around singular entities, each of which potentially has a digital twin and is interconnected when necessary. This architecture is more of an edge computing architecture than a cloud architecture, and data should be owned and managed as close as possible to its source. To achieve this, it is suggested that villages should have their own servers, and potentially each home could have instances of them too. The idea is to have a private space and several shared spaces for each entity eligible for a digital twin, and all interconnections and information exchanges between entities are to be controlled and available in these controlled shared information spaces. Thus, the major challenge is to design the architecture's engines to work in this highly distributed context of nodes, from individuals to the different layers of organization. It appears that an organic approach is necessary to be able to construct a smart territory that can scale.

4.2.2 *Orchestration of distributed services*

It is essential that the digital twin surpasses the boundaries of a single community, particularly in rural areas. The orchestration of distributed services will be provided by the digital twin workflows that can anticipate from user needs the building of the appropriate service to be proposed. The service can be based on internal service and external services. One major lock here is to provide generic service engines that operate within the systems of interdependent and composable digital twins. What is envisioned are internal service engines that belong to singular digital twins and external ones that a digital twin can subscribe to. Note that these external engines are owned and made available by other DT.

4.2.3 *Twins and proxies*

This research paper aims to explore the potential of creating a participatory Digital Twin of the territory, aggregating individual, and collective information views. The focus is on the holistic view brought by the interconnection of autonomous information spaces that compose the territory, rather than designing full-fledged digital twins of the components. It is proposed that a digital twin has in its system proxies of other entities with which it is in interaction, and these proxies may not always be connected to ensure resilience and scalability. To illustrate this approach, a village may have its own digital twins and proxies on inhabitants, which could be linked with the genuine digital twins of each person who has created their own. Each entity develops its own independent world, where some nodes may exchange information in a controlled way from the ground up. The future digital twins will be made of aggregated data from different sources in a loose coupling manner, so that it can be federated for specific goals and then unleashed after the building and delivery of the service through the digital twin and smart village.

4.3 3D representation for the territorial digital twin

Support for smart-cities and smart-villages to reach their full potential is enabled using interfaces that facilitate interaction. These interfaces are often Geographic Information Systems (GIS) that use maps and cartography to represent the territory, with additional information of interest typically presented in two-dimensional form. However, the advancement of ICT has made it possible to integrate three dimensions into these representations, providing a more natural virtual representation that is closer to the reality perceived by users. It is hypothesized that providing 3D representations of cities, villages, and territories will help to increase acceptance and usability, as it more closely resembles the real thing. Furthermore, progress in virtual, augmented, and extended reality offer promising directions for interfacing the real world with the virtual world, and to achieve this level of immersion, 3D models of the environment must be considered. To explore the feasibility of this in a rural context, low-cost and participatory reality capture is examined.

4.3.1 The Connection (3D gathering device)

3D Reality Capture Methods, such as LiDAR, Photogrammetry, and Structure from Motion (SfM), are widely used to generate point clouds for Smart Cities (Nys et al., 2020). These methods involve collecting data from various sources, such as sensors, cameras, and drones, and using this data to create a detailed 3D model of the city and its infrastructure (Poux et al., 2017) (Fig. 16.6).

LiDAR (Light Detection and Ranging) technology uses laser beams to measure the distance between the sensor and the target, providing high accuracy and precision in the data collected. LiDAR has been widely used in Smart City applications, such as mapping, 3D modeling, and change detection (Gao, 2018). Photogrammetry involves using photographs to

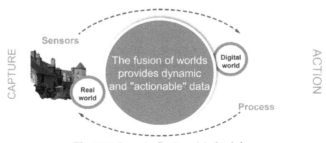

Fig. 16.6 Images Capture Methodology.

measure the location and properties of physical objects. It is also used in Smart City applications such as building façade and cultural heritage mapping (Besl, et al., 1992). SfM (Structure from Motion) is a technique that uses a sequence of images to reconstruct a 3D model of a scene. It is widely used in creating 3D models of cities, towns, and villages. (Poux et al., 2020). These methods have proven to be effective in generating highly accurate and detailed point clouds for Smart Cities. However, they may not be as applicable to Smart Villages due to the smaller scale and simpler set of systems present in a village. Furthermore, the cost of acquiring and maintaining the necessary equipment and technology for these methods are prohibitive for smaller, rural communities (VanDerHorn et al., 2021). Indeed, acquiring datasets for entire cities often relies on dynamic systems such as Mobile Mapping or Aerial Configurations. Mobile Mapping Systems (MMS) and Aerial LiDAR are widely used for creating digital twins of smart cities. MMS involves mounting LiDAR sensors and cameras on a vehicle and driving around the city to collect data. Aerial LiDAR, on the other hand, involves using a laser scanner mounted on an aircraft to collect data from above. Both methods provide highly accurate and detailed point clouds of the city and its infrastructure. However, there are several limitations to the use of MMS and Aerial LiDAR for smart cities. One major limitation is the high cost of these systems. The cost of acquiring and maintaining the necessary equipment and technology for MMS and Aerial LiDAR can be prohibitive. Additionally, the scalability of these methods is limited in tight environments such as narrow alleys, urban canyons, and crowded city centers where the vehicle or aircraft may not have enough space to maneuver or a poor global navigation satellite system (GNSS) signal. Furthermore, MMS and Aerial LiDAR are not well-suited for capturing data of buildings and other structures with complex geometries, due to the limitations of the sensors used (Zhou, et al., 2020). Also, MMS and Aerial LiDAR may not be able to capture data of buildings and other structures that are obscured by trees, buildings, or other obstacles (Zhao, et al., 2021). Therefore, MMS and aerial LiDAR are widely used for creating digital twins of smart cities, but they have their limitations. These systems are high-cost and have limited scalability in tight environments, also they have difficulty capturing data of buildings and other structures with complex geometries and obscured buildings. On top, these methods may not be as applicable to Smart Villages due to the smaller scale, simpler set of systems and the cost of acquiring and maintaining the necessary equipment and technology for these methods.

4.3.2 *Toward sustainable 3D Data Capture*

4.3.2.1 Low-cost Reality Capture

Low-cost terrestrial photogrammetry is a method of generating accurate and up-to-date point clouds of small villages that has several strengths.

This method involves using a consumer-grade camera, such as a DSLR or a smartphone camera, mounted on a tripod or handheld, to capture a series of overlapping images of the village. These images are then processed using photogrammetry principles to obtain such exhaustive 3D spatial information (Zhang et al., 2017) primarily as a point cloud. It is a {X, Y, Z} (+ attributes) spatial ensemble which digitally represents the recorded environment w.r.t the sensor strengths and limitations. The landscape of these instruments and acquisition methodologies is mature enough to allow digital replicas of the real world ranging from the object scale to the country scale (Narayanan et al., 2011). One of the main strengths of low-cost terrestrial photogrammetry is its affordability. Compared to other 3D Reality Capture Methods, such as Aerial LiDAR or MMS, this method requires minimal investment in equipment and technology. This makes it accessible to small villages that may not have the resources to invest in more expensive equipment. Another strength of this method is its ability to capture high-resolution and up-to-date images of the village. The use of consumer-grade cameras allows for capturing high-resolution images, which can be used to create detailed point clouds of the village. Additionally, the method can be easily repeated to capture updated images of the village, providing an up-to-date point cloud. Furthermore, low-cost terrestrial photogrammetry does not require special permissions or clearance, which makes it more accessible to small villages with limited resources. Therefore, the gain in affordability, high-resolution and up-to-date image capture that provides photogrammetry can be a valuable tool for small communities that wish to create a digital twin of their village.

4.3.2.2 Crowdsourced photogrammetry

Low-cost and crowdsourced photogrammetry have emerged as valuable techniques for creating up-to-date digital twins for Smart Villages. The use of low-cost equipment and crowdsourcing techniques, where the data is collected by members of the community, allows for the creation of digital twins that are both cost-effective and reflective of the local community's needs and perspectives. One of the key advantages of low-cost and crowdsourced photogrammetry is its ability to provide up-to-date digital twins of the village. The use of consumer-grade cameras and smartphones allows for the frequent capture of images of the village, which can be used to create updated point clouds. Additionally, the participation of members of the community in the data collection process ensures that the digital twin reflects the current state of the village and the community's needs and perspectives. Especially, it can be useful for the analysis of public opinion and sentiment regarding the village and its development (Zhou, et al., 2020).

Crowdsourced photogrammetry also has the potential to engage and empower members of the community, particularly youth and women, in the data collection process and decision-making related to the village's development. It can provide a platform for community participation and co-creation in the smart village development. This is particularly important in rural areas where community participation and empowerment are key to sustainable development (Hossain, et al., 2020). Furthermore, low-cost, and crowdsourced photogrammetry can be integrated with other data sources such as satellite images, citizen-generated data, and open data to provide a more comprehensive picture of the village and its systems. For example, integrating data from social media platforms, such as Twitter and Facebook.

In conclusion, low-cost and crowdsourced photogrammetry bring important perspectives and valuable tools for creating up-to-date digital twins for Smart villages. These methods allow for the creation of cost-effective digital twins that are reflective of the local community's needs and perspectives. Additionally, crowdsourced photogrammetry has the potential to engage and empower members of the community in data collection. Therefore, this method is very promising to establish an efficient Digital Twin for monitoring, planning and decision-making related to the village's development and resource management (Khan et al., 2020).

Fig. 16.7 Example of 3D reconstruction from crowd-sourced photogrammetry.

4.4 Simulation for large complex system of systems

In the previous sections, it has been demonstrated that simulation can be a powerful tool for validating the behavior of a system and can be used in the context of a digital twin to analyze different scenarios and anticipate the future. However, the works presented thus far have been largely limited to specific sectors, failing to take a holistic, global view. Therefore, there is a need for a global architecture that can gather, aggregate, and process data and information from sensors, considering uncertainty, spatial, and temporal dimensions, and enable communication between officials and citizens. The feedback from previous projects suggests that modeling and simulation play an important role in digital twins for smart cities. However, the models are often heterogeneous and not always formal and unambiguous. Building a model is a specific process that involves selecting a theoretical framework, such as a formalism, to accurately describe the source system or object of study, while also ensuring that the whole process is suitable for the problem at hand. As Robinson (2015) notes, "conceptual modeling is more of an art than a science".

In the case of large and complex systems digital twins, such as a village, city, or region, it is necessary to use multi-modeling approaches to select the most appropriate methods for modeling the various phenomena, such as water networks, circulation, electricity production/consumption, consumers, and inhabitants.

Multi-modeling, as defined by Vangheluwe et al. (2002), is the method of combining multiple perspectives of a study system into a coherent representation. This involves managing phenomena at different levels (micro and macro), at different scales (temporal and spatial), and from different perspectives, by using multiple formalisms or modeling paradigms for the same problem. Other approaches, such as co-simulation (Gomes et al., 2018; Camus et al., 2018), can also be used in a concurrent or complementary manner.

According to Gomes et al. (2018), "co-simulation is a theory and set of techniques for simulating a coupled system, by combining individual simulators. Each simulator is generally considered as a black box that can represent behavior, consume inputs, and produce outputs."

These approaches focus on model composition and aim at selecting and connecting the most appropriate modeling techniques. However, they can be limited when the spatial or geographical representation dimension also needs to be considered in the model. This is particularly the case when it is necessary to stack layers of models representing phenomena at different scales, such as wastewater flow, car traffic, weather, etc. (Broutin et al., 2008; Broutin et al., 2010). Spatial representation is an essential element for any territory-scale modeling (Roche, 2014), and as stated by Williamson et al. (2010), a smart city must be equipped with spatial capabilities. Indeed, a "spatially active society is an evolving concept in which location, place, and other spatial information are available to governments, citizens, and businesses as a means to organize their activities and information".

Multilevel modeling can be used to solve several modeling problems (Morvan, 2013) that are related to the concerns of multi-paradigm modeling: (1) modeling the interaction between levels, (2) coupling heterogeneous models, and (3), dynamically adapting the abstraction level (Franceschini et al., 2019). There are several approaches to model this spatial dimension, some of which even allow modeling and simulation:

Geographic information systems (GIS) (Fischer et al., 1992) (Bernhardsen T., 2002) can display spatially indexed data from various sources, modifying these data, storing them, retrieving them, and manipulating them for analysis. Their great strength lies in their ability to manage large, heterogeneous, multi-layered databases and to interactively query the existence, location, and properties of a wide range of spatial objects. However, the lack of analysis, modeling and simulation capabilities is widely recognized as a major shortcoming.

Nested graph structured representations (NGSR) (Macedo et al., 1998) are described by the nodes of the graph and a set of links between them, represented by the edges, with the particularity that each of the nodes of the graph can integrate another graph, and so on recursively.

Multilayered cellular automata (Bandini S., 1999) is based on the definition of a nested graph, it allows to reproduce multiscale processes with a fine spatial/temporal resolution, and especially to model and simulate transition rules between the states of the different layers.

Multi-Level Agent-based Modeling Modeling (ML-ABM) (Morvan, 2013) is characterized by the following definitions: (1) A level is a point of view on a system, embedded in a model as a specific abstraction. This name refers to levels of the organization, observation, analysis, etc. (2) An agent-based multilevel model integrates heterogeneous (agent-based) models, representing complementary views, called levels, of the same system. (3) Integration means that the ABMs in an ML-ABM can interact and share entities such as environments and agents. (4) Heterogeneity means that the ABMs integrated into an ML-ABM can be based on different modelling paradigms (differential equations, cellular automata, etc.), use different temporal representations (discrete events, stepwise, continuous system), and represent processes at different spatiotemporal scales. (5) The points of view are complementary for a given problem because they cannot be taken in isolation to treat it.

Tools such as Gama (Drogoul et al.,2013;Drogoul et al, 2013) and netlogo (Hjorth et al., 2020) incorporate these considerations. There is also formal work based on the DEVS formalism (Zeigler et al., 2018) that addresses these considerations (Broutin et al., 2008; Foures et al., 2018).

Capturing the globality of the points of view on the studied system is an essential prerequisite for its good modeling (Stanton et al., 2006). In this context, the vision advocated in some works on digital twins (augmented model) seems very relevant at the scale of a territory, and to set up processes to help decision-making based on data and models (Ketzler et al., 2020; Xia et al., 2022).

5. Conclusion

This chapter demonstrated that the digital world is one lever to facilitate life of population in the small cities and more rural territories. It focused on the contribution of digital to Smart villages and smart territories. The current developments in relation to smart villages provided interesting results as well as direction to get a larger scale territory that goes beyond the frontier of a given city or village. The first part that was dealing with recalls on experiences and feedback from the proof of concept in Cozzano showed that digitalized approach to the territory can make it possible to imagine different scenarios to resolve crises in the countryside. Then territorial services in large rural areas including medical and air quality cases presented the interest of modeling simulation to provide decision support to citizen, patient, and professionals. Finally, this study revealed that major required concepts of a territorial System of digital twins must include humans, system of systems and 3D environments. In the future Smart Village, the implementation of Digital Twin may be less complex and will require less advanced technology due to the smaller scale and simpler set of systems present in a village. The Digital Twin of a Smart Village may focus on creating an efficient virtual model with data collected from various sources, such as sensors and IoT devices with the use of simulation. However, the digital twin will exist only by and with humans. It will be made as a system of systems, based on an ecosystem, distributed, at the edge. One must consider not only cities and villages but the territory to achieve smartification of the territory.

References

Antoine-Santoni, Thierry and Bastien Poggi. (2022). Towards an edge infrastructure evolution for rural Mediterranean activities in a Smart Village. In: Workshops at 18th International Conference on Intelligent Environments (IE2022)., *Biarritz: IOS Press*, pp. 80–89. https://doi.org/0.3233/AISE220025.

Antoine-Santoni, Thierry, Bastien Poggi, Evelyne Vittori, Ho Van Hieux, Marielle Delhom, and Antoine Aiello. (2019). Smart Entity:How to Build DEVS Models from Large Amount of Data and Small Amount of Knowledge? In: International Conference on Simulation Tools and Techniques. *Springer*, pp. 615–626.

Antoine-Santoni, Thierry, Oumaya Baala, Manuele Kirsch Pinheiro, Fabien Mieyeville, Bertrand Mocquet, and Luiz-Angelo Steffenel. (2022). A Smart Territory, the key to resilient territory. In: Resilient and Sustainable Cities, Resilient and Sustainable Cities. Elsevier.

Bandini, S. and Mauri, G. (1999). Multilayered cellular automata. *Theoretical Computer Science*, 217(1): 99–113.

Béler, C. and Grabot, B. (2015). Secured and efficient information exchanges in collaborative networks: The singular information system (October). Pro-VE 15.

Bernhardsen, T. (2002). Geographic information systems: An Introduction. John Wiley & Sons.

Besl, P.J. and McKay, N.D. (1992). A method for registration of 3-D shapes. *IEEE Transactions on Pattern Analysis and Machine Intelligence*, 14(2): 239–256.

Boardman, J. and Sauser, B. (2006, April). System of Systems –the meaning of. In: 2006 IEEE/ SMC International Conference on System of Systems Engineering IEEE, 6 pp).

Boubekri, M. (2017). Smart Cities: Definition, dimensions, performance, and initiatives. *Sustainability*, 9(11): 2082.

Broutin, E., Bisgambiglia, P.A. and Santucci, J.F. (2008). Multilayered and heterogeneous modelling and simulation of natural complex systems. In European Simulation and Modelling Conference 2008, pp. 338. SCS.

Broutin, E., Bisgambiglia, P.A. and Santucci, J.F. (2010, July). Time management in a multilayered architecture. In Proceeding of the SCS Summersim 2010 Conference, pp. 6. SCS.

Boy, G.A. (2020). Human–Systems Integration: From Virtual to Tangible (1st Edn.). CRC Press. https://doi.org/10.1201/9780429351686.

Camara Dit Pinto, S. et al. (2021). Digital twin design requirements in downgraded situations management. *IFAC-PapersOnLine*, 54(1): 869–873. doi: 10.1016/j.ifacol.2021.08.102.

Camus, B., Paris, T., Vaubourg, J., Presse, Y., Bourjot, C., Ciarletta, L., and Chevrier, V. (2018). Co-simulation of cyber-physical systems using a DEVS wrapping strategy in the MECSYCO middleware. *Simulation*, 94(12): 1099–1127.

Cathelat, B. et al. (2019). Smart Cities: Shaping the Society of 2030. UNESCO publishing, ISBN: 978-92-3-100317-2.

Deng, T., Zhang, K. and Shen, Z.J.M. (2021). A systematic review of a digital twin city: A new pattern of urban governance toward smart cities. *Journal of Management Science and Engineering*, 6(2): 125–134.

Drogoul, A., Amouroux, E., Caillou, P., Gaudou, B., Grignard, A., Marilleau, N. ... and Zucker, J. D. (2013, May). Gama: A spatially explicit, multi-level, agent-based modeling and simulation platform. In: International Conference on Practical Applications of Agents and Multi-Agent Systems. Berlin, Heidelberg: *Springer*, pp. 271–274.

Drogoul, A., Amouroux, E., Caillou, P., Gaudou, B., Grignard, A., Marilleau, N. ... and Zucker, J. D. (2013, May). Gama: Multilevel and complex environment for agent-based models and simulations. In: AAMAS'13: 12th International Conference on Autonomous Agents and Multi-Agent Systems. *IFAAMAS: International Foundation for Autonomous Agents and Multiagent Systems, Richland, SC*, pp. 1361–1362.

Endsley, M. et al. (2003). Designing for Situation Awareness. CRC Press.

Eshuis, P., Pas Sopheon, E. ten, Rutten Sopheon, H., and Verlinden Zorggemak, J.-M. (2010). State of the Art Clinical Pathway Definition: Gap Analysis. https://itea3.org/project/ workpackage/document/download/249/07011-Edafmis-WP-1-D11.pdf.

Fischer, M.M. and Nijkamp, P. (1992). Geographic information systems and spatial analysis. *The Annals of Regional Science*, 26(1): 3–17.

Foures, D., Franceschini, R., Bisgambiglia, P.A. and Zeigler, B.P. (2018). multiPDEVS: A parallel multicomponent system specification formalism. Complexity, 2018.

Franceschini, R., Van Mierlo, S. and Vangheluwe, H. (2019, December). Towards adaptive abstraction in agent based simulation. In 2019 Winter Simulation Conference (WSC) (pp. 2725–2736). IEEE.

Gao, F. (2018). Digital twin technology: A review. *CIRP Annals-Manufacturing Technology*, 67(1): 367–386.

Gil-Garcia, J. Ramon, Theresa A. Pardo and Taewoo Nam. 2015. What Makes a City Smart? Identifying Core Components and Proposing an Integrative and Comprehensive Conceptualization. *IOS Press, Information Polity*, 20(1): 61–87. https://doi.org/10.3233/IP-150354.

Gomes, C., Thule, C., Broman, D., Larsen, P.G, and Vangheluwe, H. (2018). Co-simulation: A survey. *ACM Computing Surveys (CSUR)*, 51(3): 1–33.

Hjorth, A., Head, B., Brady, C. and Wilensky, U. (2020). Levelspace: A netlogo extension for multi-level agent-based modeling. Journal of Artificial Societies and Social Simulation, 23(1).

Hossain, M., Islam, M. and Luitel, B. (2020). Community participation and empowerment in smart village development: A case study from Bangladesh. *Sustainability*, 12(14): 5155.

Jiang, X., Li, Y., Sun, Y. and Wang, Y. (2019). A comprehensive review of digital twin: Definition, architecture, and applications. *Journal of Manufacturing Systems*, 49: 1–12.

Ketzler, B., Naserentin, V., Latino, F., Zangelidis, C., Thuvander, L. and Logg, A. (2020). Digital twins for cities: A state of the art review. *Built Environment*, 46(4): 547–573.

Khan, M., Ali, M. and Khan, M. (2020). Low-cost photogrammetry for monitoring and planning of smart villages. *Sustainability*, 12(2): 522.

Komorowski, L. and Stanny, M. (2020). Smart Villages: Where Can They Happen? *Land 9*, no. 5: 151. https://doi.org/10.3390/land9050151.

Macedo, L. and Cardoso, A. (1998, September). Nested graph-structured representations for cases. In: European Workshop on Advances in Case-Based Reasoning. Berlin, Heidelberg: Springer, pp. 1–12.

Morvan, G. (2012). Multi-level agent-based modeling: A literature survey. arXiv preprint arXiv: 1205.0561.

Narayanan, S.G. and Quattrochi, D.A. (2011). A review of image-based 3D modeling techniques. *Journal of Spatial Science*, 56(2): 121–142.

Nys, G.A., Poux, F. and Billen, R. (2020). CityJSON Building generation from airborne LiDAR 3D point clouds. *ISPRS International Journal of Geo-Information*, 9(9): 521.

Oliveira, Ana Paula, Cavalcante de Oliveira, Mariana Gabriel, Mario Roberto Dal Poz, and Gilles Dussault. 2017. Challenges for ensuring availability and accessibility to health care services under Brazil's Unified Health System (SUS). Ciência & amp; Saúde Coletiva, 22 (4): 1165–1180. https://doi.org/10.1590/1413-81232017224.31382016.

Poupry et al. (2022). Contribution to the design and implementation of a reflexive cyber-physical system: Application to air quality prediction in the Vallées des Gaves. *In*: Proceedings of the European Conference of the PHM Society 2022.

Poux, F. et al. (2020). Initial User-Centered Design of a Virtual Reality Heritage System: Applications for Digital Tourism. *Remote Sensing*, 12(16): 2583. https://doi.org/10.3390/rs12162583.

Robinson, S. (2015). A tutorial on conceptual modeling for simulation. In: Proceedings of the 2015 Winter Simulation Conference. *IEEE Press*, pp. 1820–1834.

Roche, S. (2014). Geographic Information Science I: Why does a smart city need to be spatially enabled? *Progress in Human Geography*, 38(5): 703–711. https://doi.org/10.1177/0309132513517365.

Sbayou, Mariem, Gregory Zacharewicz, Youssef Bouanan, and Bruno Vallespir. (2019). BPMN Coordination and Devs Network Architecture for Healthcare Organizations . *International Journal of Privacy and Health Information Management (IJPHIM)*, 7(1): 103–115. https://doi.org/10.4018/IJPHIM.2019010106.

Stanton, N.A. et al. (2006). Distributed Situation Awareness in Dynamic Systems: Theoretical Development and Application of an Ergonomics Methodology. *Ergonomics*, 49(12–13): 1288–1311. https://doi.org/10.1080/00140130600612762.

Starkiene, Liudvika. 2013. Inequitable geographic distribution of physicians: Systematic review of international experience. *Sveikatos Politika ir Valdymas*, 5(1).

Sylvain Poupry, Cédrick Béler and Kamal Medjaher (2022). Development of a reliable measurement station for air quality monitoring based on low-cost sensors and active redundancy. *IFAC-PapersOnLine*, 55(5): 7–12. ISSN 2405-8963, https://doi.org/10.1016/j.ifacol.2022.07.631.

VanDerHorn, E. and Mahadevan, S. (2021). Digital Twin: Generalization, characterization and implementation. *Decision Support Systems*, 145: 113524.

Vangheluwe, H., De Lara, J. and Mosterman, P.J. (2002). An introduction to multi-paradigm modelling and simulation. *In*: Proceedings of the AIS'2002 Conference (AI, Simulation and Planning in High Autonomy Systems), *Lisboa, Portugal*, pp. 9–20.

World Health Organization. (2021). WHO global air quality guidelines: Particulate matter (PM2.5 and PM10), ozone, nitrogen dioxide, sulfur dioxide and carbon monoxide: Executive summary. World Health Organization. https://apps.who.int/iris/handle/10665/345334.

Williamson, I., Rajabifard, A., and Holland, P. (2010) Spatially enabled society. *In:* Proceedings of the FIGURE Congress 2010, Facing the Challenges – Building the Capacity, Sydney. Available at:http://www.fig.net/pub/fig2010/papers/inv03%5Cinv03_williamson_rajabifard_et_al_4134.pdf.

Xia, H., Liu, Z., Maria, E., Liu, X. and Lin, C. (2022). Study on city digital twin technologies for sustainable smart city design: A review and bibliometric analysis of geographic information system and building information modeling integration. *Sustainable Cities and Society*, 104009.

Zeigler, B., Traoré, M.K., Zacharewicz, G. and Duboz, R. (2019). IET Healthcare Technologies Series, 15. London: Institution of Engineering and Technology, 376.

Zeigler, B.P., Muzy, A. and Kofman, E. (2018). Theory of Modeling and Simulation: Discrete Event & Iterative System Computational Foundations. Academic Press.

Zhang, L., Liu, Y. and Liu, Y. (2017). A review of structure from motion techniques. *Journal of Applied Remote Sensing*, 11(2).

Zhou, Y. and Liu, Y. (2020). A review of mobile mapping systems: data acquisition, processing, and applications. *International Journal of Geographical Information Science*, 34(2): 235–254.

Zhao, J., Liu, Y. and Gao, F. (2021). A review of airborne LiDAR data processing: Algorithms, methods and applications. *ISPRS Journal of Photogrammetry and Remote Sensing*, 166: 1–18.

Zubizarreta, Iker, Alessandro Seravalli and Saioa Arrizabalaga. (2015). Smart City Concept: What It Is and What It Should Be. *Journal of Urban Planning and Development*, 142(1). https://doi.org/10.1061/(ASCE)UP.1943-5444.0000282.

PART IV
Super Smart Society and Sustainability

Conclusions and Outlook

At the end of this journey, we can conclude that Society 5.0 is the ideal continuation of Sociaty 4.0 in which digitalization and a 360° innovation (i.e., technology, materials, business models, etc.) constitute the basis for an economy based almost exclusively on sustainability and data security. From Society 1.0 to Society 5.0, the characteristics of each "evolutionary era" of society led to the modern concept of Society 5.0, a data-driven society that sees emerging technologies as its main enabler. Society 5.0 represents an evolution from various points of view, primarily technological. The digital transformation underway is supported by AI, capable of supporting humans in analyzing the enormous amount of data that that the internet of things (IoT) systems continuously acquire in bridging the real world and the virtual world. To successfully respond to the sustainability challenge that awaits us in the near future, Company 5.0 is based on three fundamental pillars:

1. The transition to the Society 5.0 model and Productivity Revolution, through the use of IoT, Big Data, and Artificial Intelligence (AI) technologies;

2. The creation of resilient, environmentally friendly and attractive communities, through the Future City Initiatives for the realization of the United Nations Sustainable Development Goals (SDGs);

3. The empowerment of future generations through a revolution in the development of human resources to make the most of the rich creative and communication skills, focusing on the gender objectives of the SDGs.

The forecasts on a global scale suggest a series of almost certain dynamics, such as the global aging of the population. To think effectively about the ultra-aging society that awaits us. Towards a more sustainable, resilient, and human-centric future, it will be necessary to act in two directions: maximizing the length of a healthy life and favoring environments that allow fragile individuals to live independently. The language of innovation should no longer be limited exclusively to the pursuit of maximum profit, but should include terms such as equality, equity, solidarity, sustainability, inclusion, and change among its keywords. The challenge to global aging is only one of the fronts that sees the technologies of the 5.0 paradigm deployed at the forefront.

In addition to issues related specifically to economic and environmental sustainability, we will see some technologies take on a key role in the development of the 5.0 Company. Connectivity, and in particular 5G technology, is seen as a fundamental enabler to guarantee devices and applications to communicate with each other with optimal latency. Equally important will be the role of super computers (HPC - High Performance Computing), with the arrival of the long-awaited exascale class and large data storage systems, in the dual direction of cloud computing and edge computing. Therefore, Society 5.0 will be a data-driven society capable of finding its application in areas such as precision medicine, domestic robotics, industrial robotics, smart grids, autonomous driving, smart cities, decentralized finance, digital twin, and many others, which will contribute to creating the metaverses of a society where technology will have to take on an instrumental and human-centric value. The ultimate goal is a society for *'everyone'* and *"of each of us"*, without exclusion, where technology is at the service of human well-being in all its forms.

Editors

Chaudhery Mustansar Hussain
Department of Chemistry and Environmental Science
New Jersey Institute of Technology, Newark, N J 07102, USA
chaudhery.m.hussain@njit.edu

Antonella Petrillo
Department of Engineering
University of Naples Parthenope, CDN 80143 Napoli, ITALY
antonella.petrillo@uniparthenope.it

Shahid Ul Islam, PhD
Department of Biological and Agricultural Engineering
University of California Davis, United States
shads.jmi@gmail.com

Index

About the Editors

Chaudhery Mustansar Hussain, PhD is an Adjunct Professor, Academic Advisor and Director of Chemistry & EVSc Labs in the Department of Chemistry & Environmental Sciences at the New Jersey Institute of Technology (NJIT), Newark, New Jersey, USA. His research is focused on the applications of Nanotechnology & Advanced Materials in Environment, Analytical Chemistry and Various Industries. Dr. Hussain is the author of numerous papers in peer-reviewed journals as well as prolific author and editor of several (around 50 books) scientific monographs and handbooks in his research areas published with ELSEVIER, Royal Society of Chemistry, Wiley, and Springer, etc.

Antonella Petrillo, PhD is a Mechanical Engineer. Currently she is Professor in DII, Department of Engineering of University of Napoli "Parthenope" where she teaches Facility Management, Quality Management and Safety at workplace. She serves as International Reviewer for over 50 International Journals and she is member of Scientific Committee for PhD Programs, International Conferences and Proposal Evaluations. She has over 160 Scientific Publications on International Journals and Conferences and she is author of 5 books on Innovation and Decision Making in Industrial Applications and Engineering.

Shahid Ul Islam is currently a Fulbright researcher and Cultural Ambassador at the University of California, Davis, United States. Before joining UCD, he was a Principal Project Scientist at the Indian Institute of Technology, New Delhi, India, where he led a number of projects funded by the Science and Engineering Research Board (DST-SERB) and Council of Scientific and Industrial Research (CSIR) agencies, Govt. of India. His areas of interest include smart functional materials, green chemistry, and sustainable engineering practices in different application sectors. He is a member of many groups, including the American Chemical Society (USA) and is a life member of the Asian Polymer Association. He is the author of numerous scientific articles, including peer-reviewed journal papers, reviews, chapters, and books published by the American Chemical Society, Royal Society of Chemistry, Elsevier, Wiley, and Springer, etc.